Food Engineering Series

For further volumes:
http://www.springer.com/series/5996

Enrique Ortega-Rivas

Non-thermal Food Engineering Operations

Springer

Enrique Ortega-Rivas
School of Chemical Sciences
Autonomous University of Chihuahua
Chihuahua, Mexico

ISSN 1571-0297
ISBN 978-1-4899-8926-0   ISBN 978-1-4614-2038-5 (eBook)
DOI 10.1007/978-1-4614-2038-5
Springer New York Dordrecht Heidelberg London

Printed on acid-free paper

Springer is part of Springer Science+Business Media (www.springer.com)

To my three girls who brought me love, happiness, understanding, and support during the time spent completing this long-cherished dream

Preface

Food process engineering comprises a series of unit operations traditionally applied in the food industry. One major component of these operations relates to the use of heat, directly or indirectly, to provide foods free from pathogenic microorganisms. Thermal processes are able to control microbial populations, but can also affect the biochemical composition of many foods, resulting in losses of quality, both sensory and nutritional. The last three decades have witnessed the advent and adaptation of several operations, processes, and techniques aimed at producing microbiologically safe foods, but with minimum alteration of sensory and nutritive properties. Some of these techniques have eliminated totally the thermal component in food processing, so they are considered to be alterative methods.

Most of the above-mentioned operations not relying on heat to preserve foods have received different denominations. Many terms, such as "emerging technologies," "novel processes," "cold pasteurization techniques," and "non-thermal processing," have been used to refer to them. Some of these terms are limited or inaccurate. For example, emerging technologies once exploited on a commercial scale may become established, whereas cold pasteurization or sterilization may be interpreted as being conducted at temperatures well below room temperature. The two common features that properly describe all these technologies are their application under room (or ambient) conditions and their elimination of the heat component to preserve or convert foods. Thus, the most generic terms encompassing the technologies under discussion are "ambient-temperature processes" and "non-thermal food processes." Since there is also the matter of convention within disciplines, a suitable term to describe alternative technologies in food processing is necessary. Food scientists seem to agree on the ambiguity of the terms "ambient temperature" and "room temperature," and so they prefer to simply define non-thermal food processing as those technological alternatives aimed at preserving the quality of treated foods with the absence of heat treatments.

In terms of education, some programs of study in chemical engineering, agricultural engineering, biosystems engineering, food science and technology, and so on, include food process engineering as a subject. This theme or topic is offered at both the undergraduate and the graduate level, and sometimes a distinction is made between preserving operations and transformation operations, or between traditional thermal processing and alternative non-thermal processing. The Graduate Program in Food Science and Technology of the Autonomous University of Chihuahua has combined these criteria for some time, offering courses on conventional food processing technologies, alternative food processing technologies, and conversion operations. This book has been written as a textbook for a course covering alternative food processing technologies and conversion operations, i.e., the most important non-thermal food processing operations in the food industry, as currently taught at the Autonomous University of Chihuahua. It is divided into three parts: an introductory part which covers handling of food materials along with preliminary operations such as cleaning and sorting, a second part dealing with processing or conversion operations, and a third part devoted to the study of preservation operations, where the most relevant recent alternatives such as ultrahigh hydrostatic pressure and high-voltage pulsed electric fields are included.

The idea of writing the book originated from experiences accumulated over the years in researching and teaching both food conversion operations and alternative food processing technologies. The author reckoned on the need for a compilation of information on the subject of non-thermal technologies, whether used as a preserving method or not, in a single volume. The book is intended to be used as a textbook for different food-processing-related courses, or as a book to be consulted by practicing engineers dealing with these subjects in the food industry. The project took several years of intensive work to collect, analyze, and refine information on the subject from varied and valuable sources. One of these sources was the feedback from students of the above-mentioned course on alternative food processing technologies and conversion operations. Many contributions from students were helpful in organizing the work. Particularly relevant for preparing the third part of the book were analytical reviews on the corresponding themes provided by Maria Antonieta Anaya-Castro, Arcely Córdova-Muñoz, and Edmundo Juárez-Enríquez.

Sincere and fondest appreciation is given to my wife Sylvia and my daughters Samantha and Christina, who suffered the inevitable reduction of time shared with them owing to the extra investment in time needed to write this book.

Chihuahua, Mexico   Enrique Ortega-Rivas

Contents

Part II   Processing Operations

Part I
Introductory Aspects

Chapter 1
Classification of Food Processing and Preservation Operations

1.1 Introduction

The origins of food processing may be placed in ancient Egypt in that period when diverse developments seemed to symbolize the history of the culture of mankind. Crude processing methods that included slaughtering, fermenting, sun-drying, and preserving with salt are evidenced, not only in the writings of the ancient Egyptians, but also in those attributed to Greek, Chaldean, and Roman civilizations. All these cultures tried and tested processing methods, which remained essentially the same until the Industrial Revolution. Examples of ready-to-eat meals existed even prior to the Industrial Revolution, for instance, the Cornish pasty and haggis, a traditional Scottish dish. Modern food processing technology was largely developed in the nineteenth and twentieth centuries to serve mainly military needs. A typical example is the invention of a vacuum bottling technique by Nicolas Appert in 1809 to supply food for French troops, which contributed to the development of tinning and then canning by Peter Durand in 1810. Although initially expensive and somewhat hazardous owing to the lead used in cans, canned goods later became a staple around the world. Pasteurization, discovered by Louis Pasteur in 1862, was a significant advance in ensuring the microbiological safety of food. During the twentieth century, World War II, the space race, and the rising consumer society in developed countries contributed significantly to the growth of food processing with such advances as spray-drying, juice concentrates, and freeze-drying, as well as the introduction of artificial sweeteners, coloring agents, and food preservatives. By the late twentieth century, products such as dried instant soups, reconstituted fruits and juices, and self-cooking meals were developed. In Western Europe and North America, the second half of the twentieth century witnessed a rise in the pursuit of convenience, as food processors marketed their products especially to middle-class working wives and mothers, using the perceived value of time to appeal to the postwar population. Nowadays, the processed foods that are on the shelves of grocery shops and supermarkets are modern processed foods or traditional ones, but their manufacturing technology, process control, and packaging

E. Ortega-Rivas, *Non-thermal Food Engineering Operations*,
Food Engineering Series, DOI 10.1007/978-1-4614-2038-5_1,
© Springer Science+Business Media, LLC 2012

methods have been advanced and rationalized to an incomparable extent in recent years. As a result, products with high quality and uniformity are now being manufactured, based on the advancement of food science, applied microbiology, mechanical engineering, chemical engineering, and some other interrelated disciplines.

The range of the food manufacturing industry is wide, so classification differs regionally. The overall food processing industry is generally termed the "food and drink industry," and the classification is usually based on grouping the raw products being transformed as being of animal origin, vegetal origin, and so on. Further classification can be accomplished in the sense of considering the specific nature of the raw products from the primary sector as, for example, livestock compared with fish, or grains compared with pulses, or fruits compared with vegetables. Diverse criteria are used worldwide to classify industrial activity, including the activity of the food industry. In the USA, for example, the North American Industry Classification System (NAICS) has replaced the Standard Industrial Classification (SIC) system. The major NAICS categories for the food industry are food manufacturing, beverage manufacturing, food and beverage stores, and food services and drinking places. A summary of a broad food manufacturing industry classification is presented in Table 1.1.

The food processing industry is one of the largest manufacturing industries worldwide and possesses, undoubtedly, global strategic importance. It has a critical need of growth based on future research directions detected by an integrated interdisciplinary approach to problems in food process engineering. In this sense, food process engineers all around the world are faced with the common problem of finding a compromise between the severity of the processes, mainly used as preservation techniques, and the quality of the final products. Normally, the severity of processing and overall quality is inversely proportional, so finding the above-mentioned compromise is a difficult task. The food industry has relied on chemical engineering principles, using them in modeling and problem-solving of routine food processing operations. Applying chemical engineering principles directly in food processing operations does not represent, generally, the best option since the raw materials used in the chemical industry and the food industry differ considerably in properties. The chemical industry generally employs inert raw materials whose composition is definable and their changes in processing are fairly predictable on the basis of the kinetics of chemical reactions. On the other hand, the raw materials used in the food industry have a very complex composition, so they are not as easily characterized as inert materials and the changes they undergo on processing are hard to describe by chemical reactions kinetics. There is, therefore, an urgent need to define whether the food industry will be capable of coping with challenges in growth and development using the somewhat traditional approach of adapting chemical engineering principles, in order to provide nutritive, safe, and premium-quality food products to an increasingly demanding population.

Table 1.1 A general classification of the food industry

Food manufacturing
  Grain and oilseed milling
    Milling of flour and manufacturing of malt
    Manufacturing of starch and vegetable fats and oils
    Manufacturing of breakfast cereals
  Sugar and confectionery product manufacturing
    Manufacturing of sugar
    Manufacturing of chocolate and confectionery from cacao beans
    Manufacturing of confectionery from purchased chocolate
    Manufacturing of non-chocolate confectionery
  Fruit and vegetable preservation and specialty food manufacturing
    Manufacturing of frozen foods
    Canning, pickling, and drying of fruits and vegetables
  Dairy product manufacturing
    Manufacturing of dairy products (except frozen dairy products)
    Manufacturing of ice cream and frozen desserts
  Animal slaughtering and processing
    Slaughtering of animals (except poultry)
    Processing of meat from carcasses
    Rendering and processing of meat by-products
    Processing of poultry
  Seafood product preparation and packaging
    Canning of seafood
    Processing of fresh and frozen seafood
  Bakery product and tortilla manufacturing
    Manufacturing of bread and bakery products
    Manufacturing of cookies, crackers, and pasta
    Manufacturing of tortilla
  Other food manufacturing
    Manufacturing of snack foods
    Manufacturing of roasted nuts and peanut butter
    Manufacturing of coffee and tea
    Manufacturing of flavoring syrup and concentrates
    Manufacturing of seasoning and dressings
    All other food manufacturing
Beverage manufacturing
  Manufacturing of soft drinks and ice
  Breweries
  Wineries
  Distilleries

1.2   The Disciplines of Food Engineering and Food Science

There has been an ongoing argument on whether food engineering can be considered an independent engineering discipline, or should part of chemical engineering. Although chemical engineering principles have traditionally been employed in the design and control of food processing operations in a direct manner, there is an upsurge in activities aimed at adapting such principles to food processes in order to find better practical results. Theoretical principles of chemical engineering focus on the writing of mass and energy balance equations for individual pieces of equipment to scale them up to perform a given unit operation, or to determine and control the performance of a given unit. The application of these principles has been suitable to the chemical processing industry, but it has not been possible to find such suitability when dealing with processes in the food industry. As a consequence, the many benefits achieved by the chemical industry by applying the above-mentioned principles are not equally advantageous for the food industry.

The principal factor discouraging the widespread use of chemical engineering principles for food processing systems is the complex composition of food materials, which makes their properties poorly understood as compared with those of well-defined chemical systems. Essentially, food materials consist of complicated matrices comprising a wide variety of chemical compounds often including diverse macromolecules. Chemical reactions involving different materials are governed by the laws of kinetics and thermodynamics. For most inert materials, such as non-metallic minerals, the reactants and products are in well-defined thermodynamic states, i.e., the reactants can be transformed into the products and vice versa. In contrast, biological substances such as many food materials cannot be precisely defined in a thermodynamic sense, as many of the various stages that a food material undergoes as it is processed cannot be described as transitions between thermodynamic states. This means that the changes in food materials induced by processing are inevitably irreversible. For instance, dough can be baked to make bread, but neither the dough nor the bread can be thermodynamically characterized with any degree of precision. The bread, of course, cannot be changed back to dough. Furthermore, process control strategies extensively applied in chemical processing cannot be easily applied to food processing. For example, a simple feedback control system where the feed parameters are controlled to yield a product with the desired properties would be extremely difficult to apply to a food system. It may not be possible to establish the desired properties in food processing using measurable physical parameters, because the food industry often relies on sensory panel assessment to define a number of quality attributes. Also, if the desired properties of the product are quantifiable through physical variables, appropriate online sensors are not normally available for food processes. To complicate matters even further, food materials act as suitable substrates for microbial growth, so the kinetics of reproduction (as well as eradication, of course) of microorganisms have to be taken into account in the processing of foodstuffs. Finally, understanding the effects of process alterations on the food product is grossly inadequate for any form of control strategy to be implemented.

The established discipline dealing with the application of fundamental principles of physics, chemistry, and mathematics, along with transport phenomena criteria, to transformation of raw materials into finished products is, undoubtedly, chemical engineering. When these raw materials are of biological origin and used for human consumption, as stated above, chemical engineering principles fail to an extent in producing satisfactory practical results for the operation of food processing plants. It is, therefore, necessary to understand deeply the effects of different degrees of alterations caused by the processing of food materials. The detailed study of biological materials used as foodstuffs corresponds to another established discipline known generally as food science. Both chemical engineering and food science can be considered mature disciplines, counting on programs of study in higher-education institutions and well-established professional organizations supporting them. It may thus be fair to consider food engineering or food process engineering as an applied engineering discipline that may equally fit within the various chemical engineering topics, or as one of the major components of the food science discipline (along, mainly, with food microbiology and food chemistry/biochemistry). In any case, food engineering as such has become an established engineering discipline when it is considered part of chemical engineering, or a mature applied science when it is considered within the context of food science. The relevance of food engineering lies in the fact that, compared with either chemical engineering of food science, it is also supported by programs of study, professional associations, periodical publications, and classic texts and reference books.

1.3   Chemical Engineering Principles Involved in Food Processing

As stated before, a number of chemical engineering principles can be applied to the processing of food materials, bearing in mind theoretical aspects of food science. Interest in chemical engineering in the food industry is increasing rapidly because the quantity of processed foods is increasing worldwide. The demand for a wider variety of foods will also increase, and the complexity of the products from the food industry may include "designed" foods, functional foods, and custom-oriented foods. Nowadays, the consumer has become cultivated in diverse topics, and food is not an exception. Consumers are mainly concerned with safety, nutritional quality, and sensory aspects of processed foods. They are also demanding functional foods which, apart from the nourishing aspect, may contribute to raising standards of quality of life. In this sense, another increasing concern has to do with possible long-term effects of preservatives and additives used by food processors, as well as potential side effects of novel and unconventional manners of processing foodstuffs. It has, therefore, become imperative for food process engineers to become knowledgeable in the different food processing technologies that have been applied over the years. This knowledge will contribute to developing expertise in different food processes in order to modify traditional techniques, or suggest new directions in the technology of food processing as a way of meeting continually increasing consumer demands.

The vested interest in food processing by the discipline of chemical engineering may be related to some deep understanding of the main physicochemical aspects of food materials through food science principles. The acquisition of such knowledge has benefited from advances in instrumental methods of chemical analyses, increases in the sophistication of technology, widespread exchange of information, and interaction between scientific disciplines. Current instrumental and analytical techniques have revealed structures and intimate configurations of food materials that needed to be somewhat "guessed" in the recent past. Simulations are possible to predict the physical properties of foods under various conditions, such as temperature and moisture levels, in terms of the mass fraction of water, protein content, or fat levels. The traditional principles of chemical engineering known collectively as transport phenomena, coupled with simulation and analytical methods to predict food properties, have allowed more appropriate application of such principles in numerous processes in the food industry.

In the food industry, many processes have relied on typical unit operations based on heat transfer and mass transfer phenomena. The ultimate goal in the processing of food materials has been traditionally related to the need to control microbial populations that may become a threat to public safety. This is why heat transfer, normally coupled with mass transfer, still represents the main principle on which unit operations used for preservation of foods is based. The severity of thermal processing, normally expressed as temperature and holding time, is directly proportional to safety, but indirectly proportional to quality as a whole. Principles of chemical kinetics dictate that, in general, the reaction velocity roughly doubles with a temperature increase of 10°C. A number of biochemical reactions triggered by an increase in temperature in food processing will cause irreversible effects on the nutritional and sensory aspects of the foods being processed. The first upsurge of alternative technologies aimed at maintaining food quality of processed foods, while rendering them safe for human consumption, was based on manipulation of the temperature–time relationship. A second stage focused on seeking a solution for the safety–quality dilemma in food processing is represented by a number of proposed technologies relying on antimicrobial factors different from heat. Satisfactory evaluation of novel preservation technologies depends on reliable estimation of many aspects, including microbiological, physicochemical, nutritive, and sensory characteristics. The evaluation of processing effects on the safety and quality of preserved food materials is a task needed to determine the viability of novel techniques as feasible alternatives to traditional processes in order to meet consumer demands.

1.4　Categorizing Food Processing and Preservation

The systematic study of food processing and preservation can be focused on dealing with the different technologies for food manufacturing, such as those listed in Table 1.1, or treating the common unit operations used in different food industries, regardless of the raw materials being used. Programs of study in higher-education institutions normally offer courses on the different food processing technologies, e.g. grain processing, sugar and confectionary technology, and fruit and vegetable technology, but also include courses on the different food engineering operations,

Table 1.2 Main categories of food processing and preservation operations

Unit operation	Example
Preliminary processes	
Classification	Sorting, grading
Depuration	Cleaning, peeling, deskinning, dehusking
Preparation	Coring, filleting, trimming, stemming
Physical conversions	
Size reduction	Cutting, slicing, dicing, shredding, milling, grinding, pulverizing
Size enlargement	Agglomeration, granulation, instantization, coagulation, flocculation
Shaping and reshaping	Sheeting, tabletting, flaking, puffing
Mixing and emulsification	Blending, dispersing, aerating, kneading, homogenization
Mass transfer operations	Extraction, leaching, rehydration
Others	Encapsulation, coating
Chemical conversions	
Biochemical	Fermentations, enzymic conversions
Purely chemical	Caramelization, alkalinization, acidification (pickling)
Preservation techniques	
Elevated temperature	Blanching, pasteurization, sterilization, canning, dehydration
Ambient temperature	Salting/brining, irradiation, high-pressure processes, high-intensity-field processes
Low temperature	Chilling, freezing, freeze-concentration, freeze-drying
Others	Packaging, coating, waxing
Separation techniques	
Solid–solid	Screening, leaching
Solid–gas	Cyclonic separation, air filtration, air classification
Solid–liquid	Sedimentation, centrifugation, filtration, membrane separations
Liquid–liquid	Distillation, extraction

such as freezing and drying. The term "unit operation" is used to describe a physical and/or mechanical procedure occurring parallel to a chemical reaction known as a unit process, which happens in diverse material processing industries. In many industrial plants that process materials, a fundamental aspect of their operation is a chemical reaction known as a unit process, such as oxygenation, hydrogenation, or polymerization. In order for a particular reaction to be performed, a series of controlled steps to create optimal conditions are required. These steps or maneuvers are physical and/or mechanical processes, such as evaporation, distillation, and pulverization, and are known as unit operations. About 20 unit operations are commonly employed in the chemical processing industry (Green and Perry 2008; McCabe et al. 2005), whereas in the food manufacturing industry that number is at least twice as large (Brennan et al. 1990; Fellows 2000; Zeuthen and Bogh-Sorensen 2003). Owing to the complexities of food processes discussed above, a standardized classification of unit operations for the food industry would be difficult to establish. A list of unit operations as normally utilized in the food industry is given in Table 1.2. These operations may be broadly divided into the following four categories: preliminary processes, conversion operations, preservation techniques, and separation techniques. Several of the food operations listed could appear in more than one category, so the classification is not exhaustive. Whereas some of the listed unit operations are common to either chemical or food processes (e.g., distillation and extraction), differing only in the raw materials used, some others can be considered specific to the food industry, such as puffing. The specificity lies mainly in the fact that, for some food-oriented unit operations, chemical engineering principles and concepts, or even idealized models, have not even been proposed. A number of unit operations that have been well identified in food processing are all those based on heat treatment. Pasteurization, sterilization, or evaporation can be considered unit operations of wide application in different industries and, as such, they rely on chemical engineering fundamentals.

In the attempt to substitute heat in food processing, at least partially, the so-called emerging technologies or non-thermal processes constitute some of those unit operations in the food industry that are still at the stage of developing theoretical fundamentals. All these types of technologies have appeared in the food technology arena in the recent decades and they have shaped and modified the classifications or categorizations previously described. Many of these emerging technologies would be better described as non-predominant, non-conventional, or simply, ambient (room) temperature food processing technologies. Whereas some ambient temperature food processing technologies are really traditional operations, such as mixing and comminution, some others are definitely novel as applied to food materials. Among these, some specific technologies such as ultrahigh hydrostatic pressure and high-voltage pulsed electric fields have been extensively studied in recent years, and are considered potential methods for industrialization and wide commercialization. It is not foreseen, of course, that they may displace thermal processing technologies totally. It is more likely that they will compete with or complement the traditional food preservation technologies, in order to provide consumers with fresh-like, safe, and nutritious food products. This book reviews and discusses all the unit operations used for food processing and preservation, whether novel or not, with the common feature of not relying on heat as the food preservation factor or the food transformation medium.

References

Brennan JG, Butters JR, Cowell ND, Lilly AEV (1990) Food Engineering Operations. Elsevier Science Publishers, London.

Fellows PJ (2000) Food Processing Technology: Principles and Practice. Woodhead Publishing Ltd, Cambridge.

Green DW, Perry RH (2008) Perry's Chemical Engineers' Handbook. 8th Ed. McGraw-Hill, New York.

McCabe WL, Smith JC, Harriot P (2005) Unit Operations in Chemical Engineering. 7th Ed. McGraw-Hill, New York.

Zeuthen P, Bogh-Sorensen L (2003) Food Preservation Techniques. Woodhead Publishing Ltd, Cambridge.

Chapter 2
Common Preliminary Operations: Cleaning, Sorting, Grading

2.1 Main Characteristics of Raw Materials in the Food Industry

The raw materials for the food industry are usually collected or harvested in farms and open land, and the means of transporting them to the processing plants take many forms. Raw materials arriving at processing factories are, therefore, exposed to various contamination sources, so the logical order of preliminary operations is cleaning, sorting, and grading. Transporting of raw materials includes such varied operations as hand and mechanical harvesting on the farm, refrigerated trucking of perishable produce, box car transportation of live cattle, and pneumatic conveying of flour from a rail car to bakery storage bins. Throughout such operations, emphasis must be given to maintaining sanitary conditions, avoiding product losses, maintaining raw material quality, minimizing bacterial growth, and timing all transfers and deliveries so as to shorten the holdup time, which can be costly as well as detrimental to product quality.

The suitability of a raw food material for preliminary treatment is determined by a thorough assessment of its main characteristics, such as availability, physical properties, and functionality. Food processing is seasonal in nature as the availability of raw materials is a function of well-established harvest times of specific crops, and the demand for some processed products varies depending on the time of the year, for example, ice cream and salads are in greater demand during the summer months. As just stated, an important number of raw materials used in the food industry are seasonal harvested crops, such as fruits, vegetables, cereals, and legumes, so the most important physical properties to consider for further processing are those related to solid pieces of fairly regular size and shape. Some of the relevant physical properties of solid foods are listed in Table 2.1; many of these properties are influenced by the chemical composition of the food, particularly its moisture content.

Food pieces with regular geometry are easier to handle and better suited to high-speed mechanization. Also, the more uniform the geometry, the less rejection and

E. Ortega-Rivas, *Non-thermal Food Engineering Operations*,
Food Engineering Series, DOI 10.1007/978-1-4614-2038-5_2,
© Springer Science+Business Media, LLC 2012

Table 2.1 Physical properties of solids

Appearance, size, shape, color
Specific gravity, density
Rheological properties: texture, plasticity, elasticity, hardness
Thermal properties: specific heat, latent heat, thermal conductivity
Electrical conductance or resistance, dielectric constant
Diffusion and mass transfer characteristics

wastage in cleaning operations such as automatic peeling, trimming, and slicing. For example, potatoes of smooth shape and with shallow eyes are preferred for mechanical peeling and washing, and smooth-skinned fruits or vegetables are more easily washed than ribbed varieties and less likely to have insects and fungi on their surface. Size and shape are inseparable properties, but are very difficult to define mathematically in solid food materials. Mohsenin (1989) discussed numerous approaches by which the size and shape of irregular food units may be described. These include the development of statistical techniques based on a limited number of measurements and more subjective approaches, including visual comparison of units with charted standards. Uniformity of size and shape is vital for most operations and processes; for example, the uniformity of wheat kernel size is crucial for flour milling. Specific surface may be considered a relevant expression of geometry especially for surface-related matters, such as the economics of fruit peeling. Surface defects and irregularities complicate accurate quantification of geometrical properties and represent cleaning and handling problems, apart from possible yield loss. The cost of labor and energy can be improved considerably by adding a step for imperfection removal, so selection of specific varieties with a minimum defect level is normally advisable.

Color and color uniformity are important attributes in fresh foods and play an important role in consumer choice, but may be less important in selecting raw materials for further processing, especially severe thermal processing. For example, for green vegetables such as peas, spinach, and green beans, heating results in a change color from bright green to dull olive green owing to conversion of chlorophyll to pheophytin, and potatoes and other vegetables are subject to browning during heat processing owing to the Maillard reaction. Some treatments may be suggested to diminish color alteration caused by processing of different foods of vegetal origin, but texture changes, mainly softening, may be a consequence of such treatments. Therefore, adding colorants to different thermally processed vegetal foods may be considered a more practical solution. For low-temperature processing, such as chilling or freezing, color changes will be minimal, and color uniformity of the raw material will represent a reasonable index of its suitability.

Texture represents, perhaps, the most changed attribute of raw food materials in any processing step in the food industry, and such changes can be caused by a great variety of effects. Water loss along with protein denaturation, which may result in loss of water-holding capacity, or coagulation, hydrolysis, and solubilization of proteins, are two of the most important factors accounting for textural damage in processed foods. The raw material should, therefore, be robust enough to withstand the mechanical stresses in preliminary operations, such as abrasion during cleaning of fruits and vegetables. Fruit and vegetable varieties with improved mechanical strength have been developed, including tough-skinned peaches and tomatoes suited to mechanical peeling, washing, and sorting. Some other examples are the selection of blackcurrant varieties suitable for mechanical strigging and pea and bean varieties able to withstand mechanical podding. Texture depends not only on the variety of the raw material, but also on its degree of maturity, and may be assessed by sensory panels or commercial instruments. Prior to preliminary or further treatment, raw materials should be tested daily and harvested or taken for processing when at their optimum consistency.

The functionality of a raw material can be interpreted as the combination of properties that will determine product quality and process effectiveness. Such properties may differ greatly for different raw materials and processes, and may be measured by chemical analysis or analytical testing. There are many examples of varieties developed for special purposes, such as cattle bred for meat or milk and sheep raised for wool or meat. Wheat for different purposes may be chosen according to the protein content, with hard wheat (11.5–14.0% protein) being suitable for white bread, whereas wheat with higher protein content (14.0–16.0% protein) is required for whole-wheat bread (Chung and Pomeranz 2000). In contrast, soft or weak flours with lower protein levels (8.0–11.0% protein) are best suited for chemical leavened products with a lighter or tender structure, such as biscuits, cakes, and pastries. Selection of raw materials on a functional basis usually involves process testing of varieties to assess function by chemical or physical testing, or a combination of both. Examples are the above-mentioned case of wheat varieties, which can be tested for their suitability to produce flours for bread, cake, or pastry manufacturing, and the analyses for fat and protein in milk for suitability in manufacturing cheese, yogurt, or cream. In practice, food processors will define their requirements in terms of raw material specifications for any process for the raw material that arrives at the factory gate. Acceptance and the price paid for the raw material may depend upon the results of specific testing. Milk deliveries will be routinely tested for hygienic quality, somatic cells, antibiotic residues, extraneous water, fat level, and protein content. For fruits and vegetables, processors may issue specifications to account for the size of units, the presence of extraneous vegetable matter, or the level of specific defects.

2.2 Cleaning of Raw Materials

All raw materials in the food industry must be cleaned before processing. The main purpose is to remove contaminants, which range from innocuous to hazardous. Removal of contaminants is a necessity, not only for the sake of the final consumer, but also to protect processing equipment. For example, is vital to eliminate sand, stones, or metallic objects from grains and cereals prior to milling, in order to avoid serious damage to machinery such as roller mills. Contaminants on raw food materials can be of different origins:

- Mineral: soil, stones, sand, metal, oil
- Plant: twigs, leaves, husks, skins
- Animal: feces, hair, insects, eggs
- Chemical: pesticides, fertilizers, other contaminants
- Microbial: yeasts, molds, bacteria, metabolic by-products

Removal of all these types of contaminants should be guaranteed before raw food materials are sent for further processing. Important considerations in the cleaning stage are the higher possible efficiency of contaminant removal, the minimal loss or damage to the raw materials being cleaned, and the precautions needed to avoid recontamination after cleaning. Dry and wet cleaning are the two main methods for removal of contaminants and preparation for further processing in a food manufacturing plant. Dry cleaning methods include screening, abrasion, brushing, aspiration, magnetism, and electricity, whereas wet cleaning methods comprise soaking, spraying, fluming, and flotation. Combination of these methods is possible to obtain the best attainable results.

Dry cleaning methods have the advantages of relatively low cost and convenience while leaving the cleaned surface in a dry condition. A great disadvantage is, however, the spread of dust with risk of recontamination and potential dust explosion hazard. Careful control of dust is important in the food industry generally, but especially in dry cleaning since dusty conditions may arise in conjunction with potential spark generators, such as stones and tramp metal within the raw material as extraneous matter. Measures that may be taken to minimize explosion risk and hazard include dust-proofing of equipment, removal of dust by gas–solid separators such as cyclones, rigorous housekeeping, and spark-proofing fittings.

Screening is widely used for removing contaminants considerably different in size from the foods being treated. Primary screens are basically sorting machines that may also be used for cleaning purposes. As shown in Fig. 2.1, flatbed screens consist of one or more flat sieve decks fixed together in a dust-tight casing with the assembly shaken by different devices and normally including hard rubber tapper balls between the decks to minimize blinding of the sieve aperture. Flatbed screens constitute an excellent choice for cleaning of fine materials, such as flour and ground spices, since they are not easily blinded, although the units may be easily accessible for frequent cleaning and maintaining. Rotary drum screens, often referred to as trommels, centrifugal screens, or reels, are continuous units with numerous applications in the food industry, including cleaning of raw materials. Cleaning may be done so as to retain oversize extraneous material from flour, salt, or sugar, while discharging a clean product. Alternatively, the discharge material may be the contaminant and the oversize clean material the retained material as, for example, in the removal of weed seeds, grit, and small stones from cereals. Screens of this kind are relatively inexpensive to install and operate but, unless carefully designed, they are difficult to clean, with the consequent risk of recontamination.

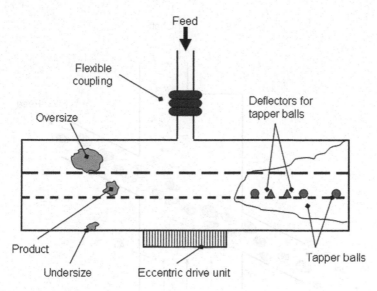

Fig. 2.1   A flatbed screening unit

Abrasion cleaning represents another dry cleaning alternative in the food industry. The technique consists in promoting abrasion between pieces of food, or between the food and moving parts of cleaning machinery, in order to loosen or remove adhering contaminants. Diverse equipment, such as trommels, tumbling drums, vibrators, abrasive discs, and rotating brushes, have been adapted and used for abrasion cleaning of raw materials in different food processing industries. Special care must be taken when operating abrasion cleaning machinery, so as to avoid recontamination, protect operators, and prevent dust explosions.

Aspiration cleaning, also known as winnowing, is a cleaning alternative for removing debris differing in buoyancy from that of a raw food material. Aspiration cleaning is, in fact, an application of air classification where separation is based mainly upon particle size, although other particle properties such as shape and density may play a role. The general principle of air classification involves a particulate system subjected to an upward air stream at constant and uniform velocity. Some of the particles will become fluidized, some will be conveyed and carried away by the air stream, and others will remain stationary. A degree of separation is achieved, in consequence, and the remaining material can then be subjected to a higher velocity, removing another fraction, and this process can be continued. The forces participating in this simple process are the drag forces acting on the particles due to the air stream, which counteract the force due to gravity. The same separating effect would be achieved if a mixture of particles approximately equal in size, but with significantly different density, were subjected to the air stream. If, the air stream is fed with cereals grains and their accompanying chaff instead of with particles of different size (or different density), the separation effect will be the same as that illustrated in Fig. 2.2. Winnowing has been, therefore,

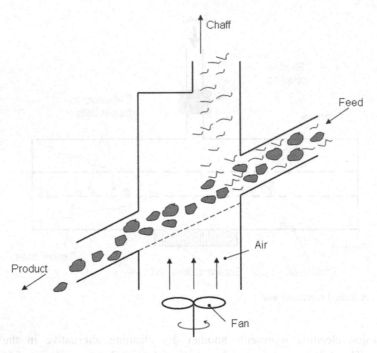

Fig. 2.2 Air classifier used for aspiration cleaning

traditionally used to separate chaff from grains after threshing, and represents one of the simplest forms of air classification. Aspiration cleaning is extensively used in combine harvesters, in pea viners, and in similar machines. The aspiration principle is also used to clean onions, melons, eggs, and some other foods not suitable for wet cleaning. The soil is loosened by abrasion using brushing or contacting with rotating pintles, and the debris is then selectively removed by air streams (Ryall and Pentzer 1982). Aspirator cleaners are capable of very precise adjustment, other important applications are the removal of bran particles from flour and, even, the discrimination between protein and starch fragments in the production of protein-enriched flours.

Magnetic materials can be removed by powerful magnets, whereas non-ferrous metals, such as aluminum, are detected by passing the material through a strong electromagnetic field which is distorted and sends a warning signal. X-rays have also been used for products in sealed containers, such as cocoa, prone to contamination by iron from hammer mills, impact mills, and the agitator blades of rotating mills. Magnets are usually located in the conveyor trunking, but may also take the form of stationary or rotating magnetic drums, magnetized belts, magnets located over belts carrying the food, or staggered magnetized grids through which the food is passed. Electrostatic methods for cleaning materials are also available, taking advantage of the differences in the electrostatic charge of the material under controlled-humidity conditions. The raw food material is fed from a hopper onto a rotating drum, which is either charged to a potential of 5–20 kV or earthed, and the oppositely charged particles are separated as they are more strongly attracted to the drum. They may then be removed from the drum by a scraper. This electrostatic method can be used to eliminate dust and stalk from tea fannings, as well as some unwanted seeds from cereals and oilseeds.

Wet cleaning methods are also extensively used as the first stage of processing of raw materials in the food industry. Wet cleaning can be highly efficient for heavily soiled raw materials, mainly fruits, vegetables, and tubers. A disadvantage of wet cleaning is the large quantity of high-purity water required and the same quantity of dirty effluent produced. Detergents and sanitizing chemicals, such as chlorine, citric acid, and ozone, are commonly used in wash waters, especially in association with peeling, when reducing enzymic browning may also be of interest (Ahvenian 2000). Soaking, spray washing, and flotation washing are the main technologies employed in wet washing of raw food materials.

Heavily soiled fruits and vegetables can be simply presoaked in water, using agitation to assist the process. Soak tanks are constructed out of metal, smooth concrete, or other construction materials suitable for regular cleaning and disinfection. Gridded bottom outlets are provided for the removal of heavy soil, whereas side outlets may be fitted to allow the removal of light debris, which would otherwise drain back into the cleaned material, causing recontamination.

A more efficient wet cleaning procedure is spray washing, which uses high-pressure sprays and requires smaller volumes of water. Spray belt washers and spray drum washers are the two most commonly employed designs of spray washers. The former are, simply, conveyor perforated belts, which carry the food pieces beneath banks of water sprays. The latter consist of reels made of metal slats or rods, spaced in order to retain the food while debris is washed through, and equipped with a central spray rod fitted with jets or slots to spray washing water radially onto the food pieces tumbling inside the reel. The reel rotates slowly and is inclined to the horizontal to allow the food pieces to be sprayed continuously. In this design the abrasion occurring may be useful in loosening of dirt, but may also cause textural damage to some foods.

Flotation washing is based on the difference in buoyancy between the desired and undesired parts of the raw food materials being treated. For example, if harvested fruit is immersed in a large volume of water, heavy pieces and particles such as soil, metal, glass, and rotten fruit will sink, whereas the sound fruit pieces are naturally buoyant and will float. Fluming fruit in water over a series of weirs will, therefore, result in very effective cleaning not only of fruit, but also of some other foods such as peas and beans. A diagram of a flotation separation system is shown in Fig. 2.3.

Wet cleaning normally leaves the treated product with an excessive amount of water, so removal of this is frequently required. Dewatering may be accomplished by passage of the cleaned raw food material over vibrating screens, or by the use of dewatering reels, such as specially designed rotary screens. Dewatering centrifuges may also be used and, occasionally, it may be necessary to resort to drying

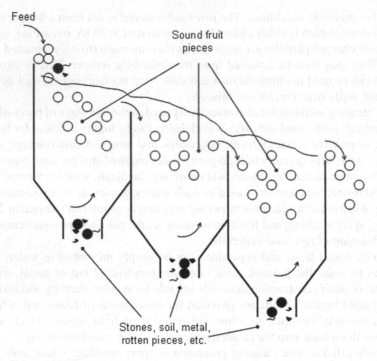

Fig. 2.3 Principle of flotation washing

procedures, like in the case of washed cereals or wet-cleaned fruit, which is to be stored or sold as a finished product.

As previously mentioned, cleaning methods are often used in combination and many pieces of cleaning equipment involve several stages mounted as a single unit. For example, bean or pea washers consist, usually, of a soak tank linked to a spray drum washer followed by a dewatering screen. Cleaning screens may normally be combined with an aspiration device and with a magnetic separator. A typical example of a combined cleaning procedure is the pretreatment of wheat for milling into flour, a diagram of which is presented in Fig. 2.4. The stages of the process include a magnet, a screen, an aspirator, a disc separator, a washer, a centrifuge, and a drier. In combined cleaning procedures, interstage handling methods require careful design and selection in order to control material damage, effluent volumes, and handling costs.

2.3 Dehulling and Peeling

Legumes are sometimes initially processed by removing the seed coat or hull and splitting the seed into its dicotyledonous components in order to reduce fiber and tannin content, as well as to improve appearance, cooking quality, texture,

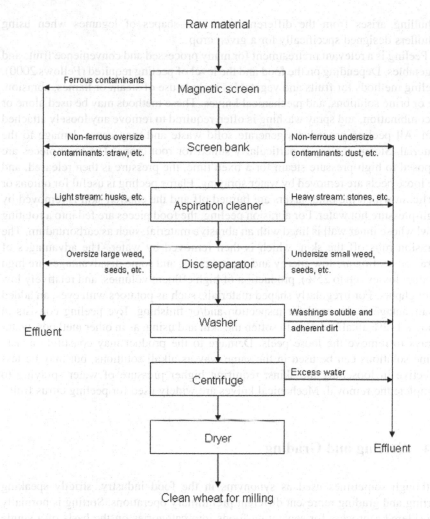

Fig. 2.4 Cleaning procedure for wheat prior to milling

palatability, and digestibility. Removing hulls from many legumes is awkward and tedious work and is often performed manually by previous soaking and subsequent drying. Totally removing the hulls from legumes can only be achieved manually, but equipment can be used to facilitate the operation. For most commercial applications, dehulling can be approached by employing abrasion and attrition mills. Attrition mills are normally used for legumes where the hull is not firmly attached to the seed. For abrasion mills the yields are much lower than the theoretical yields, and losses are higher as cotyledon material (on the order of 30%) is lost with the hull. Soaking methods and residual mechanical hull removal are methods still used to evaluate the efficiency of hull removal. A typical problem with dehulling arises from the different sizes and shapes of legumes when using dehullers designed specifically for a given crop.

Peeling is a relevant pretreatment for many processed and convenience fruits and vegetables. Depending on the food and the level of peeling required (Fellows 2000), peeling methods for fruits and vegetables include use of steam or flames, abrasion, lye or brine solutions, and mechanical knives. These methods may be used alone or in combination, and spray washing is often required to remove any loosely attached peel. All peeling procedures generate solid waste and may cause damage to the material. Steam peeling is particularly suited for root crops. The food pieces are exposed to high-pressure steam for a fixed time, the pressure is then released, and the loose peels are removed by water spraying. Flame peeling is useful for onions or garlic, in which the outer layers are burned off and the charred skin is removed by high-pressure hot water. For abrasion peeling, the food pieces are fed into a rotating bowl whose inner wall is lined with an abrasive material, such as carborundum. The abrasion rubs off the skin, which is then removed by water. The advantages of abrasion peeling are low energy and capital costs, and some disadvantages are high product losses (up to 25%), production of high effluent volumes, and relatively low throughputs. For irregularly shaped materials, such as potatoes with eyes, an added disadvantage is the need for inspection and/or finishing. Lye peeling consists in using a 1–2% alkali solution to soften the skin and using, as in other methods, water sprays to remove the loose peels. Damage to the product may constitute a risk. Brine solutions can be used in the same way as alkali solutions, but may be less effective in loosening the skins, requiring higher pressure of water spraying to complete the removal. Mechanical knives are widely used for peeling citrus fruits.

2.4 Sorting and Grading

Although sometimes used as synonyms in the food industry, strictly speaking sorting and grading represent different preliminary operations. Sorting is normally considered a process for separating foods into categories on the basis of a single physical property, such as size, shape, weight or color. Grading, on the other hand, is a separation with a number of factors to be taken into account to accomplish it. Some examples are color, absence of blemishes, flavor, and texture. Food grading is usually performed manually, by trained personnel, because it is not normally possible to link quality with one particular physical attribute.

Before any step is taken for sorting or grading, care should be taken to avoid any sort of damage, which may result in substantial economic loss to the processor. The emptying of containers onto sorting belts and falls from sorters can cause extensive product damage. Different devices have been used to minimize damage on emptying raw materials onto sorting equipment, ranging from simple padded collection chutes to more sophisticated arrangements, such as pivoted bins. Air cushioning is another alternative suggested to minimize drop damage of raw food materials, and investigations aimed at controlling drop damage aerodynamically showed typical

Table 2.2 Terminal settling velocities of different fruits

Fruit	Terminal velocity (m/s)
Apples	44
Apricots	35
Blackberries	19
Cherries	25
Peaches	44
Plumbs	35

Table 2.3 Types of screens used for food sorting

Fixed aperture	Variable aperture
Stationary	Roller
Vibratory	Cable or belt
Rotary	Belt and roller
Gyratory	Screw
Reciprocating	

values for terminal settling velocities of some fruits as given in Table 2.2. The data shown suggest that falls of only a few centimeters are sufficient to damage many raw materials, so, therefore, to be fully effective air cushioning requires upward air streams of similar magnitudes to the terminal velocities.

Sorting is conducted on the basis of physical properties, and details of the principles and equipment were given by Sarvacos and Kostaropolous (2002), as well as by Peleg (1985). No sorting system is absolutely precise, and a balance between precision and flow rate is often pursued. Means of sorting include weight, size, shape, and color sorting.

Weight sorting is usually a very precise method since is not dependent on the geometry of the product. Diverse types of foods, such as eggs, fruits, and vegetables, can be weight-sorted using spring-loaded, strain gauge or electronic weighing devices incorporated into conveying systems. By utilization of tipping or compressed air mechanisms set to trigger at progressively lesser weights, heavier items are removed first, followed by the next lighter category, and ending with the lightest portion of a particular load. An alternative for weight sorting uses a catapult principle, in which units are thrown into different collecting chutes, depending on their weight, by spring-loaded catapult arms. A general disadvantage of weight-sorting methods is the relatively long time required per sorted item.

Size sorting is not as precise as weight sorting, but is considerably cheaper. Size categories include different physical dimensions, such as diameter, length, and projected area. The diameter of spheroidal units, such as tomatoes and citrus fruits, is conventionally considered to be orthogonal to the fruit stem, whereas the length is considered to be coaxial. Rotating the units in a conveyor can, therefore, make size sorting more precise. Sorting into size categories can be conveniently performed using some type of screen, many designs of which were discussed by Slade (1967). The screen designs most commonly used for food sorting are listed in Table 2.3.

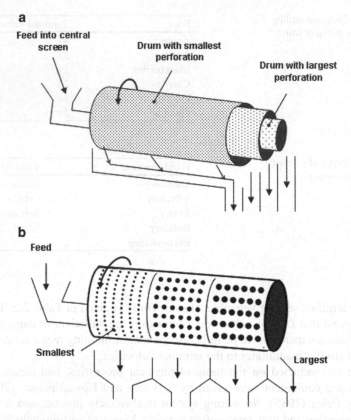

Fig. 2.5 Drum screens for sorting of foods: (**a**) concentric drum screen, (**b**) series-type, consecutive drum screen

Sorting by size of fruits and vegetables is extensively performed in flatbed screens as well as in trommels or drum screens. Single-deck flatbed screens are used for preliminary sorting of potatoes, carrots, and turnips, whereas multideck screens of this type find extensive use in the size-sorting of raw materials such as cereals and nuts, as well as in part processes and finished foods such as flour, sugar, salt, herbs, and ground spices. Drum screens are used extensively as size sorters for peas, beans, and similar foods which will withstand the tumbling action produced by rotation of the drum. Drum sorters are usually required to separate the feedstock into more than two streams and, thus, two or more screening stages are needed. To achieve this, the screens may be arranged to operate concentrically or consecutively. The concentric drum screen illustrated in Fig. 2.5a has the advantage of compactness but, because it is fed at the center, the highest product loading goes through the smallest screen area. The series-type consecutive drum screen shown in Fig. 2.5b has the disadvantage of requiring a large floor area.

Also, since the feed enters at the end which has the smallest aperture screen, the whole screen tends to become overloaded at the inlet end, resulting in inefficient sorting. There is a parallel-type consecutive drum screen arrangement which overcomes the disadvantages of the previously described designs by first contacting the inlet material with the large-aperture screen, leaving the following smaller-aperture screens to deal with a reduced quantity of undersize material. Another type of drum screen, which is reported to reduce damage during pea sorting, uses spaced, circumferential, wedge-section rods instead of perforated-screen drums. The spacing of these rods increases in steps from the inlet to the outlet, giving a series-type consecutive system. Built-in flights ensure smooth transfer of the peas through the sorter.

Shape sorting is useful in cases with the raw food pieces are contaminated with particles of similar size and weight. Such a case is particularly common for cereal grains, which may contain different seeds. The principle of shape sorters for this type of grain is that discs or cylinders with accurate shape indentations will pick up seeds of the correct shape when they are rotated through a stock, whereas seeds with different shapes will remain in the feed.

Foods can be sorted on the basis of their color, for example, when removing discolored baked beans prior to a blanching process. One of the most common applications is to pick out wrongly colored pieces, and the simplest method is by visual inspection as the food on conveyor belts passes by trained operators. Manual color sorting is also performed on sorting tables, but is an expensive procedure. The process can be automated using highly accurate photocells, which compare the reflectance of food pieces with preset standards and can eject defective or wrongly colored pieces by using a blast of compressed air, as illustrated in Fig. 2.6. This system can be used for small particulate foods, such as navy beans and maize kernels for canning, or nuts, rice, and small fruits. Extremely high throughputs (up to 16 t/h) have been reported (Fellows 2000), and if more than one photocell positioned at different angles is used, blemishes on larger food pieces, such as potatoes, can be detected. Color sorting can also be used to separate materials to be processed separately, such as red and green tomatoes. It is also feasible to use transmittance as a basis for sorting although, as most foods are completely opaque, very few opportunities are available. The principle has been used for sorting cherries with and without stones, and for internal examination of eggs.

Grading refers to classification based on quality and incorporates commercial value, end use, and official standards. As previously mentioned, grading is mainly performed by trained operators who provide judgment, making use of sets of charted standards or manufactured models. For example, a fruit grader can judge simultaneously shape, color intensity, color uniformity, and degree of russeting in apples. Egg candling involves inspection of eggs spun in front of a light so as to a detect flaw, such as shell cracks, diseases, blood spots, or fertilization. Experienced candlers can, apparently, grade thousands of eggs per hour. Grading by use of instruments may only be possible when the quality of food can be linked to a single

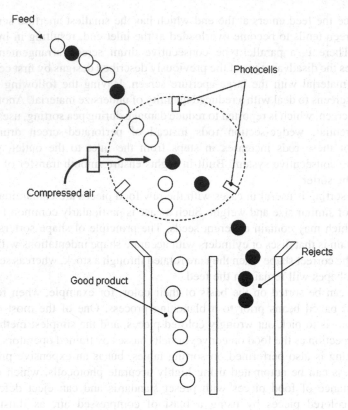

Fig. 2.6 Principle of color sorting equipment

physical property and, thus, a sorting operation will result in different grades of the food material being obtained. As an example, the size of specific varieties of peas is related to tenderness and sweetness, so size sorting results in different quality grades. Grading of foods can also be related to the quality of a batch by systematic sampling. Again, trained personnel, or human graders, may assess the quality of random samples of foods, such as cheese and butter, for a number of criteria. Some quick analytical testing can aid the inspector's decision on sampling grading. There is vested interest in developing rapid, non-destructive methods to assess quality aspects of foods for sorting and grading. Cubeddu et al. (2002) described the potential application of advanced optical techniques to provide information on surface and internal properties of fruits, including textural attributes and chemical composition, to allow classification in terms of maturity, firmness, and presence of defects. Another approach is represented by the potential use of sonic techniques to evaluate the texture of fruits and vegetables (Abbot et al. 1992). Similar applications of X-rays, lasers, infrared rays, and microwaves have also been discussed (Sarvacos and Kostaropolous 2002).

References

Abbot JA, Affeldt HA, Liljedahl LA (1992) Firmness measurement in stored "Delicious" apples by sensory methods, Magness-Taylor, and sonic transmission. J Am Soc Hort Sci 117: 590–595.

Ahvenian R (2000) Ready-to-use Fruit and Vegetable. Fair-Flow Europe Technical Manual F-FE 376A/00. Fair-Flow, London.

Chung OK, Pomeranz Y (2000) Cereal processing. In: Nakai S, Modler HW (eds) Food Proteins: Processing Applications, pp 243–307. Wiley-VCH, Chichester, UK.

Cubeddu R, Pifferi A, Taroni P, Torricelli A (2002) Measuring food and vegetable quality: advanced optical methods. In: Jongen W (ed) Fruit and Vegetable Processing: Improving Quality, pp 150–169. Woodhead Publishing, Cambridge.

Fellows PJ (2000) Food Processing Technology: Principles and Practice. Woodhead Publishing, Cambridge.

Mohsenin NN (1989) Physical Properties of Food and Agricultural Materials. Gordon and Breach Science Publishers, New York.

Peleg K (1985) Produce Handling, Packaging and Distribution. AVI Publishing Company, Westport, CT.

Ryall AL, Pentzer WT (1982) Handling, Transportation and Storage of Fruits and Vegetables. AVI Publishing Company, Westport, CT.

Sarvacos GD, Kostaropolous AE (2002) Handbook of Food Processing Equipment. Kluwer Academic, London.

Slade FH (1967) Food Processing Plant, Volume 1. Leonard Hill, London.

Chapter 3
Handling of Materials in the Food Industry

3.1 Handling of Liquids

The liquid state is, undoubtedly, of utmost importance in the material processing industries, including the food industry. Numerous raw materials and products are in liquid form in a variety of processes. Liquid foods, such as milk, honey, fruit juices, beverages, and vegetable oils, are handled at different stages in food processing lines. More consistent food materials, such as salad dressings and mayonnaise, may also behave somewhat like liquids. There are many important properties of liquids to deal with, such as thermal conductivity and specific heat. However, the most relevant are those related to flow since most liquid foods are mainly conveyed and pumped, so their flow behavior properties are important for determining the power requirements for pumping or the sizing of pipes. The transport of liquid foods by pumps and other relevant applications in processing are directly related to their properties, especially density and viscosity.

An appropriate understanding of the flow behavior of the liquid state is important for the food process engineer, in order to design properly processes and operate them efficiently in different food-related industries. Rheology is the discipline focused on the study of deformation and flow. Flow behavior of liquids, semisolids, and even solids is described by rheological models. Rheological properties are also used as a means of controlling or monitoring processes in the food industry.

3.1.1 Classification of Fluids

According to rheology, which embraces the study of flow behavior in a very general way, there are two main types of flow: viscous and elastic. Whereas the first occur in fluids, the second is common in solids. An intermediate flow behavior is also found. Thus, perhaps the most useful classification that can be made in terms of flow behavior is by means of a spectrum extending from elastic solids at one

E. Ortega-Rivas, *Non-thermal Food Engineering Operations*,
Food Engineering Series, DOI 10.1007/978-1-4614-2038-5_3,
© Springer Science+Business Media, LLC 2012

Table 3.1 Classification of fluids

Fluid	Characteristic function	Examples
Newtonian	$\tau = \mu\dot{\gamma}$	Air, water, steam, all gases, milk, vegetable oil, honey
Non-Newtonian		
Time-independent		
Bingham plastic	$\tau = \tau_o + \mu_o\dot{\gamma}$	Toothpaste, peanut butter, butter, potter's clay, mustard, mayonnaise
Shear thickening (dilatant)	$\tau = K\dot{\gamma}^n$ $n > 1$	Quicksand, thick starch solutions, wet beach sand, fine powders in suspension
Shear thinning (pseudoplastic)	$\tau = K\dot{\gamma}^n$ $n < 1$	Paper pulp, paint, apple sauce, banana purée, orange juice concentrate
Time-dependent non-elastic		
Rheopectic	No unique function	Bentonite clay
Thixotropic	No unique function	Paints, printing inks, tomato ketchup, oil drilling mud
Viscoelastic	No unique function	Saliva, nearly all biological fluids, concentrated tomato soup, bread dough, many polymeric solutions

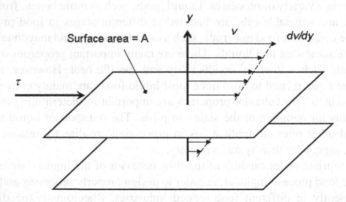

Fig. 3.1 Definition of viscosity: τ is shear stress, v is fluid velocity, and $\mathrm{d}v/\mathrm{d}y = \dot{\gamma}$ is shear rate

extreme to viscous flow at the other. Because the laws governing flow of solids differ totally from those describing flow of fluids, viscous and viscoelastic flow are normally studied together. Table 3.1 presents a summary of the main types of fluids.

Newtonian theory provides the simplest case of viscous behavior, that in which the stresses are related to the velocity gradients existing at the time of the observation. For an incompressible fluid in a simple shear, such a theory may be stated, according to Fig. 3.1, by the first function in Table 3.1. In the most general sense, any fluid response not explainable by Newtonian theory may be termed non-Newtonian. In practice, however, non-Newtonian fluids, as opposed to Newtonian ones, are those for which the viscosity is not a constant value, but is a function of the imposed shear rate.

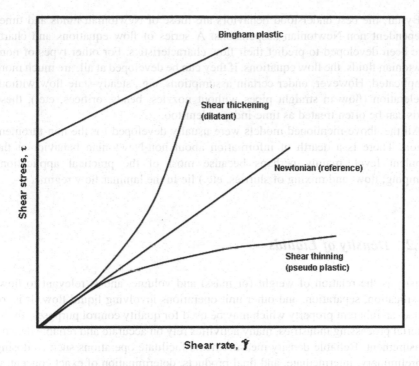

Fig. 3.2 Rheograms for different types of fluids

A wide variety of non-linear relationships between stress and shear rate have been used. Possibly, the most common one is the power law, also appearing in Table 3.1, representing the shear-thickening and shear-thinning fluids as described there. If the exponent n is less than 1, the material flows more easily the faster it is sheared, and the apparent viscosity decreases with increasing shear rate. If, one the other hand, n is greater than 1, the apparent viscosity increases with the increasing shear rate. Because they follow the power law, dilatant and pseudoplastic fluids are also known as power-law fluids. Another case of flow behavior can be considered when a solid-like material does not flow initially at all (or hardly at all), until some critical stress is reached, and then it flows in a Newtonian manner. Such a behavior characterizes the so-called Bingham plastic fluids, as can be described by the second equation in Table 3.1. Graphically, all the above-mentioned models can be represented as shown in Fig. 3.2.

There is a further approach to describe the flow behavior of systems in which the viscosity varies with time as well as with the shear rate. A material whose viscosity reduces with time is said to be thixotropic, whereas one whose viscosity increases with time is said to be rheopectic. Finally, viscoelastic behavior refers to a wide variety of effects which are somewhere between solid and fluid responses. With viscoelastic materials, the τ versus $\dot{\gamma}$ diagram only covers part of the whole behavior, so transient experiments are needed to characterize their elastic properties.

By far, the best understood behaviors are those of Newtonian fluids and time-independent non-Newtonian suspensions. A series of flow equations and charts have been developed to predict their flow characteristics. For other types of non-Newtonian fluids, the flow equations, if they can be developed at all, are much more complicated. However, under certain assumptions, e.g., steady-state flow without acceleration (flow in straight pipes without nozzles, bends, orifices, etc.), these fluids can be often treated as time-independent too.

All the above-mentioned models were usually developed for the non-turbulent region. There is a dearth of information about non-Newtonian behavior at the turbulent level, mainly perhaps because most of the practical applications (pumping, flow, and mixing of slurries, etc.) lie in the laminar flow regime.

3.1.2   Density of Liquids

Density is the relation of weight (or mass) and volume, and is relevant to flow, classification, separation, and other unit operations involving liquid flow. It is, of course, an inherent property which may be used for quality control purposes. In the material processing industries, many activities rely on accurate and reliable density measurement. Reliable density measurements facilitate operations such as dosing of preliminary, intermediate, and final products, determination of exact concentration of ingredients, monitoring of quality, and controlling of processes. Food processing often requires the combination of multiple ingredients and progressively monitoring the density enables operators to determine the exact amount of ingredient to be added.

As stated above, the density of a substance is the relation of its mass to its volume and is customary expressed in SI units in kilograms per cubic meter. This definition applies for solid, liquid, gaseous, and disperse systems, such as foams, bulk goods, and powders. The ratio of the absolute density of a material to the density of a reference material is known as specific gravity. Water at 4°C is often used as a reference material for this purpose.

Laboratory measurements of liquid density can be performed by different techniques. Pycnometry and hydrometry are the two methods most often employed to quantify density of liquids, but apparatuses such as the hydrostatic balance and the Mohr–Westphal balance are also used for specific purposes. For liquids of thick consistency, such as heavy oils and syrups, a submersion balance can be used for accurate density determination.

When pycnometry is used, by weighing of a known volume of liquid, one can measure the density of that liquid in a simple way. Glass bulbs with precisely known volume used for this purpose are termed "pycnometers," and have a marker to which the liquid sample must be carefully filled. The density of the liquid can be then calculated by

$$\rho = \frac{m_F - m_0}{V},$$
(3.1)

where ρ is the density of the liquid, m_F is the weight of the pycnometer filled with the sample, m_0 is the weight of the empty pycnometer, and V is the volume of the sample (volume of the empty pycnometer to the mark). Because of thermal expansion of the glass, it is important to note that the pycnometer volume is known only for the temperature at which it was calibrated. Careful control of the temperature for the whole procedure of determining the liquid density is, therefore, a crucial factor for accuracy of measurement.

To measure the specific gravity, the pycnometer is weighed with the sample liquid and again with the reference liquid. The ratio of both weights gives the specific gravity as

$$SG = \frac{m_F}{m_R} = \frac{m_F V}{V m_R} = \frac{\rho}{\rho_R}, \tag{3.2}$$

where SG is the specific gravity of the sample, m_R is the weight of the pycnometer filled with the reference liquid, and ρ_R is the density of the reference liquid. As can be inferred from Eq. 3.2, the advantage of this determination is that neither the weight of the empty pycnometer nor its volume needs to be known, provided that both weights are measured at the same temperature. Also, if the density of the reference material is known from the literature, the absolute density of the sample can be found by, simply, transposing it from Eq. 3.2, i.e.,

$$\rho = SG \rho_R. \tag{3.3}$$

Hydrometers are hollow glass bodies with the shape of a buoy designed with a volume to mass ratio in such a way that the glass body will float at a certain depth in the liquid whose density is being measured. Depending on the density of the liquid being measured, the hydrometer will float at a higher or lower position. The upper part of the hydrometer has a scale for reading the non-submerged part of the floating glass body. The floating depth position of the hydrometer depends on the weight force F_G and the interfacial force F_σ, which are related by

$$F_A = F_G + F_\sigma, \tag{3.4}$$

where F_A is the buoyancy force.

Substituting the definitions of the three forces, we obtain

$$V_S \rho g = mg + \sigma \pi d, \tag{3.5}$$

where V_S is the submersed volume of the hydrometer, ρ is the density of the liquid, g is the acceleration due to gravity, m is the weight of the hydrometer, σ is the interfacial tension, and d is the diameter of the neck of the hydrometer.

If the submersed volume of the hydrometer V_S is subtracted from the total volume of the hydrometer V, Eq. 3.5 will transform to

$$\left(V - \frac{\pi d^2}{4}h\right)\rho g = mg + \sigma\pi d. \tag{3.6}$$

The non-submersed length of the hydrometer will, therefore, be represented by the height h in Eq. 3.6, and can be transposed as

$$h = \frac{4}{\pi d^2}\left(V - \frac{mg + \sigma\pi d}{\rho g}\right). \tag{3.7}$$

If interfacial tension effects are not considered, the hydrometer equation simplifies to

$$h = \frac{4}{\pi d^2}\left(V - \frac{m}{\rho}\right). \tag{3.8}$$

The non-submersed length of the hydrometer h can be read with the aid of a scale on the upper part of the hydrometer. A weight at the bottom of the hydrometer acts like a keel in a sailboat to ensure that it will float in the liquid and maintain a vertical orientation.

The scale on the hydrometer can be calibrated directly into units of density. Hydrometers with special scales are available for specific applications, such as sugar solutions (saccharimeter), alcohol solutions (alcoholometer), acids (acid hydrometer), salt solutions (Baumé hydrometer), and milk (Quevenne lactometer). Figure 3.3 shows a diagram of a typical hydrometer.

For highly viscous liquids, measurement of density can be done using a submersion balance, such as that illustrated in Fig. 3.4. A beaker with the viscous liquid sample is set on a balance platform. The display value is either recorded or set to zero, and then a test body with known volume is submerged in the sample in the beaker. The buoyancy force caused by the submerged test body is transferred to the balance and appears on the display as an apparent increased weight Δm. This increased weight force equals the buoyancy force, being the weight of the displaced liquid, which equals the volume of the submersed solid body. On the basis of this principle, the following relations can be established:

$$\Delta G = F_A \tag{3.9}$$

and

$$\Delta mg = \rho V_K g, \tag{3.10}$$

where ΔG is the apparent weight force increase and V_K is the volume of the test body.

Fig. 3.3 A hydrometer

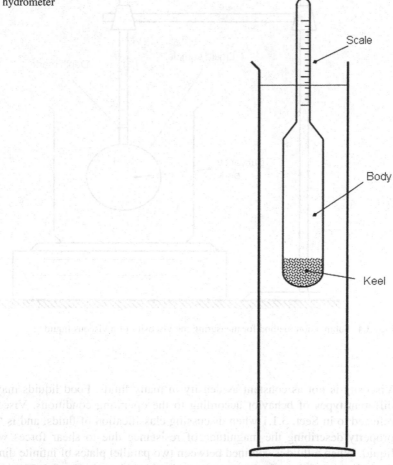

Scale

Body

Keel

Transposing from Eq. 3.10, one can determine the density of the viscous liquid ρ as

$$\rho = \frac{\Delta m}{V_K}. \tag{3.11}$$

For accuracy in measurement, the buoyancy body can be a hollow metal sphere with calibrated volume. To avoid errors from buoyancy of the mounting rod, there is normally a depth mark on it, which indicates the right position for immersion to account for the submerged section of the rod in the calibrated volume.

3.1.3 Viscosity: Definitions and Measurement

Viscosity is another important property in processing of liquid foods. It is not as easily defined as density and is more affected by variables such as temperature.

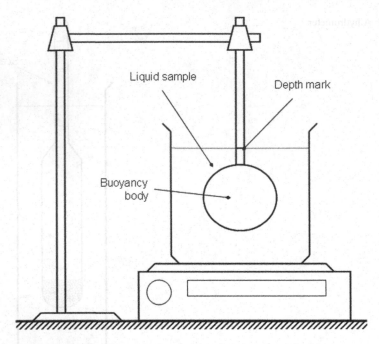

Fig. 3.4 Submersion method for measuring the viscosity of a viscous liquid

Viscosity is not as constant as density in many fluids. Food liquids may present different types of behavior according to the operating conditions. Viscosity was referred to in Sect. 3.1.1 when discussing classification of fluids, and is the liquid property describing the magnitude of resistance due to shear forces within the liquid. When a fluid is confined between two parallel plates of infinite dimensions, the influence of shear force can be conceived as shown in Fig. 3.1. The lower plate is held stationary while a force F is applied on the upper plate to produce a velocity of displacement of the plates of surface area A. The velocity near the stationary plate is zero, whereas the liquid near the top plate will be displaced at a velocity v, resulting in a velocity profile. The shear force τ on the plate surface area A will have a shear stress represented mathematically by

$$\tau = \frac{F}{A}. \tag{3.12}$$

As the distance between plates is y, the velocity gradient can be described as dv/dy (Fig. 3.1). This gradient is a measure of the shear rate applied to the fluid and can also be represented by $\dot{\gamma}$ (Table 3.1).

As listed in Table 3.1 and shown in Fig. 3.2, for an ideal Newtonian fluid, the shear stress is a linear function of the shear rate, and the proportionality constant μ is called the dynamic viscosity:

$$\mu = \frac{\tau}{\dot{\gamma}}. \tag{3.13}$$

Many liquid foods, such as milk, fruit juices, wine, and beer, exhibit Newtonian behavior. For Newtonian fluids the viscosity can be determined by applying a single shear rate and measuring the corresponding shear stress. It is normally recommended to express the dynamic viscosity in the SI unit of newton seconds per square meter, which is equivalent to pascal seconds. The corresponding unit of viscosity in the cgs system is dyne seconds per square centimeter, commonly referred to as poise.

An important number of food materials, such as fruit nectars, cream, honey, salad dressings, and syrups, follow non-Newtonian behavior; the common types of non-Newtonian fluids have already been described and some examples are listed in Table 3.1.

Another definition of viscosity is the kinematic viscosity v, represented as the relation between the dynamic viscosity and the density, i.e.,

$$v = \frac{\mu}{\rho}. \tag{3.14}$$

A general relationship to describe the behavior of non-Newtonian fluids, often referred to as the Herschel–Bulkley model, is

$$\tau = K\dot{\gamma}^{n} + \tau_0. \tag{3.15}$$

As can be seen, when τ_0 crosses the origin, the Herschel–Bulkley model describes the so-called power law fluids, i.e., the shear-thickening-type and shear-thinning-type of fluids.

The relationship given by Eq. 3.13 is also known as the apparent viscosity η, and may be represented by

$$\eta = f\dot{\gamma} = \frac{\tau}{\dot{\gamma}}. \tag{3.16}$$

From inspection of Eq. 3.16, it is clear that for Newtonian fluids the apparent viscosity and the true viscosity (also termed Newtonian viscosity) are the same. The definition of apparent viscosity is useful for non-Newtonian fluids, since by substituting Eq. 3.13 into Eq. 3.16, we obtain

$$\eta = \frac{K\dot{\gamma}^{n}}{\dot{\gamma}} = K\dot{\gamma}^{n-1}. \tag{3.17}$$

An important number of power-law liquids, such as apple sauce and orange juice concentrate (Table 3.1), are handled in the food industry. Their flow behavior can, thus, be described by the power law in order to perform calculations for pipe sizing and pump energy requirements. Under the assumptions that purely viscous behavior prevails and no slip occurs at a tube wall, the power-law equation for laminar flow has the form (Geankoplis 2003)

$$\tau_w = \frac{D\Delta P}{4L} = K'\left(\frac{8v}{D}\right)^{n'},$$

(3.18)

where τ_w is the shear stress at the wall of the tube, D is the tube diameter, ΔP is the pressure drop, L is the length of the tube, v the mean velocity, and n' is the slope of the line when the data are plotted with logarithmic coordinates. For $n' = 1$, the fluid is Newtonian; for $n' < 1$, it is pseudoplastic, or Bingham plastic if the curve does not go through the origin; and for $n' > 1$, it is dilatant. K' is the consistency index; as the name suggest, the larger its value, the thicker or more viscous the fluid.

It may be shown (Dodge and Metzner 1959) that K' is related to the analogous power-law constant K (Table 3.1) as follows:

$$K' = K\left(\frac{3n' + 1}{4n'}\right)^n \left(\frac{8v}{D}\right)^{n-n'}.$$

(3.19a)

If the fluid obeys the power law, $n = n'$ and

$$K' = K\left(\frac{3n + 1}{4n}\right)^n.$$

(3.19b)

Since Eq. 3.18 rigorously portrays the laminar flow behavior of the fluid (provided n' and K' are evaluated at the correct shear stress), it may be used to define a Reynolds number applicable to all purely viscous fluids under laminar flow conditions. This dimensionless group can be derived simply by the substitution of $D\Delta P/4L$ from Eq. 3.18 into the usual definition of the fanning friction factor, i.e.,

$$f = \frac{D\Delta P/4L}{\rho v^2/2}.$$

(3.20)

Such substitution leads to

$$f = \frac{16\gamma}{D^{n'}v^{2-n'}\rho},$$

(3.21)

where $\gamma = K'8^{n'-1}$ and all the remaining components are as already defined.

By letting $f = 16/\text{Re}$ as for Newtonian fluids in laminar flow, we can obtain the above-mentioned generalized Reynolds number as

$$\text{Re}^* = \frac{D^{n'} v^{2-n'} \rho}{\gamma}. \tag{3.22}$$

If the equation is desired in terms of K instead of K', Eq. 3.19b may be substituted into Eq. 3.22 and we obtain

$$\text{Re}^* = \frac{D^n v^{2-n} \rho}{K8^{n-1} \left(\frac{3n+1}{4n}\right)^n}. \tag{3.23}$$

For Newtonian fluids $n' = 1$ and $K' = \mu$, so γ reduces to μ and Re^* in Eq. 3.23 transforms to the familiar $Dv\rho/\mu$, showing that this traditional dimensionless group is merely a special restricted form of the more general one described here.

A plot or graph showing the relationship between shear stress and shear rate, like the one given in Fig. 3.2, is known as a rheogram. It is used to determine, in a graphical form, the values of the parameters K and n of non-Newtonian fluids, and it is the best way to characterize fluids of this type. A rheological measurement is taken by imposing a well-defined shear stress and measuring the resulting shear rate. The most commonly used instruments for achieving steady shear flow in liquids are the capillary tube viscometer and the rotational viscometer. The use of narrow-gap rheometers, such as the cone-and-plate rheometer, is limited to relatively small shear rates. At high shear rates, end effects arising from the inertia greatly complicate measurement. The edge end effects are caused mainly by the finite dimensions of the system, the shape of the free surface, the related surface tension, and fracture of the samples.

For Newtonian fluids, whose viscosities range from 0.4 to 20,000 mPa s, capillary tube viscometers also known as U-tube viscometers can be used as standard instruments for measuring viscosity. In a capillary tube viscometer, such as the one shown in Fig. 3.5, by measuring the time t for the flow of fluid of a constant volume V of a liquid of known density ρ, one can determine the viscosity from the following equation (Steffe 1996):

$$\mu = \left(\frac{\pi \rho g h R^4}{8LV}\right) t, \tag{3.24}$$

where g is the acceleration due to gravity, h is the liquid head, R is the inner tube radius, and L is the capillary section length.

Kinematic viscosity can also be calculated using a capillary tube viscometer, by measuring the time t for the liquid to drain between two etched marks on a

Fig. 3.5 Ostwald capillary
tube viscometer

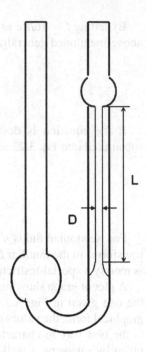

capillary tube bulb. Since all the terms within the parentheses in Eq. 3.24 are
constant for a capillary viscometer, the kinematic viscosity can be calculated as

$$v = ct. \tag{3.25}$$

The capillary viscosity constant c can be easily determined by measuring the
draining time of a fluid of known kinematic viscosity and, then, calculating the
kinematic viscosity of the test liquid by (Kawata et al. 1991)

$$v = c_1 t - \frac{c_2}{t}, \tag{3.26}$$

where c_1 and c_2 are constants for a specific capillary tube. When the kinematic
viscosities v_1 and v_2 and the flow times t_1 and t_2 of known liquids are known, the
instrument constants can be calculated by

$$c_1 = \frac{v_1 t_1 - v_2 t_2}{t_1^2 - t_2^2} \tag{3.27}$$

and

$$c_2 = \frac{(v_1 t_1 - v_2 t_2) t_1 t_2}{t_1^2 - t_2^2}. \tag{3.28}$$

Fig. 3.6 A coaxial cylinder
rotational viscometer

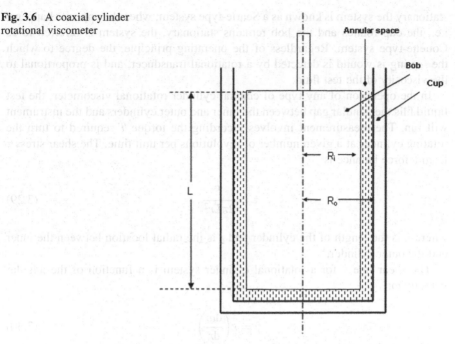

Basic diagram of a co-axial cylinder rotational viscometer

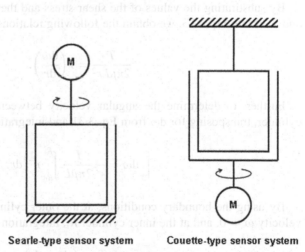

Searle-type sensor system   Couette-type sensor system

The rheological parameters of non-Newtonian test liquids are best estimated
from the shear stress and shear rate relationship data generated by the use of coaxial
cylinder viscometers (Fig. 3.6). In rotational coaxial cylinder viscometers, the
typical mounting consists of a spindle or bob immersed in a liquid sample within
a cup. Two operating possibilities arise: when the bob rotates and the cup remains

stationary the system is known as a Searle-type system; when the opposite happens, i.e., the cup rotates and the bob remains stationary, the system is referred to as Couette-type system. Regardless of the operating principle, the degree to which the spring is wound is detected by a rotational transducer, and is proportional to the viscosity of the test fluid.

In the operation of any type of coaxial cylinder rotational viscometer, the test liquid fills the annular gap between the inner and outer cylinders and the instrument will run. The measurement involves recording the torque T required to turn the rotating cylinder at a given number of revolutions per unit time. The shear stress τ from a force balance is

$$\tau = \frac{T}{2\pi L r^2},$$ (3.29)

where L is the length of the cylinder and r is the radial location between the inner and the outer cylinder.

The shear rate $\dot{\gamma}$ for a rotational cylinder system is a function of the angular velocity ω:

$$\dot{\gamma} = r\left(\frac{d\omega}{dr}\right).$$ (3.30)

By substituting the values of the shear stress and the shear rate from Eqs. 3.29 and 3.30 into Eq. 3.13, we obtain the following relationship:

$$\frac{T}{2\pi\mu L r^2} = r\left(\frac{d\omega}{dr}\right).$$ (3.31)

Further, to determine the angular velocity between the inner and the outer cylinder, transposing for $d\omega$ from Eq. 3.31 and integrating, we obtain

$$\int_0^i d\omega = -\frac{T}{2\pi\mu L}\int_{R_0}^{R_i} r^{-3} dr.$$ (3.32)

By using the boundary conditions, at the outer cylinder radius R_0, the angular velocity $\omega = 0$, and at the inner cylinder R_i, integration leads to

$$\mu = \frac{T}{8\pi^2 NL}\left(\frac{1}{R_i^2} - \frac{1}{R_0^2}\right),$$ (3.33)

where N is the rotational velocity ($\omega = 2\pi N$).

Table 3.2 lists densities and viscosities of some common liquid foods, and Table 3.3 gives details of parameters characterizing non-Newtonian food systems.

Table 3.2 Densities and viscosities of common liquid foods[a]

Material	Density (kg/m³)	Viscosity (mPa·s)
Honey	1,360	4.8
Condensed milk (75% solids)	1,300	2,160
Sucrose (20% solution)	1,070	1.92
Single-strength orange juice	1,048	2
Apple juice (60°Brix)	1,044	25
Milk (skim)	1,040	1.4
Milk (whole)	1,030	2.2
Cream (20% fat)	1,010	6.2
Cream (30% fat)	1,000	13.8
Olive oil	910	84
Soybean oil	910	40
Rapeseed oil	900	118

[a]At approximately room temperature, except for cream (at approximately 3°C)

Table 3.3 Parameters characterizing some non-Newtonian food products

Product	Temperature (°C)	K (Pa·s^n)	n	τ_0 (Pa)
Apple sauce	20	2.6	0.30	0
Banana purée (1.17°Brix)	24	6.1	0.43	0
Blueberry pie filling	20	6.1	0.43	0
Comminuted batter meat (15% fat)	15	693.3	0.16	1.53
Corn starch suspension (53% solids)	25	0.131	1.72	0
Mayonnaise	25	6.4	0.55	0
Melted chocolate	46	0.57	0.57	1.16
Mustard	25	19.1	0.39	0
Orange juice concentrate (65°Brix)	25	0.4	0.76	0
Peach purée (20% solids)	27	13.4	0.4	0
Protein concentrate from milk	5	20	1.13	0
Tomato juice concentrate (25% solids)	32	12.9	0.41	0
Tomato ketchup	25	18.7	0.27	32.0
Tomato paste (35°Brix)	25	120.0	0.30	0

3.2 Handling of Solids

3.2.1 Introductory Aspects

The properties of the solid state may be approached from different points of view. The properties of solids in large pieces differ from the properties of particulate solids. In the food industry a number of raw materials, such as fruits and vegetables, are examples of solids in discrete large pieces. Particulate solids, on the other hand, include a range of systems from coarse, intermediately sized, fine, and ultrafine powders such as flour, cocoa and icing sugar to granular materials such as cereal grains and oilseeds. The most important properties of pieces of solids include density, hardness, fragility, and tenacity (Brown 2005). The properties of particulate materials can be broadly divided into two categories (Barbosa-Cánovas et al. 2005): primary properties (those inherent to the intimate composition of the material) and secondary properties (those relevant when considering the systems as assemblies of discrete particles whose internal surfaces interact with a gas, generally air).

3.2.2 Solids in Pieces

3.2.2.1 Description of Main Properties

Apart from the density, some other relevant properties of solids in pieces, including consistency, hardness, and tenacity, can be summarized using the concept of texture. The density of solids in pieces, such as fruits and vegetables, can be determined by a setup somewhat similar to the submersion method for measuring viscous liquid viscosity described in Sect. 3.1.2. For this case, the solid piece whose density is to be determined replaces the buoyancy body, and the test liquid is some kind of liquid which will not dissolve the solid. The weight of the beaker with the liquid in it is recorded and the solid object is completely immersed and suspended at the same time, using a string, so that it does not touch either the sides or the bottom of the beaker. The total weight of this arrangement is recorded again, and the volume of the solid V_s can be calculated by

$$V_s = \frac{m_{\text{LCS}} - m_{\text{LC}}}{\rho}, \tag{3.34}$$

where m_{LCS} is the weight of the container with the liquid and the submerged solid, m_{LC} is the weight of the container partially filled with the liquid, and ρ is the density of the liquid. Knowing the volume of the piece and using its weight, one can easily compute the density.

3.2.2.2 Texture: Definitions and Measurement

The textural properties of a food material are related to its deformation, disintegration, and flow under a force. In this context, food texture can be considered to be equivalent to viscosity, as applied to solid foods instead of fluid foods. In a broader sense, food texture may be defined as a manifestation of the rheological features of a food (Pomeranz and Meloan 1994), and its different components have been proposed by Szczesniak (1983) as listed in Table 3.4. According to Szczesniak (1983), all texture-measuring devices comprise the following five main elements: (1) a driving mechanism which can range from a simple weight and pulley

Table 3.4 Texture terms and their definitions (From Szczesniak 1983)

Term	Definition
Hardness	The force necessary to attain a given deformation
Cohesiveness	The strength of the internal bonds making up the body of the product
Viscosity	The rate of flow per unit force
Springiness	The rate at which a deformed material returns to its original condition after the deforming force is removed
Adhesiveness	The work necessary to overcome the attractive forces between the surface of the food and the surface of other materials (e.g., tongue, teeth, palate)
Fracturability	The force with which the material fractures. It is related to the primary parameters of hardness and cohesiveness. In fracturable materials, cohesiveness is low and hardness can range from low to high
Chewiness	The energy required to masticate a solid food product to a state ready for swallowing. It is related to the primary parameters of hardness, cohesiveness, and springiness
Gumminess	The energy required to disintegrate a semisolid food product to a state ready for swallowing. It is related to the primary parameters of hardness and cohesiveness

arrangement to a more sophisticated variable-drive electrical motor or hydraulic system; (2) a probe element such as a cutting blade, spindle, or plunger, to come in contact with the food sample; (3) a force applied vertically, horizontally, or leveraged to cut, pierce, puncture, compress, grind, shear, or pull the food sample; (4) a sensing element; and (5) a display to show the results of the measurement.

In terms of actual instruments, Bourne (2002) classified the main methods as a function of which quantity or relation is measured, i.e., force, distance, energy, ratios, multiple units, or multiple variables (Table 3.5). He also listed chemical tests for texture and miscellaneous devices. Force-measuring instruments are commonly used in the food industry and they operate by varying the force while keeping the time and distance constant, in order to assess a number of characteristics such as puncture strength, extrusion, shear, crushing, tensile, torque, and snapping strength. Distance-measuring instruments evaluate length, area, or volume, keeping force and time constant. Time-measuring devices calculate the time required for a standard volume of fluid to flow through a restricted opening or the time required for a ball to fall a given distance through a liquid. Multiple-measurement devices determine any of various forces, distances, areas, and time, and record the results. Multiple-measurement instruments, such as the Instron universal testing machine and a number of texture analyzers, use an assortment of attachments to study a wide range of foods and their textural properties.

Stress is the force applied to an object, and which is distributed entirely throughout it. If at any point within the object a plane is drawn at right angles to this internal force, the definition of stress σ at that point is the magnitude of the force F per unit cross-sectional area A, i.e.,

$$\sigma = \frac{F}{A}. \tag{3.35}$$

Table 3.5 Classification of texture measurement methods (From Bourne 2002)

Method	Measured variable	Dimensional units[a]	Examples
Force-measuring	Force	MLT^{-2}	Tenderometer
Distance-measuring	Distance	L	Penetrometer
	Area	L^2	
	Volume	L^3	
Time-measuring	Time	T	Ostwald viscometer
Energy-measuring	Work (force × distance)		Farinograph
Ratio-measuring	Force, or distance, or time, or work, measured twice	Dimensionless	Cohesiveness
Multiple-measuring	Force, distance, and time, and work	MLT^{-2}, L, T ML^2T^{-2}	Instron
Multiple-variable	Force, or distance, or time, all vary	Unclear	Durometer
Chemical analysis	Concentration	Dimensionless (% or ppm)	Alcohol-insoluble solids
Miscellaneous	Anything	Anything	Optical density

[a]Dimensions are given in terms of mass (M), length (L), and time (T)

The different types of stresses that can be applied to materials are compressive, tensile, shear, and isotropic.

Compressive stress results when an object is placed between a pair of opposing forces pointing toward each other, as shown in Fig. 3.7a. Tensile stress, as illustrated in Fig. 3.7b, is exerted when an object is held by a pair of opposing forces pulling away from each other, so as to stretch the object. Shear stress is applied by opposite forces pulling away from each other and holding an object between them, just like tensile strength, except that they do not occur along a common axis, so the effect is to skew the object rather than stretch it. For example, it the top of a rectangular object is pulled to the right while the bottom is pulled to the left, as shown in Fig. 3.7c, the object will deform into a parallelogram. Isotropic stress comes equally from all directions, as with hydrostatic pressure. Isotropic stress is illustrated in Fig. 3.7d, and is identical to the pressure on the surface of the object. Shear stress is represented by the symbol τ, as previously described by Eq. 3.12.

Further categorization of stresses is related to the direction of the applied forces and the calculations performed to interpret them in a proper manner. In both compressive and tensile stresses, the pair of applied forces exist along a common axis, so they are classified as axial stresses. When computing the compressive or tensile stresses on an object, one normally divides the applied force by the cross-sectional area of the object, which is perpendicular to the axis of the force. Since the area is normal to the force, the stress is referred to as a normal stress. Also, when computing shear stress, one divides the magnitude of the forces by the

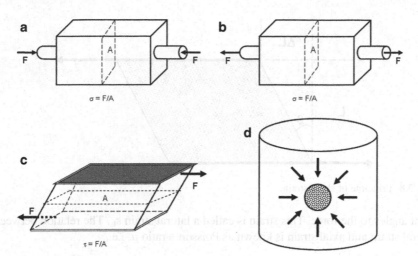

Fig. 3.7 Types of stresses applied on objects: (**a**) compressive stress, (**b**) tensile stress, (**c**) shear stress, (**d**) isotropic stress

cross-sectional area of the object that is parallel to these forces. Since the forces are tangential, and not perpendicular to the area, this stress is known as tangential stress.

Strain is another property relevant for characterization of pieces of solid food materials, and is related to the possible changes in dimensions arising from applying stresses and forces on specific objects. If an object is subjected to stress, one or more dimensions, for example, length, may be changed. The magnitude of this dimensional change is known as deformation. The strain ε is defined as the magnitude of the change ΔL divided by the original dimension L_0:

$$\varepsilon = \frac{\Delta L}{L_0}. \tag{3.36}$$

The strain defined in Eq. 3.36 is often referred to as engineering strain. True strain is given by the relationship

$$\varepsilon = \ln\left(\frac{L}{L_0}\right), \tag{3.37}$$

where L is the stressed length after elongation or compression.

When an object is subjected to compressive or tensile strain, its length will decrease or increase along the axis of the stress, and the change is known as the axial strain ε_A. If an object is stretched, it may become thinner, whereas if it is compressed, it may become thicker. Thus, for any axial strain there is usually a compensating strain at

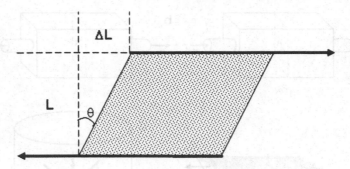

Fig. 3.8 Principle of shear strain

right angles to the force. This strain is called a lateral strain ε_L. The relation between lateral strain and axial strain is known as Poisson's ratio μ, i.e.,

$$\mu = \frac{\varepsilon_L}{\varepsilon_A}. \tag{3.38}$$

The strain can be volumetric ε_{vol} when it refers to a change in volume divided by the initial volume at gauge or absolute pressure, or can be shear strain γ when distortion in a material takes place with the opposite forces not in line with each other. Shear strain γ equals the tangent of the angle of deformation θ caused by the shear stress (Fig. 3.8):

$$\gamma = \frac{\Delta L}{L} = \tan \theta. \tag{3.39}$$

Stress and strain can be related in a number of ways. For example, an object subjected to stress and then released from it may or may not return to its original dimensions. An object that does return to its original dimensions is said to be ideal elastic. Hooke's law establishes that there is a linear response between stress and strain, and may be represented by

$$\sigma = E\varepsilon. \tag{3.40}$$

The proportionality constant E in Eq. 3.40 is called Young's modulus and is a measure of the stiffness or resistance to deformation of materials.

For volumetric changes the relationship between hydrostatic pressure and volumetric strain is known as the bulk modulus K. The ratio of shear stress to shear strain is called the shear modulus G.

Testing the properties of a solid material, including solid food materials of course, may be characterized by the following relationship between the above-described constants μ, E, K, and G:

$$E = 3K(1 - 2\mu) = 2(1 + \mu)G. \tag{3.41}$$

Compression, tensile, and flex tests are standard tests performed on solid food materials using instruments such as the universal testing machine or the texture analyzer previously described.

In the compression test, the main aim is to determine Young's modulus of the material. This is considered the typical test performed to determine the firmness of diverse solid raw materials in the food industry, such as fruits and vegetables, and this can be related to quality attributes such as degree of maturity and handling damage. In the tensile test, the material is stretched instead of compressed, and the tensile test is useful for describing the properties of fibrous or elastomeric materials. In the flex test, a rectangular piece of material is suspended across two parallel cylindrical rods and a third rod, parallel to the first two, is lowered into the middle part of the piece between the supports in order to flex it.

Characterization of solid food materials still relies on sensory testing for verifying how ingredient and process modifications may affect some attributes such as color, taste, odor, and texture. Some instrumental tests have not yet been able to displace the judging made by sensory human perception. Matching instrumental measurements and sensory evaluation to achieve proper characterization of solid food materials remains a subject requiring considerable research effort.

3.2.3 Particulate Solids

3.2.3.1 Background

Taking into account all the sizes and shapes that may be found in solids, the most important from the standpoint of material process engineering is the small particle. An understanding of the characteristics and features of populations of minute solids is needed to design processes and equipment dealing with streams containing such solids. As opposed to chunky solid pieces, finely divided solids can be referred to by denominations as varied as particulate solids, pulverized solids, granular materials, etc. In the process engineering literature (chemical engineering, mechanical engineering, chemistry, physics, materials science, and so on), a generic term that has been used to describe a huge population of small particles forming a defined material, is simply "powder" or "industrial powder." Powders have some characteristics so distinctive that they make them practically different, by comparison, from any other state of matter. Intrinsically, molecules forming a solid piece are joined by chemical bonds and interactions stronger than those forming liquids or gases. Externally, however, there are no interactions of the chemical bond type keeping together the small particles forming a powder batch. It has been suggested that the individual entities within masses of particulate solids are kept so only by geometrical accommodation.

Some of the distinctive characteristics of powders that have made them the subject of research efforts to understand their behavior are:

- Powders are not solids, but may deform under compression.
- Powders are not liquids, but they may flow under certain circumstances.
- Powders are not gases, but they may compressed to a degree.

From the features of powders listed above, it is clear that it would be difficult to establish even a definition of a powder. A first approach defines powder as a group of solid particles not filling completely the space they occupy; the space not filled by particles is filled by gas. This definition is somewhat inappropriate since nobody would call a pile of rocks a powder. In the search for a more appropriate definition, it has been observed for quite a while that dry powders posses many properties common to fluids, such as exerting pressure on container vessels and flowing through channels or orifices. There are, nonetheless, important differences, such as:

- A powder does not exert uniform pressure on all directions in confinement. The exerted pressure is minimal in the perpendicular direction of the applied pressure.
- An applied shear force on the surface of a mass of powder is transmitted through all the static mass, unless a fracture occurs.
- The density of a mass of powder varies depending on its degree of packaging; it increases if the powder is compacted by vibration, shaking, tapping, etc.

The characteristics listed above resulted in, from the 1960s, special attention being focused on the fluid-related properties that a powder may acquire under certain conditions. If powders are able to follow some of the typical behavior of fluids, it would be possible to study them by making use of modified fluid mechanics theoretical models. A suitable way of making a powder behave as a fluid is to suspend its individual particles within a stream of gas. When a stable suspension of solid particles of a powder in a gas stream is achieved, it is said that the powder has been fluidized. A powder that is able to be fluidized would be most properly defined as any disperse two-phase system in which the disperse phase consists of particles of a finely divided solid and the continuous phase consists of a gas. The solid particles form a sort of mechanical network owing to certain interparticle forces. The network can be expanded under some conditions, but will reassume its packed state under the influence of gravity. The packed, stationary state of a mass of powder exhibits mechanical resistance and some degree of elasticity. The particle–gas interaction is fundamental, and determines primarily the flowing capacity of the powder.

3.2.3.2   Primary Properties

Particle characterization, i.e., description of the primary properties of food powders in a particulate system, underlies all work in particle technology. The primary particle properties, such as particle shape and particle density, together with the

Table 3.6 Terms recommended by the British Pharmacopoeia for use with powdered materials

Powder type	British standard meshes	
	All passes	Not more than 40% passes
Coarse	10	44
Moderately coarse	22	60
Moderately fine	44	85
Fine	85	–
Very fine	120	–

primary properties of a fluid (viscosity and density) and also with the concentration and state of dispersion, govern the secondary properties, such as settling velocity of particles, rehydration rate of powders, and resistance of filter cakes. On can argued that it is simpler, and more reliable, to measure the secondary properties directly without reference to the primary ones. Direct measurement of secondary properties can be done in practice, but the ultimate aim is to predict them from the primary ones, as when determining pipe resistance to flow from known relationships, feeding in data from primary properties of a given liquid (viscosity and density), as well as properties of a pipeline (roughness). As many relationships in powder technology are rather complex and often not yet available in many areas, such as food powder processing, particle properties are mainly used for qualitative assessment of the behavior of suspensions and powders, for example, as an equipment selection guide. Since a powder is considered to be a dispersed two-phase system consisting of a dispersed phase of solid particles of different sizes and a gas as the continuous phase, complete characterization of powdered materials is dependent on the properties of a particle as an individual entity, the properties of the assembly of particles, and the interactions between those assemblies and a fluid.

The "size" of a powder or particulate material is very relative. The term "size" is often used to classify, categorize, or characterize a powder, but even the term "powder" is not clearly defined and the common convention considers that for a particulate material to be considered a powder its approximate median size (50% of the material is smaller than the median size and 50% is larger) should be less than 1 mm. It is also common practice to talk of "fine" and "coarse" powders; several attempts have been made to standardize particle nomenclature in certain fields. For example, Table 3.6 shows the terms recommended by the British Pharmacopoeia referred to standard sieve apertures. Also, by convention, particle sizes may be expressed in different units depending on the size range involved. Coarse particles may be measured in centimeters or millimeters, fine particles in terms of screen size, and very fine particles in micrometers or nanometers. However, because recommendations of the International Organization for Standardization (ISO), SI units have been adopted in many countries and, thus, particle size may be expressed in meters when doing engineering calculations, or in micrometers by virtue of the small range normally covered or when constructing graphs. An important number

Table 3.7 Approximate ranges of the median sizes of some common food powders

Commodity	British standard mesh	Microns
Rice and barley grains	6–8	2,800–2,000
Granulated sugar	30–34	500–355
Table salt	52–72	300–210
Cocoa	200–300	75–53
Icing sugar	350	45

of food powders may be considered in the fine size range. Some median sizes of common food commodities are presented in Table 3.7.

Particle size, as an independent property, is useless because there is not a particulate material having a single particle size. Any powder will consist of a population of particles of the same chemical composition, but with a wide range of individual sizes. Particle size distribution measurement is a common method in any physical, mechanical, or chemical process because the particle size distribution is directly related to the behavior of the material and/or the physical properties of products. Foods are frequently in the form of fine particles during processing and marketing (Schubert 1987). The bulk density, compressibility, and flowability of a food powder are highly dependent on the particle size and its distribution (Barbosa-Cánovas et al. 1987). Segregation will happen in a free-flowing powder mixture because of the differences in particle sizes (Barbosa-Cánovas et al. 1985). Size distribution is also one of the factors affecting the flowability of food powders (Peleg 1977). For quality control or description of system properties, the need to represent the particle size distribution of food powders is paramount as are proper descriptors in the analysis of the handling, processing, and functionality of each food powder. Many different types of instruments are available for measuring the particle size distribution but most of them fall into four general categories: sieving, microscope counting techniques, sedimentation, and stream scanning. In particle size measurement two most important decisions have to be made before a technique is selected for the analysis; these are concerned with the two variables measured, the particle size, and the occurrence of such a size. It is important to bear in mind that great care must be taken when selecting the particle size, as an equivalent diameter, in order to choose the size most relevant to the property or process which is to be controlled. The occurrence of an amount of particle matter which belongs to specified size classes may be classified or arranged by diverse criteria so as to obtain tables or graphs. In powder technology the use of graphs is convenient and customary for a number of reasons. For example, a particular size which is to be used as the main reference for a given material is easily read from a specific type of plot. Particle sizing was thoroughly discussed by Allen (1997).

It practice, the particles forming a powder will rarely have a spherical shape. Many industrial powders are of mineral (metallic or non-metallic) origin and have been derived from hard materials by a size-reduction process. In such a case, the comminuted particles resemble polyhedrons with nearly plane faces

Table 3.8 General definitions of particle shape

Shape name	Shape description
Acicular	Needle shape
Angular	Roughly polyhedral shape
Crystalline	Freely developed geometrical shape in a fluid medium
Dentritic	Branched crystalline shape
Fibrous	Regularly or irregular threadlike
Flaky	Platelike
Granular	Approximately equidimensional irregular shape
Irregular	Lacking any symmetry
Modular	Rounded irregular shape
Spherical	Global shape

(from four to seven) and sharp edges and corners. The particles may be compact, with length, breadth, and thickness nearly equal, but sometimes they may be platelike or needlelike. As particles become smaller, and because of the influence of attrition resulting from handling, their edges may become smoother, and thus they can be considered to be spherical. The term "diameter" is, therefore, often used to refer to the characteristic linear dimension. All these geometrical features of an important number of industrial powders, such as cement, clay, and chalk, are related to the intimate structure of the elements forming them, whose arrangements are normally symmetrical with definite shapes such as cubes and octahedrons. On the other hand, particulate food materials are mostly organic in origin, and their individual grain shapes can have a great diversity of structures, since their chemical compositions are more complex than those of inorganic industrial powders. The shape variations in food powders are enormous, ranging from extreme degrees of irregularity (ground materials such as spices and sugar) to an approximate sphericity (starch and dry yeast) or well-defined crystalline shapes (granulated sugar and salt).

General definitions of particle shapes are listed in Table 3.8. It is obvious that such simple definitions are not enough to compare particle sizes measured by different methods or to incorporate them as parameters into equations where particle shapes are not the same (Allen 1997). Shape, in its broadest meaning, is very important in particle behavior and just looking at the particle shapes, with no attempts at quantification, can be beneficial. The earliest methods of describing the shape of particle outlines used length L, breadth B, and thickness T, in expressions such as the elongation ratio (L/B) and the flakiness ratio (B/T). The drawback with simple, one-number shape measurements is the possibility of ambiguity; the same single number may be obtained from more than one shape. Nevertheless, a measurement of this type, which has been successfully employed for many years, is the so-called sphericity Φ_s defined by the relation

$$\Phi_s = \frac{6V_p}{x_p s_p},$$

(3.42)

Table 3.9 Densities of common food powders

Powder	Density (kg/m^3)
Glucose	1,560
Sucrose	1,590
Starch	1,500
Cellulose	1,270–1,610
Protein (globular)	~1,400
Fat	900–950
Salt	2,160
Citric acid	1,540

where x_p is the equivalent diameter of the particle, s_p is the surface area of one particle, and V_p is the volume of one particle. For spherical particles Φ_s is 1, whereas for many crushed materials it lies between 0.6 and 0.7.

The density of a particle is defined as its total mass divided by its total volume. It is considered quite relevant for determining other particle properties such as bulk powder structure and particle size, so it requires careful definition (Okuyama and Kousaka 2006). Depending on how the total volume is measured, different definitions of particle density can be given: the true particle density, the apparent particle density, and the effective (or aerodynamic) particle density. Since particles usually contain cracks, flaws, hollows, and closed pores, it follows that all these definitions may be different. The true particle density represents the mass of the particle divided by its volume excluding open and closed pores, and is the density of the solid material of which the particle is made. For pure chemical substances, organic or inorganic, this is the density quoted in reference books of physical/chemical data. Since most inorganic materials consist of rigid particles, whereas most organic substances are normally soft, porous particles, the true density of many food powders will be considerably lower than that of mineral and metallic powders. Typical non-metallic minerals will have true particle densities well over 2,000 kg/m^3, whereas some metallic powders can have true densities on the order of 7,000 kg/m^3. By contrast, most food particles have densities considerably lower, about 1,000–1,500 kg/m^3. Table 3.9 lists typical densities for some food powders. As can be observed, salt (which is of inorganic origin) has a notably higher density than the other substances listed. The apparent particle density is defined as the mass of a particle divided by its volume excluding only the open pores, and is measured by a gas or liquid displacement method such as liquid or air pycnometry. The effective particle density is referred to as the mass of a particle divided by its volume including both open and closed pores. In this case, the volume is within an aerodynamic envelope as "seen" by a gas flowing past the particle and, as such, this density is of primary importance in applications involving flow around particles such as in fluidization, sedimentation, or flow through packed beds.

The three particle densities defined above should not be confused with the bulk density of materials, which includes the voids between the particles in the volume measured. The different values of particle density can also be expressed

Fig. 3.9 Density
determination of solids by
liquid pycnometry: (a)
description of the
pycnometer, (b) weighing, (c)
filling to about half full with
powder, (d) adding liquid to
almost full, (e) eliminating
bubbles, (f) topping and final
weighing

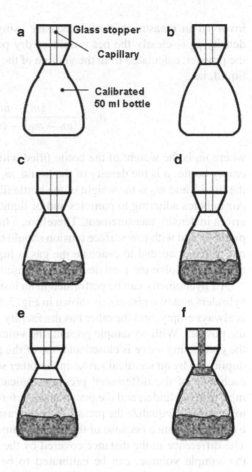

in a dimensionless form, as relative density or specific gravity, which is simply the ratio of the density of the particle to the density of water. It is easy to determine the mass of particles accurately but it is difficult to evaluate their volume because they have irregular shapes and voids between them.

The apparent particle density, or if the particles have no closed pores also the true density, can be measured by fluid displacement methods, i.e., pycnometry, which are in common use in industry. The displacement can be performed using either a liquid or a gas, with the gas employed normally being air. Thus, the two techniques to determine true or apparent density, when applicable, are liquid pycnometry and air pycnometry.

Similar to the case of liquid density, liquid pycnometry can be used to determine the particle density of fine and coarse materials depending on the volume of the pycnometer bottle used. For fine powders a pycnometer bottle of 50-mL volume is normally employed, whereas coarse materials may require larger calibrated containers. Figure 3.9 shows a schematic diagram of the sequence of events involved in measuring particle density using a liquid pycnometer. The particle density ρ_s is clearly the net weight of dry powder divided by the net volume of the powder, calculated from the volume of the bottle minus the volume of the added liquid, i.e.,

$$\rho_s = \frac{(m_s - m_0)\rho}{(m_1 - m_0) - (m_{sl} - m_s)}, \tag{3.43}$$

where m_s is the weight of the bottle filled with the powder, m_0 is the weight of the empty bottle, ρ is the density of the liquid, m_1 is the weight of the bottle filled with the liquid, and m_{sl} is the weight of the bottle filled with both the solid and the liquid. Air bubbles adhering to particles and/or liquid absorbed by the particles can cause errors in density measurement. Therefore, a liquid which is slowly absorbed by the particles and with low surface tension should be selected. Sometimes, when heating or boiling is needed to evacuate the gas, a liquid that has a high boiling point and does not dissolve the particles should be used (Okuyama and Kousaka 2006).

Air pycnometry can be performed in an instrument which usually consists of two cylinders and two pistons, as shown in Fig. 3.10. One is a reference cylinder, which is always empty, and the other has the facility for inserting a cup with the sample of the powder. With no sample present, the volume in each cylinder is the same, so if the connecting valve is closed and one of the pistons is moved, the change must be duplicated by an identical stroke in the other so as to maintain the same pressure on each side of the differential pressure indicator. If a sample is introduced in the measuring cylinder, and the piston in the reference cylinder is advanced all the way to the stop, to equalize the pressures, the measuring piston will have to be moved by a smaller distance because of the extra volume occupied by the sample (Fig. 3.10). The difference in the distance covered by the two pistons, which is proportional to the sample volume, can be calibrated to be read directly in cubic centimeters, usually with a digital counter. The method will measure the true particle density if the particles have no closed pores or the apparent particle density if there are any closed pores, because the volume measured normally excludes any open pores. If, however, the open pores are filled either by wax impregnation or by adding water, the method will also measure the envelope volume. From the difference between the two volumes measured, the open pore volume can be obtained, and can be used as a measure of porosity.

The bulk density of food powders is so fundamental to their storage, processing, and distribution that it merits particular consideration. When a powder just fills a vessel of known volume V and the mass of the powder is m, then the bulk density of the powder is m/V. However, if the vessel is tapped, it will be found in most cases that the powder will settle and more powder needs to be added for the vessel to be completely filled. If the mass now filling the vessel is m', the bulk density is $m'/V > m/V$. Clearly, this change in density was caused by the influence of the fraction of the volume not occupied by particles, known as porosity. The bulk density is, therefore, the mass of the particles that occupy a unit volume of a bed,

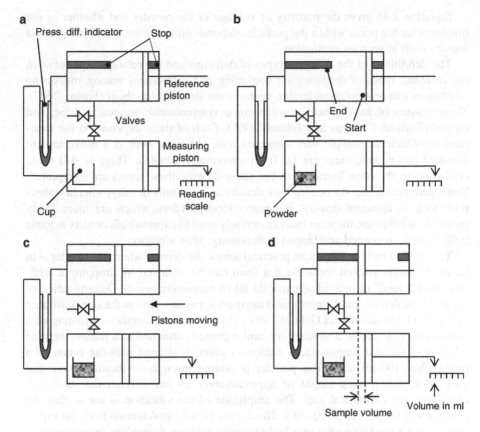

Fig. 3.10 Density determination of solids by using an air pycnometer: (**a**) description of the instrument, (**b**) filling of the cup, (**c**) displacement of pistons, (**d**) reading

whereas porosity or voidage is defined as the volume of the voids within the bed divided by the total volume of the bed. These two properties are in fact related via the particle density in that, for a unit volume of the bulk powder, there must be the following mass balance:

$$\rho_b = \rho_s(1 - \varepsilon) + \rho_a\varepsilon, \tag{3.44}$$

where ρ_b is the powder bulk density, ρ_s is the particle density, ε is the porosity, and ρ_a is the air density. As the air density is low relative to the powder density, it can be disregarded and the porosity can thus be calculated simply as

$$\varepsilon = \frac{(\rho_s - \rho_b)}{\rho_s}. \tag{3.45}$$

Equation 3.45 gives the porosity or voidage of the powder and whether or not this includes the pores within the particles depends on the definition of the particle density used in such an evaluation.

The definitions of the different types of densities and the relationships between the different types of densities are confusing and differences among measuring techniques can lead to considerable errors when determining them (Fasina 2007). Three classes of bulk density have become conventional: aerated, poured, and tapped (Barbosa-Cánovas and Juliano 2005). Each of these depends on the treatment to which the sample was subjected and, although there is a move toward standard procedures, these are far from universally adopted. There is still some confusion in the open literature in the sense of how these terms are interpreted. Some people consider the poured bulk density as loose bulk density, whereas others refer to it as apparent density. For many food powders, which are more likely cohesive in behavior, the term most commonly used to express bulk density is loose bulk density, as poured and tapped bulk density, after vibration.

The aerated bulk density is, in practical terms, the density when the powder is in its most loosely packed form. Such a form can be achieved by dropping a well-dispersed "cloud" of individual particles into a measuring vessel. Determination of aerated bulk density can be performed using an apparatus such as the one illustrated in Fig. 3.11 (Abdullah and Geldart 1999). As shown, an assembly consisting of a screen cover, a screen, a spacer ring, and a chute is attached to a mains-operated vibrator of variable amplitude. A stationary chute is aligned with the center of a preweighed 100-mL cup. The powder is poured through a vibrating sieve and allowed to fall a fixed height of approximately 25 cm through the stationary chute into the cylindrical cup. The amplitude of the vibration is set so that the powder will fill the cup in 20–30 s. The excess powder is skimmed from the top of the cup using the sharp edge of a knife or ruler, without disturbing, or compacting, the loosely settled powder.

Poured density is widely used, but the measurement is often performed in a manner suitable for the requirements of the individual company or industry. In some cases the volume occupied by a particular mass of powder is measured, but the elimination of operation judgment, and thus possible error, in any measurement is advisable. To achieve this, the use of a standard volume and the measurement of the mass of powder to fill it are needed. Certain precautions to be taken are clear, e.g., the measuring vessel should be fat rather than slim, the powder should always be poured from the same height, and the possibility of bias in the filling should be made as small as possible. Although measurement of poured bulk density is far form standardized, many industries use a sawn-off funnel with a trap door or stop to pour the powder into the measuring container.

The tapped bulk density, as implied by its name, is the bulk density of a powder that has been settled into a packing closer than that which existed in the poured state, by tapping, jolting, or vibrating the measuring vessel. As with poured bulk density, the volume of a particular mass of powder may be observed, but it is generally better to measure the mass of powder in a fixed volume. Although many people in industry measure the tapped density by tapping the sample manually, it is

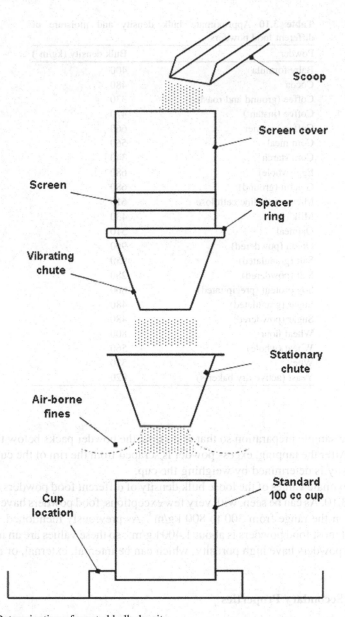

Fig. 3.11 Determination of aerated bulk density

best to use a mechanical tapping device so that the sample preparation conditions are more reproducible. A useful instrument to achieve such reproducibility is the Hosokawa powder characteristic tester, which has a standard cup (100 mL) and a cam-operated tapping device, which moves the cup upward and drops it periodically (once every 1.2 s). A cup extension piece has to be fitted and powder is added

Table 3.10 Approximate bulk density and moisture of different food powders

Powder	Bulk density (kg/m^3)
Baby formula	400
Cocoa	480
Coffee (ground and roasted)	330
Coffee (instant)	470
Coffee creamer	660
Corn meal	560
Corn starch	340
Egg (whole)	680
Gelatin (ground)	680
Microcrystalline cellulose	610
Milk	430
Oatmeal	510
Onion (powdered)	960
Salt (granulated)	950
Salt (powdered)	280
Soy protein (precipitated)	800
Sugar (granulated)	480
Sugar (powdered)	480
Wheat flour	800
Wheat (whole)	560
Whey	520
Yeast (active dry baker's)	820

during the sample preparation so that at no time the powder packs below the rim of the cup. After the tapping, excess powder is scraped from the rim of the cup and the bulk density is determined by weighing the cup.

Approximate values of the loose bulk density of different food powders are given in Table 3.10. As can be seen, with very few exceptions, food powders have apparent densities in the range from 300 to 800 kg/m^3. As previously mentioned, the solid density of most food powders is about 1,400 kg/m^3, so these values are an indication that food powders have high porosity, which can be internal, external, or both.

3.2.3.3   Secondary Properties

The requirement to get powders to flow is for their strength to be less than the load put on them, i.e., they must fail. The basic properties describing this condition are known as "failure properties" and they are the angle of wall friction, the effective angle of internal friction, the failure function, the cohesion, and the ultimate tensile strength. The failure properties take into account the state of compaction of the powder as this strongly affects its flowability unless the powder is cohesionless, like dry sand, and it gains no strength on compression. These properties may also be strongly affected by humidity and, especially in the case of food and biological materials, by temperature. The consolidation time can also have an effect on failure properties of powders. It is therefore important to test such properties under controlled conditions using sealed powder samples or air-conditioned rooms or enclosures. Also, time-consolidated samples must be tested to simulate storage conditions.

The angle of wall friction ϕ is equivalent to the angle of friction between two solid surfaces except that one of the two surfaces is a powder. It describes the friction between the powder and the construction material used to confine the powder, e.g., a hopper wall. The wall friction causes some of the weight to be supported by the walls of a hopper. The effective angle of internal friction δ is a measure of the friction between particles and depends on their size, shape, roughness, and hardness. The failure function FF is a graph showing the relationship between unconfined yield stress (or the strength of a free surface of the powder) and the consolidating stress, and gives the strength of the cohesive material in the surface of an arch as a function of the stress under which the arch was formed. The cohesion C is a function of interparticle attraction and is due to the effect of internal forces within the bulk, which tend to prevent planar sliding of one internal surface of particles upon another. The ultimate tensile strength T of a powder compact is the most fundamental strength measure, representing the minimum force required to cause separation of the bulk structure without major complications of particle disturbances within the plane of failure.

There are basically three types of shear cells available for powder testing: (1) the Jenike shear cell, also known as the translational shear box; (2) the annular or ring shear cell, also called the rotational shear box; and (3) the rotational shear cell, which is a fixture of a powder rheometer. The Jenike shear cell is circular in cross section, with an internal diameter of 95 mm. A vertical cross section through the cell is shown in Fig. 3.12. It consists of a base, a ring which can slide horizontally over the base, and a cover. The Jenike test sequence simulates the changes in stresses acting on an element of materials flowing through a bin. The ring and base are filled with the powder and a lid is placed in position. By means of a weight carrier, which hangs from a point at the center of the lid, a vertical compacting load can be applied to the powder sample. The lid carries a bracket with a projecting pin and a measured horizontal force is applied to such a bracket, causing the ring and its content, as well as the lid, to move forward at a constant speed. The test on a powder consists of three steps: preconsolidation to ensure uniformity of samples, consolidation to reproduce flow with a given stress under steady-state conditions, and shearing to measure shear stress at failure. The preconsolidation consists in filling the assembly of the base, ring, and a packing mold ring over it with a sample powder. A twisting top is placed on the sample and a force is applied to the top while giving it a number of oscillating twists. The force is released, the twisting top is removed, and the powder surface is carefully scraped level with the upper half of the cell. In the consolidation step, a shear cover is placed over the powder sample and a selected normal force is applied on it. A shear force is then applied continuously until it reaches a steady-state value indicating plastic flow. The shear force is then interrupted and the stem is retracted. The shearing step comprises replacing the normal force of consolidation by a smaller

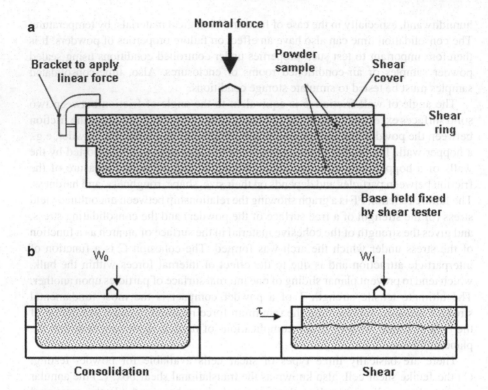

a

Normal force

Bracket to apply
linear force

Powder
sample

Shear
cover

Shear
ring

Base held fixed
by shear tester

b

W_0

W_1

τ

Consolidation   Shear

Fig. 3.12 Jenike's shear cell: (**a**) cell components, (**b**) testing steps

force and reapplying the shearing force until the stress/strain peaks and falls off, indicating a failure plane in the sample and a point in the yield locus. The procedure is repeated with five or six different vertical loads, progressively smaller, applied to a set of identical samples and the shear force needed to initiate flow is found in each case. The forces are divided by the cross-sectional area of the cell to give stresses and the shear stress is plotted against the normal stress. The resulting graph is a yield locus (Fig. 3.13), and it is a line which gives the stress conditions needed to produce flow for the powder when compacted with a fixed bulk density.

If the material being tested is cohesive, the yield locus is not a straight line and does not pass through the origin. It can be shown that the graph when extrapolated downward cuts the horizontal axis normally. As shown in Fig. 3.13, intercept T is the tensile strength of the powder compacts tested and intercept C is called the cohesion of the powder; the yield locus ends at point A. The yield locus represents the results of a series of tests on samples which have the initial bulk density. More yield loci can be obtained by changing the sample preparation procedure and, in this way a family of yield loci can be obtained. This family of yield loci contains all the information needed to characterize the flowability of a particular material; it is not, however, in a convenient form. For many powders,

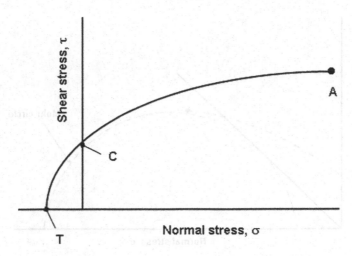

Fig. 3.13 The Jenike yield locus

yield locus curves can be described by the empirical Warren–Spring equation (Chasseray 1994):

$$\left(\frac{\tau}{C}\right)^n = \frac{\sigma}{T} + 1,$$ (3.46)

where τ is the shear stress, C is the material's cohesion, σ is the normal stress, T is the tensile stress, and n is the shear index $(1 < n < 2)$.

Apart from the cohesion and the tensile stress, the effective angle of internal friction can also be determined using the yield locus derived from testing a particular powder in a shear cell. The effective angle of internal friction is the angle resulting from drawing a tangent line to the Mohr circle inscribed in the yield locus, and passing through the origin (as shown in Fig. 3.14).

The angle of wall friction can be measured by replacing the base of a Jenike shear cell by a plate of the material of which the hopper (or any sort of container) is to be made. The ring from the shear cell is placed on the plate and filled with powder and the lid is put in position. The shear force needed to maintain uniform displacement of the ring is found for different vertical loads on the lid. The slope of the graph of shear force against normal force gives the angle of friction between the particles and the wall, or the angle of wall friction. This measure completes the testing of a particulate material using only a Jenike shear cell.

One of the major problems with using the Jenike shear cell is that during operation the shear force concentrates at the front of the shear cell. The shear force is non-uniformly applied to the sample. Similarly, the vertical force is also applied non-uniformly. At best, the results represent average stress conditions typically varying from a near zero stress to the maximum applied (about two times larger than the average).

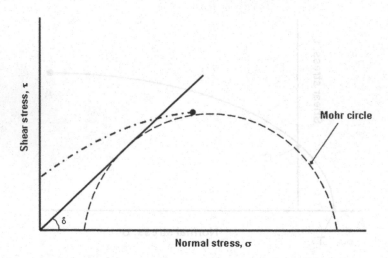

Fig. 3.14 Graphical determination of the effective angle of internal friction

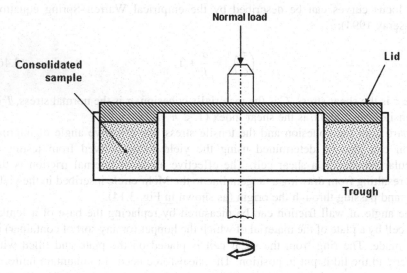

Fig. 3.15 An annular shear cell

The annular shear cell was developed by Carr and Walker (1967) and has undergone a number of modifications. In this type of cell the shear stress is applied by rotating the top portion of an annular shear, as represented in Fig. 3.15. These devices allow much larger shear distances to be covered both in sample preparation and in testing, allowing the study of flow properties after failure. The original annular shear cells tended to give lower values for the yield strength than the Jenike shear cell tester. The latest modifications to the ring cell tester have been made by Schulze and Wittmaier (2007). In the Schulze ring shear tester, the sample is placed in an outer circular channel. An angular lid attached to a crossbeam lies on top of the sample. Small bars are attached to the bottom of the lid and the bottom of the cell to prevent the powder from sliding against the lid or the cell. The movement of the cell with respect to the fixed lid causes the powder sample to shear. Load cells attached to tied rods measure the force needed to initiate this. To exert load on the sample, weights are hung from a crossbeam. This can be done during the shearing and consolidation of the powder sample. The cell can also be removed and time consolidation can be performed by placing weights on the sample outside the test device. An automatic version of this measuring cell has been developed in which a computer can automatically add loading to the sample and condition it. The instrument can be operated in a manual, semiautomatic, or totally automatic mode. Both, the original shear cells and the latest modified ones offer several important advantages over the Jenike-type shear cell. The area of shear is constant and the handling is easier, because consolidation and shear are quicker. After the sample has been consolidated, a full locus can be generated without the need to reconsolidate the sample after each load. The consolidation process becomes more automated and uniform, eliminating much of the operator variability in the measurement process characteristic of the Jenike-type tester.

Rotational shear cells, or powder rheometers, are relatively new testing instruments that can sensitively measure how flowability changes under a wide range of processing conditions, including various speeds, levels of entrapped air, and degrees of attrition. Powder rheometers can also condition the sample before testing, reducing the variability produced by differences in storage and handling. These instruments automatically rotate a blade through a cylindrical column of powder and measure the energy or force it needs to move through the sample and relate these measurements to various characteristics of the powder. The powder is placed in a circular vessel with a closed bottom. The blade is introduced into the powder and moves downward or upward in a helical motion, and the force on the blade shaft is recorded. The helical path along which the blade moves is dependent on the axial and rotational speeds, as well as on the direction of the blade rotation. The angle of approach the blade makes with the powder can be varied, so the direction and angle of the measurement allow one to measure compaction, shear, and slicing of the powder sample within the vessel. The torque data are used to determine the largest theoretical torque exerted by the whole powder column on the rotor blade for any defined test condition. Therefore, powder shear can be measured under different shear conditions, including various downward compaction modes and upward expansion modes. The blade is interchangeable, and different designs are used for different needs, including a rotating shear cell that can be used in a shallow column of powder. An arrangement such as this is shown in Fig. 3.16, and is called an open shear cell. As can be observed, the operations of this cell are analogous to those of a rotational viscometer used in characterizing highly consistent non-Newtonian fluids. In congruence with this analogy, the open shear cell will measure both the force and the torque, so these variables can be related to the shear stress and the normal stress to obtain the yield locus of a powder. Satisfactory results using rotational shear cells have been reported in different applications and

Fig. 3.16 A rotating shear
cell

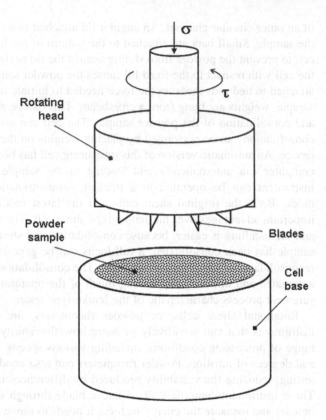

with diverse industrial powders (Freeman et al. 2009; Freeman and Cook 2006).
Powder rheometers can provide simple and sensitive testing as they are relatively
operator independent and yield quantitative values.

Another important failure property, the failure function, can be measured using a
split cylindrical die as shown in Fig. 3.17. The bore of the cylinder may be about
50 mm and its height should be just more than twice the bore. The cylinder is
clamped so that the two halves cannot separate and it is filled with the powder to be
tested, which is then scraped off level with the top face. By means of a plunger, the
specimen is subjected to a known consolidating stress. The plunger is then removed
and the two halves of the split die are separated, leaving a free-standing cylinder of
the compacted powder. A plate is then placed on top of the specimen and an
increasing vertical load is applied to it until the column collapses. The stress at
which this occurs is the unconfined yield stress, i.e., the stress that has to be applied
to the free vertical surface on the column to cause failure. If this is repeated for a
number of different compacting loads and the unconfined yield stress is plotted
against the compacting stress, the failure function of the powder will be obtained.
Although the results of this method can be used for monitoring or for comparison,
the failure function obtained will not be the same as that given by shear cell tests
because of the effect of die wall friction when forming the compact.

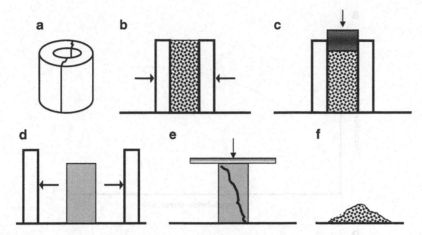

Fig. 3.17 Device for direct measurement of the failure function: (**a**) mounted measuring device, (**b**) securing halves and filling with the sample, (**c**) compacting with a plunger, (**d**) separating halves and stable column of powder, (**e**) application of normal force, (**f**) collapsing of material

The failure function, also referred as the flow function, is a measure of how the unconfined yield strength developed within the powder varies with maximum consolidation stress. This variation is illustrated in Fig. 3.18, and forms the basis of the Jenike classification of powders. The ratio of the consolidation stress to the unconfined yield strength is called the flow factor ff. Jenike (1964) proposed a classification according to the position of one point of the failure function at a fixed value of the unconfined yield stress with respect to the flow factor line. Figure 3.18b shows a schematic representation of classification of powders following this criterion. As can be seen, at a fixed value of the unconfined yield strength of 22.3 N, the straight lines through the origin at a slope 1/ff represent the categories of very cohesive (ff < 2), cohesive (2 < ff < 4), easy flow (4 < ff < 10) and free flow (10 < ff).

Failure properties are used to characterize powders, mainly in terms of powder flow. The flow characteristics of powders are of great importance in many problems encountered in bulk material handling processes in the agricultural, ceramic, food, mineral, mining, and pharmaceutical industries because the ease of powder conveying, blending, and packaging depends on flow characteristics. In designing plants involving handling and processing of food powders, diverse difficulties may arise that can be lead to severe operating problems.

Practically, for all fine powders the attractive forces between particles are large when compared with the weight of individual particles so they are said to be cohesive and normally present flow problems (Adhikari et al. 2001). Cohesion occurs when interparticle forces play a significant role in the mechanics of the powder bed. Flow problems occur in any kind of cohesive powder, but may be more serious with food powders because they are more sensitive to physical and physicochemical phenomena, which may affect their composition and properties (Fitzpatrick 2005).

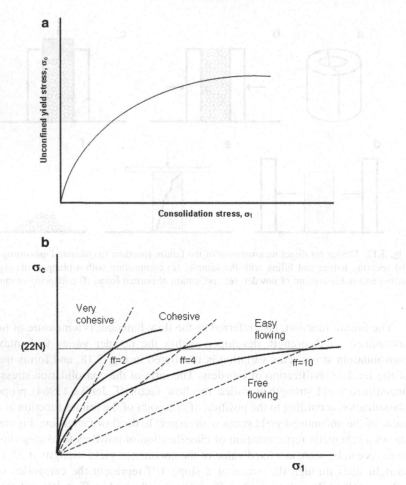

Fig. 3.18 The failure function of a food powder: (**a**) curve representing a failure function, (**b**) use of the failure function to classify powders according to Jenike

The most common effects of such phenomena in food powders lead to situations which aggravate flow properties, as they normally relate to the release of sticky substances or to the presence of hygroscopic behavior. Cohesive powders are, normally, also adhesive. In cohesion the contact surface is similar (particle–particle), whereas in adhesion the contact surface is different (particle–any surface material). For example, adhesion takes place between drying droplets and the drying wall of a spray dryer, whereas cohesion is responsible for caking of the powder during processing and storage. In food powder processing, "cohesion" is also termed "caking" and "adhesion" is also referred to as "stickiness." Table 3.11 lists cohesion values for several food powders.

Stickiness is a prevalent problem that can cause problems for operation, equipment wear, and product yield (Adhikari et al. 2001). Interaction of water with solids is the prime cause of stickiness and caking in low-moisture food powders.

Table 3.11 Moisture content and cohesion for some food powders

Material	Moisture content (%)	Cohesion (g/cm^2)
Corn starch	<11.0	4–6
Corn starch	18.5	13
Gelatin	10.0	1
Grapefruit juice	1.8	8
Grapefruit juice	2.6	10–11
Milk	1.0	7
Milk	4.4	10
Onion	<3.0	<7
Onion	3.6	8–15
Soy flour	8.0	1

Water provides the necessary plasticity to food polymeric systems, to reduce viscosity and promote molecular mobility. Chemical caking is caused by chemical reactions in which a compound has been generated or modified, as in hydration, recrystallization, or sublimation. Plastic-flow caking occurs when the particles' yield values are exceeded and they stick together or merge into a single particulate form, as in amorphous materials such as gels, lipids, and waxes. Melting and solidification of fats may also cause solid bridges and enhance caking in food powders. In this case an increase in temperature causes solid fats to melt, producing a liquid that can redistribute itself among the particles. If the temperature is subsequently reduced, the liquid fat will solidify and form solid bridges between the particles.

In general terms the condition of the surrounding atmosphere affects the secondary properties of powders, such as flowability, causing diverse types of problems. Small differences in factors such as moisture content, particle size, storage time, and temperature can result in a significant difference in flowability. If the relative humidity increases, all powders tend to absorb water, which may form liquid bridges between particles, reducing flowability. Inert powders, however, desorb water with a decrease in the surrounding relative humidity, so liquid bridges will disappear and flowability may return to normal. Food powders are predominantly soluble, so solid bridges may remain even after a decrease in humidity, causing the powder to cake and aggravating flow problems. Furthermore, because of their chemical composition, food powders are normally more sensitive to all the ambient factors previously described and the physicochemical changes largely depend on their temperature–moisture histories. For instance, increasing the temperature of a food powder increases dissolution of particles, facilitating changes in crystalline form that may also result in caking and flow problems (Teunou and Vasseur 1996).

References

Abdullah EC, Geldart D (1999) The use of bulk density measurements as flowability indicators. Powder Technol 102: 151–165.

Adhikari B, Howes T, Bhandari BR, Truong V (2001) Stickiness in foods: a review of mechanisms and test methods. Int J Food Prop 4: 1–33.

Allen T (1997) Particle Size Measurement: Surface Area and Pore Size Determination. Chapman
    & Hall, London.
Barbosa-Cánovas GV, Juliano P (2005) Physical and chemical properties of food powders.
    In: Onwulata C (ed) Encapsulated and Powdered Foods, pp 39–71. CRC Taylor & Francis,
    Boca Raton, FL.
Barbosa-Cánovas GV, Málave-López J, Peleg M (1987) Density and compressibility of selected
    food powders mixture. J Food Process Eng 10: 1–19.
Barbosa-Cánovas GV, Málave-López J, Peleg M (1985) Segregation in food powders. Biotechnol
    Prog 1: 140–146.
Barbosa-Cánovas GV, Ortega-Rivas E, Juliano P, Yan H (2005) Food Powders: Physical
    Properties, Processing, and Functionality, Kluwer Academic/Plenum Publishers, New York.
Bourne MC (2002) Food Texture and Viscosity: Concept and Measurement. Academic Press,
    London.
Brown GG (2005) Unit Operations. CBS Publishers & Distributors, New Dehli.
Carr JF, Walker DM (1967) An annular shear cell for granular materials. Powder Technol 1:
    369–373.
Chasseray P (1994) Physical characteristics of grains and their byproducts. In: Godon B, Willm C
    (eds) Primary Cereal Processing, pp 85–141. Wiley-Interscience, New York.
Dodge DW Metzner AB (1959) Turbulent flow of non-Newtonian systems. AIChE J 5: 189–204.
Fasina OO (2007) Does a pycnometer measure the true or apparent particle density of agricultural
    materials? In: 2007 ASABE Annual International Meeting, Technical Papers, Volume 13.
Fitzpatrick JJ (2005) Food powder flowability. In: Onwulata C (ed) Encapsulated and Powdered
    Foods, pp 247–260. CRC Taylor & Francis, Boca Raton, FL.
Freeman R, Cooke J (2006) Testing powders in process relevant ways. Powder Handl Process 18:
    84–87.
Freeman RE, Cooke JR, Schneider LCR (2009) Measuring shear properties and normal stresses
    generated within a rotational shear cell for consolidated and non-consolidated powders.
    Powder Technol 190: 65–69.
Geankoplis CJ (2003) Transport Processes and Separation Process Principles. Pearson Education
    Inc, Upper Saddle River, NJ.
Jenike AW (1964) Storage and Flow of Solids. Bulletin No. 123 of the Utah Engineering
    Experiment Station. University of Utah, Salt Lake City, UT.
Kawata MK, Kurase A, Nagashima A, Yoshida K (1991) Capillary viscometers. In: Wakeham
    WA, Nagashima A, Sengers JV (eds) Measurement of the Transport Properties of Fluids-
    Experimental Thermodynamics, Vol III, pp 51–75. Blackwell Scientific Publications, Oxford.
Okuyama K, Kousaka Y (2006). Particle density. In: Masuda H, Higashitani K, Yoshida H (eds)
    Powder Technology Handbook, pp 49–52. CRC Taylor & Francis, Boca Raton, FL.
Peleg M (1977) Flowability of food powders and methods for its evaluation - a review. J Food
    Process Eng 1: 303–328.
Pomeranz Y, Meloan CE (1994) Food Analysis: Theory and Practice. Chapman & Hall,
    New York.
Schubert H (1987) Food particle technology part I: properties of particles and particulate food
    systems. J Food Eng 6, 1–32.
Schulze D, Wittmaier A (2007) Measurement of flow properties of powders at very small
    consolidation stresses. Bulk Solids Powder Sci Technol 2: 47–54.
Steffe JF (1996) Rheological Methods in Food Process Engineering. Freeman Press, East Lansing, MI.
Szczesniak AS (1983) Physical properties of foods: what they are and their relations to other food
    properties. In: Peleg M, Bagley EB (eds) Physical Properties of Foods, pp 1–42. AVI Publish-
    ing Co, Westport, CN.
Teunou E, Vasseur J (1996) Time flow function: means to estimate water effect on dissoluble bulk
    materials flow. Powder Handl Process 8: 111–116.

Part II
Processing Operations

Chapter 4
Size Reduction

4.1 Principles of Size Reduction

4.1.1 Introductory Aspects

In many food processes it is frequently required to reduce the size of solid materials for different purposes. For example, size reduction may aid other processes such as expression and extraction, or may shorten heat treatments such as blanching and cooking. "Comminution" is the generic term used for size reduction and includes different operations such as crushing, grinding, milling, mincing, and dicing. Most of these terms are related to a particular application, e.g., milling of cereals, mincing of beef, dicing of tubers, or grinding of spices. The reduction mechanism consists in deforming the food piece until it breaks or tears. Breaking of hard materials along cracks or defects in their structures is achieved by applying diverse forces.

The objective of comminution is to produce small particles from larger ones. Smaller particles are the desired product either because of their large surface or because of their shape, size, and number. The energy efficiency of the operation can be related to the new surface formed by the reduction in size. The geometrical characteristics of particles, both alone and in mixtures, are important in evaluating a product from comminution. In an actual process, a given unit does not yield a uniform product, whether the feed is uniformly sized or not. The product normally consists of a mixture of particles, which may contain a wide variety of sizes and even shapes. Some types of equipment are designed to control the magnitude of the largest particles in the products, but the fine sizes are not under such control. In some machines fines are minimized, but they cannot be totally eliminated. In comminuted products, the term "diameter" is generally used to describe the characteristic dimension related to particle size. As described in Chap. 3, the shape of an individual particle is conveniently expressed in terms of the sphericity Φ_s, which is independent of particle size. For spherical particles Φ_s is 1, whereas for many crushed materials it lies between 0.6 and 0.7.

E. Ortega-Rivas, *Non-thermal Food Engineering Operations*,
Food Engineering Series, DOI 10.1007/978-1-4614-2038-5_4,
© Springer Science+Business Media, LLC 2012

Table 4.1 Types of force used in size reduction equipment

Force	Principle	Example of equipment
Compressive	Nutcracker	Crushing rolls
Impact	Hammer	Hammer mill
Attrition	File	Disc attrition mill
Cutting	Scissors	Rotary knife cutter

No single distribution applies equally well to all comminuted products, particularly for coarser particles. For finer particles, however, the most commonly found distribution follows a log-normal function (Herdan 1960), which is the most useful one among the different types of functions (Beddow and Meloy 1980).

4.1.2   Forces Used in Size Reduction

As previously mentioned, in comminution of food products the reduction mechanism consists in deforming the food piece until it breaks or tears and such breaking may be achieved by applying diverse forces. The types of forces commonly used in food processes are compressive, impact, attrition and shear, and cutting. In a comminution operation more than one type of force usually acts. Table 4.1 summarizes these type of forces in some of the mills commonly used in the food industry.

Compressive forces are used for coarse crushing of hard materials. Coarse crushing implies reduction to a size of about 3 mm. Impact forces can be regarded as general-purpose forces and may be associated with coarse, medium, and fine grinding of a variety of food materials. Shear or attrition forces are applied in fine pulverization, when the size of the products can reach the micrometer range. Sometimes the term "ultrafine grinding" is associated with processes in which the submicron range of particles is attained. Finally, cutting gives a definite particle size and may even produce a definite shape.

4.1.3   Properties of Comminuted Products

As stated earlier, the breakdown of solid material is performed through the application of mechanical forces which attack fissures present in the original structure. These stresses have been traditionally used to reduce the size of hard materials, either from inorganic origin (e.g., rocks and minerals) or from organic origin (e.g., grains and oilseeds). In both cases, the comminuted particles obtained after any size reduction operation will resemble polyhedrons with nearly plane faces and sharp edges and corners. The number of major faces may vary, but will be usually between 4 and 7. As previously mentioned, a compact grain with several nearly equal faces can be considered as spherical, so the term "diameter" is normally used to describe the particle size of these comminuted products.

The predictable shape of the products described above has to do with molecular structure since silicon and carbon, elements of the same group in the periodic table, are generally key components of the crystal units which form the solid matrix. In this sense, a good number of food materials will have the hardness associated with the rigid structure of carbon derivatives and, as such, they will fragment following the same pattern as their relatives in the inorganic world whose structure is due to the presence of silicon components. An ideal size reduction pattern to achieve a high reduction ratio of hard, brittle food materials, such as sugar crystals or dry grains, can be obtained firstly by compressing, then by using impact force, and finally by shearing or rubbing. Therefore, only these hard, brittle food materials will produce powders when subjected to different forces in a comminution operation, whereas tough, ductile food materials such as meat can only be reduced in size by applying cutting forces. In fact, cutting is considered a process totally different from comminution because its operating principles are quite different from those governing the size reduction of hard materials.

As stated earlier, in comminution of food materials more than one type of the forces described above is actually present. Regardless of the uniformity of the feed material, the product always consists of a mixture of particles covering a range of sizes. Some size reduction equipment is designed to control the size of the largest particles in the products, but the fine sizes are not under control. In spite of the hardness of the comminuted materials, the above-mentioned shape of the particles produced is subject to attrition caused by interparticle and particle–equipment contacts within the dynamics of the operation. Thus, particle angles will smooth gradually with the consequent production of fines. In practice, any feed material will possess an original particle size distribution, whereas the product obtained will end up with a new particle size distribution having the whole range finer than the feed distribution.

A product specification will commonly require a finished product not to contain particles greater than (or smaller than, depending on the application) some specified size. In comminution practice, particle size is often referred as screen aperture size. The reduction ratio, defined as the relation between the average size of the feed and the average size of the product, can be used as an estimate of the performance of a comminution operation. The values for the average size of feed and the product depend on the measurement method, but the true arithmetic mean, obtained from screen analyses on samples of the feed and product streams, is commonly used for this purpose. Reduction ratios depend on the specific type of equipment. As a general rule, the coarser the reduction, the smaller the ratio. For example, coarse crushers have size reduction ratios of below 8:1, whereas fine grinders may have ratios as high as 100:1. However, large reduction ratios, such as those obtained when dividing relatively large solid lumps into ultrafine powders, are normally attained by several stages using diverse crushing and grinding machines. A good example of this is the overall milling of wheat grain into fine flour, in which crushing rolls in a series of decreasing diameters are employed.

4.2   Energy Requirements: Comminution Laws

As previously discussed, in the breakdown of hard and brittle solid food materials two stages of breakage are recognized: (1) initial fracture along existing fissures within the structure of the material, and (2) formation of new fissures or crack tips followed by fracture along these fissures. It is also accepted that only a small percentage of the energy supplied to the grinding equipment is actually used in the breakdown operation. Figures of less than 2% efficiency have been quoted (Coulson and Richardson 1996) and, thus, grinding is a very inefficient process, perhaps the most inefficient of the traditional unit operations. Much of the input energy is lost in deforming the particles within their elastic limits and through interparticle friction. A large amount of this wasted energy is released as heat, which, in turn, may be responsible for heat damage of biological materials.

Theoretical considerations suggest that the energy required to produce a small change in the size of a unit mass of material can be expressed as a power function of the size of the material, i.e.,

$$\frac{dE}{dx} = -\frac{K}{x^n}, \tag{4.1}$$

where dE is the change in energy, dx is the change in size, K is a constant, and x is the particle size.

Equation 4.1 is often referred to as the general law of comminution and has been used by a number of workers to derive more specific laws depending on the application.

4.2.1   Rittinger's Law

Rittinger considered that for the grinding of solids, the energy required should be proportional to the new surface area produced and gave the power n the value of 2, thus giving Rittinger's law by integration of Eq. 4.1:

$$E = K\left(\frac{1}{x_2} - \frac{1}{x_1}\right), \tag{4.2}$$

where E is the energy per unit mass required for the production of a new surface by reduction, K is Rittinger's constant and is determined for a particular piece of equipment and material, x_1 is the average initial feed size, and x_2 is the average final product size. Rittinger's law has been found to hold better for fine grinding, where a large increase in surface results.

4.2.2 Kick's Law

Kick reckoned that the energy required for a given size reduction is proportional to the size reduction ratio and took the value of the power n as 1. In such a way, by integration of Eq. 4.1, the following relation, known as Kick's law, is obtained:

$$E = K \ln\left(\frac{x_1}{x_2}\right), \tag{4.3}$$

where x_1/x_2 is the size reduction ratio. Kick's law has been found to hold more accurately for coarser crushing, where most of the energy is used to cause fracture along existing cracks.

4.2.3 Bond's Law and Work Index

A third version of the comminution law is the one attributed to Bond (1963), who considered that the work necessary for reduction is inversely proportional to the square root of the size produced. In Bond's consideration, n takes the value 3/2, giving the following version (Bond's law) also by integrating Eq. 4.1:

$$E = 2K\left(\frac{1}{\sqrt{x_2}} - \frac{1}{\sqrt{x_1}}\right). \tag{4.4}$$

When x_1 and x_2 are measured in micrometers and E is in kilowatt-hours per ton, $K = 5E_i$, where E_i is Bond's work index, defined as the energy required to reduce a unit mass of material from an infinite particle size to a size such that 80% passes through a 100-µm sieve. Bond's work index is obtained from laboratory crushing tests on the feed material. Bond's law holds reasonably well for a variety of materials undergoing coarse, medium, and fine size reduction.

4.3 Size Reduction Equipment

4.3.1 Classification

As previously discussed, size reduction is a unit operation widely used in a number of processing industries. Many types of equipment are used in size reduction operations. In a broad sense, size reduction machines may be classified as crushers

Table 4.2 Size reduction machines used in food process engineering

Range of reduction	Generic name of equipment	Type of equipment
Coarse and intermediate	Crushers	Crushing rolls
Intermediate and fine	Grinders	Hammer mills
		Disc attrition mills
		Tumbling mills (rod mills)
Fine and ultrafine	Ultrafine grinders	Hammer mills
		Tumbling mills (ball mills)

used mainly for coarse reduction, grinders employed principally in intermediate and fine reduction, ultrafine grinders utilized in ultrafine reduction, and cutting machines used for exact reduction (McCabe et al. 2005). A piece of equipment is generally known as crusher when it performs coarse reduction and a mill when it is used for all other applications. The above-mentioned classification includes several categories of each type of machine so, in total, approximately 20 different designs are recognized in comminution processes. In the food industry not every one of them has important applications. For example, larger types of coarse crushers, such as jaw and gyratory crushers, are not normally encountered in the food industry. Table 4.2 lists the principal size reduction machines of used in food processing.

4.3.2   Features

Machines of various types and sizes are available for the comminution of materials in the food processing industry. The main characteristics of the most commonly employed units are discussed below.

4.3.2.1   Crushing Rolls

In this type of equipment, two or more heavy steel cylinders revolve toward each other (Fig. 4.1) so particles of feed are nipped and pulled through. The nipped particles are subjected to compressive force causing the reduction in size. In some designs differential speed is maintained so as to exert also shearing forces on the particles. The roller surface can be smooth or can have corrugations, breaker bars, or teeth, as a means to increase friction and facilitate trapping of particles between the rolls. Toothed-roll crushers can be mounted in pairs, like the smooth-roll crushers, or with only one roll working against a stationary curved breaker plate. Toothed-roll crushers are much more versatile than smooth-roll crushers but have the limitation that they cannot handle very hard solids. They operate by

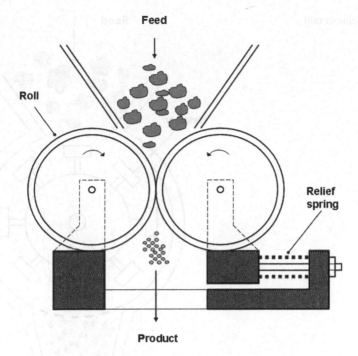

Fig. 4.1 Crushing rolls

compression, impact, and shear and not by compression alone, as do smooth-roll crushers. Crushing rolls are widely applied in the milling of wheat and in the refining of chocolate.

4.3.2.2 Hammer Mills

Figure 4.2 shows a hammer mill, equipment which contains a high-speed rotor turning inside a cylindrical case. The rotor carries a collar bearing a number of hammers around its periphery. By the rotating action, the hammers swing through a circular path inside the casing containing a toughened breaker plate. Feed passes into the action zone with the hammers driving the material against the breaker plate and forcing it to pass through a bottom-mounted screen by gravity when the particles attain a suitable size. Reduction is mainly due to impact forces, although under choke feeding conditions, attrition forces can also play a part in such reduction. The hammers may be replaced by knives, or any other devices, so as to give the mill the possibility of handling tough, ductile, or fibrous materials. The hammer mill is a very versatile piece of equipment which gives high reduction ratios and may handle a wide variety of materials from hard and abrasive to fibrous

Fig. 4.2 Hammer mill

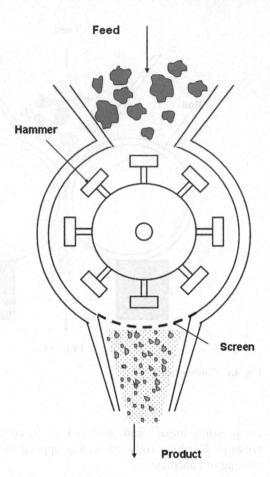

and sticky. In the food industry the applications are very varied, being extensively used for grinding spices, dried milk, sugar agglomerate, cocoa press cake, tapioca, dry fruits, dry vegetables, and extracted bones.

4.3.2.3 Disc Attrition Mills

These types of mills, such as those illustrated in Fig. 4.3, make use of shear forces for size reduction, mainly in the fine size range of particles. There are several basic designs of attrition mills. The single-disc mill shown in Fig. 4.3a has a high-speed rotating grooved disc with a narrow gap between it and its stationary casing. Intense shearing action results in comminution of the feed. The gap is adjustable, depending on the feed size and the product requirements. In the double-disc mill, like the one shown in Fig. 4.3b, the casing contains two rotating discs which rotate in opposite directions, giving a greater degree of shear compared with the single-disc mill. The pin-disc mill has pins or pegs on the rotating elements. In this case impact

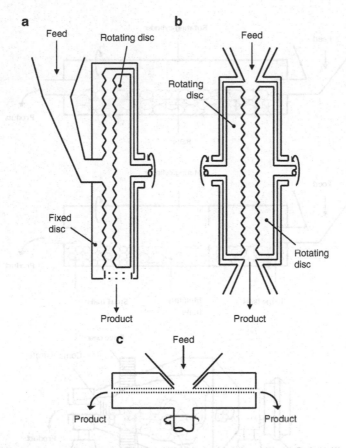

Fig. 4.3 Disc attrition mills: (**a**) single-disc mill, (**b**) double-disc mill, (**c**) Buhr mill

forces also play an important role in particle size reduction. The Buhr mill illustrated in Fig. 4.3c, which is an older type of attrition mill originally used in flour milling, consists of two circular stones mounted on a vertical axis. The upper stone is normally fixed and has a feed entry port, whereas the lower stone rotates. The product is discharged over the edge of the lower stone. The applications of attrition mills in the food industry are quite extensive. They have been employed in dry milling of wheat, as well as in wet milling of corn for the separation of starch gluten from the hulls. Other applications include breaking of cocoa kernels, preparation of cocoa powder, degermination of corn, production of fish-meal, manufacture of chocolate, and grinding of sugar, nutmeg, cloves, roasted nuts, peppers, etc.

4.3.2.4   Tumbling Mills

A tumbling mill is used in many industries for fine grinding. It basically consists of a horizontal slow-speed rotating cylinder partially filled with either balls or rods.

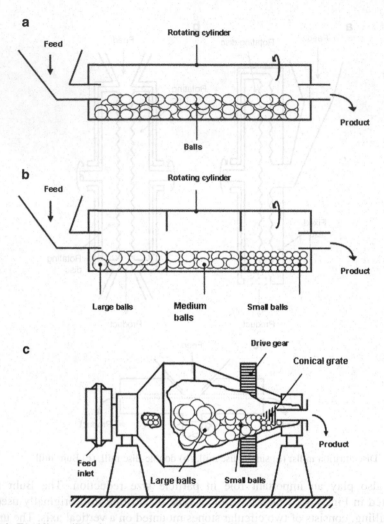

Fig. 4.4 Tumbling mill designs: (**a**) trunnion overflow mill, (**b**) compartment mill, (**c**) conical mill

The cylinder shell is usually made of steel, lined with carbon-steel plate, porcelain, silica rock, or rubber. The balls are normally made from steel or flint stones, and the rods are usually manufactured from high-carbon steel. The reduction mechanism is performed as follows. As the cylinder rotates, the grinding medium is lifted up the sides of the cylinder and drops onto the material being comminuted, which fills the void spaces between the medium. The grinding medium components also tumble over each other, exerting a shearing action on the feed material. This combination of impact and shearing forces brings about a very effective size reduction. As a tumbling mill basically operates in a batch manner, different designs have been developed to make the process continuous. As illustrated in Fig. 4.4a, in a trunnion overflow mill, the raw material is fed in through a hollow trunnion at one end of the mill and the ground product overflows at the opposite end. Putting slotted transverse partitions in a tube mill converts it into a compartment mill, as shown in Fig. 4.4b. One compartment may contain large balls, a second one small balls, and a third pebbles, thus achieving segregation of the grinding media with the consequent rationalization of energy. A very efficient way of segregating the grinding media is the use of the conical ball mill shown in Fig. 4.4c. The solid feed enters from the left into the primary grinding zone, where the diameter of the shell is a maximum, and the comminuted product leaves through the cone at the right end, where the diameter of the shell is a minimum. As the shell rotates, the large balls move toward the point of maximum diameter, and the small balls migrate toward the discharge outlet. Therefore, the initial breaking of feed particles is performed by the largest balls dropping the greatest distance, whereas the final reduction of small particles is done by small balls dropping a smaller distance. In such an arrangement the efficiency of the milling operation is greatly increased. Among the applications of tumbling mills in the food industry, the reduction of fluid cocoa mass can be named.

4.3.3 Operation

The diversity of designs of the above-described machinery implies that the operating variables differ quite considerably. Whereas the energy requirements are generally governed by the comminution laws previously discussed, some other features such as capacity and rotational velocity are particular for every particular unit and may be related, in some way, to the predominant force performing the reduction action.

In crushing rolls, the angle formed by the tangents to the roll faces at the point of contact between a particle and the rolls is called the angle of nip. It is an important variable for specifying the size of a pair of crushing rolls for a specific duty and is found as follows. Figure 4.5 shows a pair of rolls and a spherical particle just being gripped between them. The radii of the rolls and the particle are R and r, respectively. The clearance between the rolls is $2d$. Line AB passes through the centers of the left-hand roll and the particle, as well as through point C, which is the point of contact between the roll and the particle. As shown in Fig. 4.5, if α is the angle between line AB and the horizontal, line OE is a tangent to the roll at point C and it makes the same angle α with the vertical.

Disregarding gravity, two forces act at point C: the tangential frictional force F_t, with vertical component $F_t \cos\alpha$, and the radial force F_r, with vertical component $F_r \sin\alpha$. F_t is related to F_r through the coefficient of friction μ', so $F_t = \mu' F_r$. Force $F_r \sin\alpha$ tends to expel the particle from the rolls, and force $\mu' F_r \cos\alpha$ tends to pull it into the rolls. If the particle is to be crushed

$$F_r \mu' \cos\alpha \geq F_r \sin\alpha \qquad (4.5)$$

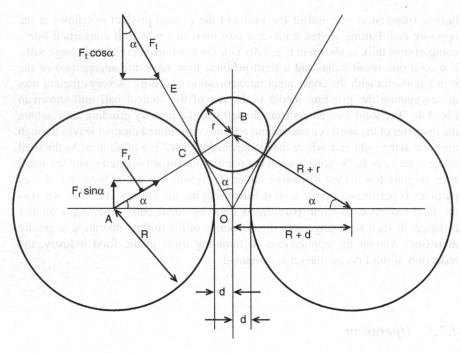

Fig. 4.5 Forces and angle of nip in crushing rolls

or

$$\mu' \geq \tan \alpha. \tag{4.6}$$

When $\mu' = \tan \alpha$, the angle α is half the angle of nip. A simple relationship exists between the radius of the rolls, the size of the feed, and the gap between the rolls. Thus, from Fig. 4.5,

$$\cos \alpha = \frac{R+d}{R+r}. \tag{4.7}$$

The largest particles in the product have diameter $2d$ and Eq. 4.7 provides a relationship between the roll diameter and the size reduction that can be expected in the mill.

From Fig. 4.5 it follows that $2R$ represents the diameter of the roll D_r, $2r$ represents the diameter of the feed D_f (when the feeding pieces have proper sphericity Φ_s), and $2d$ represents the product diameter D_p. With use of these definitions, the theoretical volumetric capacity Q of crushing rolls is the volume of the continuous ribbon of product discharged from the rolls and is given by

$$Q = \frac{N D_r D_p L}{60}, \tag{4.8}$$

where N is the roll speed in revolutions per minute and L is the length of the face in meters.

Knowing the bulk density of the discharge stream, one can estimate the approximate mass flow rate. In practice the actual capacity is found to lie between 0.1 and 0.3 of the theoretical one.

The load of balls in a tumbling mill should be such that when the mill is stopped, the balls occupy somewhat more than half the volume of the mill. In operation, the balls are picked up by the mill wall and are carried nearly to the top, from where they fall to the bottom to repeat the process. Centrifugal force maintains the balls in contact with the wall and with each other during the upward trajectory. While they remain in contact with the wall, the balls exercise a grinding action by slipping and rolling over each other. Most of the grinding occurs, however, at the zone of impact where the free-falling balls strike the bottom part of the mill.

The faster the mill rotates, the farther the balls are carried up inside the wall and the greater the power consumption. The added power is used profitably because when the balls are carried to the higher point, they will have a greater impact on the bottom and will have a better reduction capacity. When the speed is too high, however, the balls are carried over and the mill will be practically centrifuging the balls, preventing them from falling. The speed at which centrifuging occurs is called the critical speed and little or no grinding occurs when the mill operates at this, or a higher, velocity. The operating speeds must be carefully calculated for them not to be considerably lower than the critical speed, because little grinding action will occur, or considerably higher than the critical speed, because centrifuging will cancel out the grinding capacity of the mill.

The speed at which the outermost balls lose contact with the wall of the mill depends on the balance between gravitational and centrifugal forces. In Fig. 4.6, the radii of the mill and the ball at point A at the periphery of the mill are R and r, respectively. The center of the ball is, thus, $R-r$ from the axis of the mill and the radius AO forms angle α with the vertical. Two forces act on the ball: the force of gravity mg, where m is the mass of the ball, and the centrifugal force $mu^2/(R-r)$, where u is the peripheral speed of the center of the ball. The centripetal component of the force of gravity is $mg\cos\alpha$, which opposes the centrifugal force. As long as the centrifugal force exceeds the centripetal force, the particle will not lose contact with the wall. As angle α decreases, however, the centripetal force increases. If the speed does not exceed the critical value, a point is reached where the opposing forces are equal and the particle is near to falling. The angle at which this occurs is found by equating the centrifugal and centripetal forces, i.e.,

$$mg\cos\alpha = \frac{mu^2}{R-r}. \tag{4.9}$$

Transposing for $\cos\alpha$, we can transform Eq. 4.9 to

$$\cos\alpha = \frac{u^2}{(R-r)g}. \tag{4.10}$$

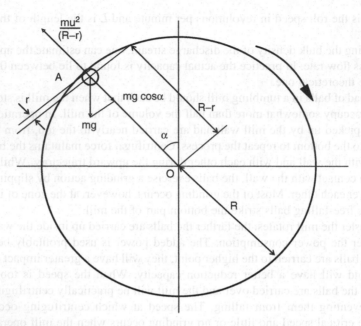

Fig. 4.6 Forces on a ball in a ball mill

The speed u is related to the speed of rotation by the equation

$$u = 2\pi N(R - r) \tag{4.11}$$

and, thus, Eq. 4.11 can be written as

$$\cos \alpha = \frac{4\pi^2 N^2 (R - r)}{g}. \tag{4.12}$$

At the critical speed, $\alpha = 0$, and consequently $\cos\alpha = 1$, and N becomes the critical speed N_c. With all these considerations, Eq. 4.12 transforms into

$$N_c = \frac{1}{2\pi} \sqrt{\frac{g}{(R - r)}}. \tag{4.13}$$

Rod mills can produce 5–200 t of material per hour reduced to about 1-mm size, whereas ball mills can give 1–50 t of fine powder per hour with 70–90% of the sizes in the range of 70 μm. The total energy requirement for a typical rod mill grinding different hard materials is about 5 hp h/t, whereas for a ball mill it is approximately 20 hp h/t. Tube mills and compartment mills normally need more power than this. As the product becomes finer, the capacity of a given mill diminishes and the energy requirement increases.

4.4 Criteria for Selection of Comminution Processes

4.4.1 General Considerations

In deciding how to crush or grind a food material, process engineers should consider factors such as the size distributions of the feed and the product, the hardness and mechanical structure of the feed, the moisture, and the temperature sensitivity of the feed. Regarding the size distributions of the materials, each type of crusher or grinder is intended for a certain size of feed and product. It is usually possible to exercise some control over the size of the feed, but sometimes it must be taken as it comes. As there is an upper limit on the size that can be accepted by a machine without jamming, if there is oversize material a guard screen is needed to keep pieces that are too large out of the crusher or grinder. In the case of too much undersize material, prescreening the feed can reduce the amount which goes through the equipment. For small-scale operation such a reduction is important as it decreases the capacity required; for large-scale equipment, though, the undersize particles simply pass through the throat, where there is always ample room, so removing them does not greatly affect the capacity. A general guide to equipment selection as a function of food material and size reduction range, is presented in Table 4.3.

Table 4.3 Application examples of size reduction machines

	Crushing rolls	Hammer mills	Attrition mills	Tumbling mills
Fineness range				
Coarse	●			
Intermediate	●	●		●
Fine and ultrafine		●	●	●
Food material				
Chocolate	●			●
Cocoa			●	●
Corn (wet)			●	
Dried fruits		●		
Dried milk		●		
Dried vegetables		●		
Grains	●		●	
Pepper		●	●	
Pulses			●	
Roasted nuts			●	
Salt		●		●
Spices		●		
Starch (wet)			●	
Sugar		●		●

4.4.2   Hardness and Abrasiveness

One of the major factors which governs the choice and design of size reduction machines is the hardness of the material to be processed. As a general rule, the hardness is defined in accordance with the Mohs scale, which is divided into ten grades of hardness. As a rule of thumb, with use of the Mohs scale, a material is considered soft if it has a value between 1 and 3, medium-hard if it has a value between 3.5 and 5, and hard if it has a value between 5 and 10. Many food materials, especially when dry, are brittle and fragile, with hardness on the Mohs scale on the order of 1–2. According to this, ball mills, hammer mills, roller mills, and attrition mills are very suitable to treat most of the solid foods commonly used in the food industry. Specific knowledge of the mechanical structure of the feed material is useful to determine the most likely force to be used in its size reduction. As mentioned above, many food materials are brittle and fragile, so compressive forces may be employed. Some other food materials have a fibrous structure and are not easily disintegrated by compressive or impact forces, so cutting may be required.

4.4.3   Mechanical Structure

Particular knowledge of the specific structure of the feed material can indicate the type of force most likely to be used in performing the size reduction. If the material is friable or has a crystalline structure, fracture may occur easily along cleavage planes, with larger particles fracturing more easily than smaller ones. In these cases, crushing using compressive forces is recommended. When few cleavage planes are present and new crack tips have to be formed, impact and shear forces may be more advisable. Many food materials have fibrous structures, so they are not easily reduced in size by compression or impact. In such cases, shredding or cutting may be the force needed to perform the desired size reduction.

4.4.4   Moisture

The presence of moisture can be either beneficial or inconvenient in comminution processes. The safety problems caused by dust formation arising during the dry milling of many solid materials are well known. The presence of small quantities of water has been found useful in the suppression of dust and, in applications where the presence of moisture is acceptable, water sprays are often used to reduce dust formation. Some other applications allow large quantities of water to be introduced in the size reduction process; wet milling of corn is a good example of this. On the other hand, in many cases a feed moisture content in excess of 2–3% can lead to clogging of the mill, with a consequent effect on throughput and efficiency. Agglomeration can also be caused by moisture, being undesirable when a free-flowing powder is needed to control the feed rate.

4.4.5 Temperature Sensitivity

As stated earlier, comminution is, possibly, the most inefficient unit operation in the food processing industry. The excessive friction that occurs in most size reduction machines causes increased heating, which can lead to a considerable rise in the temperature of the material being processed. Since food materials are normally heat-sensitive, degradation reactions can occur. The release of sticky substances caused by the temperature rise may also pose a problem. For these reasons, some crushing and grinding machinery may be equipped with cooling devices such as jackets and coils.

4.5 Applications

For the food industry size reduction is, without doubt, one of the most important processing steps. Size reduction is normally applied with an infinite variety of grinding characteristics. These range from readily grindable (sugar and salt), through tough-fibrous (dried vegetables) and very tough (gelatin), to those materials which tend to deposit (full-fat soy, full-fat milk powder). The fineness requirements may vary immensely from case to case. Many examples of applications of size reduction in food processes have been mentioned through this chapter. Summarizing, we can mentioned the milling of wheat, the refining of chocolate, the grinding of spices and dried vegetables, the breaking of cocoa kernels, the preparation of cocoa powder, the degermination of corn, the production of fish meal, the manufacture of chocolate, etc.

References

Beddow JK, Meloy TP (1980) Testing and Characterization of Powders and Fine Particles. Heyden and Son, London.

Bond FC (1963) Some recent advances in grinding theory and practice. Brit Chem Eng 8: 631–634.

Coulson JM, Richardson JF (1996) Chemical Engineering. Vol 2. Butterworth-Heinemann, Stoneham, MA.

Herdan G (1960) Small Particle Statistics. Butterworths, London.

McCabe WL, Smith JC, Harriot P (2005) Unit Operations in Chemical Engineering. 7th Ed. McGraw-Hill, New York.

Chapter 5
Size Enlargement

5.1 Introduction: Size Enlargement Processes

Size enlargement operations are used in the material process industries with different aims, such as improving handling and flowability, reducing dusting or material losses, producing useful structural forms, and enhancing appearance. Size enlargement operations are known by many names, including compaction, granulation, tabletting, briquetting, pelletizing, encapsulation, sintering, and agglomeration. Although some of these operations could be considered to be rather similar, e.g., tabletting and pelletizing, some others are relevant to a specific type of industry, e.g., sintering in metallurgical processes. In the food industry, the term "agglomeration" is applied to the process which has the main objective of controlling the porosity and density of materials in order to influence properties such as reconstitutability, dispersibility, and solubility. In this case the operation is also often referred to as instantizing, because rehydration and reconstitution are important functional properties in food processes. On the other hand, when size enlargement is used with the objective of obtaining definite shapes, the food industry takes advantage of a process which may shape and cook at the same time, known as extrusion. In a more general context, however, instantizing and extrusion of food processes are the two common categories of agglomeration—tumble/growth and pressure agglomeration—and are referred as such in the literature.

5.2 Aggregation Fundamentals: Strength of Agglomerates

Agglomeration can be defined as the process by which particles are joined or bind with one another, in a random way, ending with an aggregate of porous structure much larger in size than the original material. The term includes varied unit operations and processing techniques aimed at agglomerating particles (Green and Perry 2008). As mentioned above, agglomeration is used in food processes mainly to improve properties related to handling and reconstitution. Figure 5.1

E. Ortega-Rivas, *Non-thermal Food Engineering Operations*,
Food Engineering Series, DOI 10.1007/978-1-4614-2038-5_5,
© Springer Science+Business Media, LLC 2012

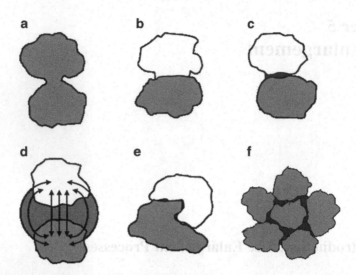

Fig. 5.1 Different binding mechanisms in agglomeration: (**a**) partial melting sinter bridges, (**b**) chemical reaction hardening binders, (**c**) liquid bridge hardening binders, (**d**) molecular and similar types of forces, (**e**) interlocking bonds, (**f**) capillary forces

shows some common binding mechanisms of agglomeration with bridges or force fields at the coordination points between particles (Pietsch 1991). The two-dimensional structure represented in such a figure is, in reality, three-dimensional, containing a large number of particles. Each particle interacts with several other particles surrounding it and the points of interaction may be characterized by contact, or by a distance small enough for the development of binder bridges. Alternatively, sufficiently strong attraction forces can be caused by one of the short-range force fields. The number of interaction sites of one particle within the agglomerate structure is called the coordination number. Particles in an agglomerate can be quite numerous, making it difficult to estimate the coordination number. Indirect measurement of the coordination number can be made as a function of other properties of the agglomerate. In regular packs of monosized spherical particles, the coordination number k and the porosity or void volume ε are related by

$$k\varepsilon \approx \pi. \tag{5.1}$$

Equation 5.1 gives a good approximation of the coordination numbers of ideal agglomerate structures. Table 5.1 lists several values of the coordination numbers calculated using Eq. 5.1 and compared with the ideal number for different structures, such as those illustrated in Fig. 5.2.

A general relation describing the tensile strength of agglomerates σ_t held together by binding mechanisms acting at the coordination points is

$$\sigma_t = \frac{1-\varepsilon}{\pi} k \frac{\sum\limits_{i=1}^{n} A_i(x, \ldots)}{x^2}, \tag{5.2}$$

Table 5.1 Geometrical arrangement, porosity, and coordination number of packings of monosized particles as shown in Fig. 5.2

Geometric arrangement	Porosity (ε)	Coordination number (π/ε)	k
Cubic	0.476	6.60	6
Orthorhombic	0.395	7.95	8
Tetragonal-spheroidal	0.302	10.40	10
Rhombohedral (pyramidal)	0.260	12.08	12
Rhombohedral (hexagonal)	0.260	12.08	12

Fig. 5.2 Packings of monosized spherical particles: (**a**) cubic, (**b, c**) orthorhombic, (**d**) tetragonal-spheroidal, (**e**) rhombohedral (*pyramidal*), (**f**) rhombohedral (*hexagonal*)

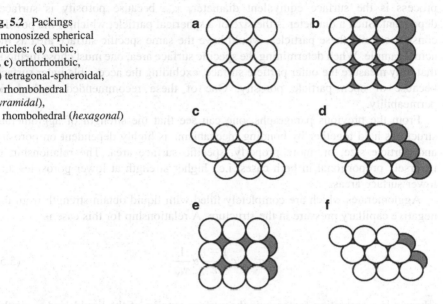

where A_i is the adhesion force caused by a particular binding mechanism and x is the representative size of the particles forming the agglomerate.

Substituting Eq. 5.1 into Eq. 5.2, we obtain the following relation:

$$\sigma_t = \frac{1-\varepsilon}{\varepsilon} \frac{\sum\limits_{i=1}^{n} A_i(x,...)}{x^2}. \qquad (5.3)$$

A further simplification results because many binding mechanisms are a function of the representative particle size x, and thus

$$\sigma_t = \frac{1-\varepsilon}{\varepsilon} \frac{\sum\limits_{i=1}^{n} A_i(x,...)}{x}. \qquad (5.4)$$

The three dots in parentheses in Eqs. 5.2, 5.3, and 5.4 indicate that A_i is also a function of other, unknown, parameters.

When liquid bridges have formed at the coordination points, A_i depends on the bridge volume and the wetting characteristics represented by the wetting angle. There are models available for predicting adhesion forces of various types (Pietsch 1991), but A_i might be of different magnitude at each of the many coordination points, owing to roughness or the microscope structure of particulates forming the agglomerates.

The representative particle size most appropriate to describe the agglomeration process is the surface equivalent diameter, x_{sv}, because porosity is surface-dependant. Such a diameter is the size of a spherical particle, which, if the powder consists of only these particles, would have the same specific surface area as the actual sample. When determining the specific surface area, one must chose methods that only measure the outer particle surface, excluding the accessible inner surface because of open particle porosity. One of these recommended methods is permeability.

From the previous paragraphs, one can see that the strength of agglomerate structures held together by bonding mechanisms is highly dependent on porosity and particle size, or, more properly, specific surface area. The relationship is inversely proportional in both cases, i.e., higher strength at lower porosities and lower surface areas.

Agglomerates which are completely filled with liquid obtain strength from the negative capillary pressure in the structure. A relationship for this case is

$$\sigma_t = c\,\frac{1-\varepsilon}{\varepsilon}\,\alpha\,\frac{1}{x_{sv}}, \tag{5.5}$$

where c is a correction factor, α is the surface tension of the liquid, and x_{sv} is the surface equivalent diameter of the particle. In order to apply Eq. 5.5, there must be a complete wetting of the solids by the liquid.

For high-pressure agglomeration and the effect of matrix binders, general formulas have not been developed yet. It can be considered, however, that the effects of variables will follow the trend described before, with porosity, particle surface, contact area, and adhesion all playing an important role. For non-metallic powders, the following equation can be used to evaluate the applied pressure p needed for agglomeration:

$$\log p = m V_R + b, \tag{5.6}$$

where V_R represents the relationship V/V_s, V being the compacted volume at a given pressure and V_s being the volume of the solid material to be compacted; m and b are constants.

5.3 Agglomeration Methods

With few exceptions, agglomeration methods can be classified into two groups: tumble/growth agglomeration and pressure agglomeration. Also, agglomerates can be obtained using binders or in a binderless manner. The tumble/growth method produces agglomerates of approximately spherical shape by buildup during tumbling of fine particulate solids. The resulting granules are at first weak and require binders to facilitate formation, and posttreatment is needed to reach the final and permanent strength. On the other hand, products from pressure agglomeration are made from particulate materials of diverse sizes, are formed without the need for binders or posttreatment, and acquire immediate strength.

5.3.1 Tumble/Growth Agglomeration

The mechanism of tumble/growth agglomeration is illustrated in Fig. 5.3. As shown, the overall growth process is complex and involves both disintegration of weaker bonds and reagglomeration by abrasion transfer and coalescence of larger units (Cardew and Oliver 1985). Coalescence occurs at the contact point when, at impact, a binding mechanism develops which is stronger than the separating forces. Additional growth of the agglomerate may proceed by further coalescence or by layering, or both. The most important and effective separation force counteracting the bonding mechanism is the weight of the solid particle. For particles below approximately 10 μm, the natural attraction forces, such as molecular, magnetic, and electrostatic forces, become significantly larger than the separation forces because of the particle mass and external influences. In such a way, natural agglomeration occurs.

The mechanism of tumble/growth agglomeration is similar to that of natural agglomeration. The particles to be agglomerated are larger, however, so the particle-to-particle adhesion needs to be increased by addition of a binder, such as water or other more viscous liquids, depending on the properties of particles being agglomerated and the required strength of the agglomerate structure. On the other hand, the collision probability may be enhanced by providing a higher particle concentration.

The conditions needed for tumble/growth agglomeration can be provided by inclined discs, rotating drums, any kind of powder mixer, and fluidized beds (Fig. 5.4). In general terms, any equipment or environment creating random movement is suitable for performing tumble/growth agglomeration. In certain applications, very simple tumbling motions, such as on the slope of storage piles or on other inclined surfaces, are sufficient for the formation of crude agglomerates. The most difficult task of tumble/growth agglomeration is to form stable nuclei because of the presence of few coordination points in small agglomerates. Also, since the masses of particles and nuclei are small, their kinetic energy is not high enough to cause microscopic deformation at the contact points, which enhances bonding.

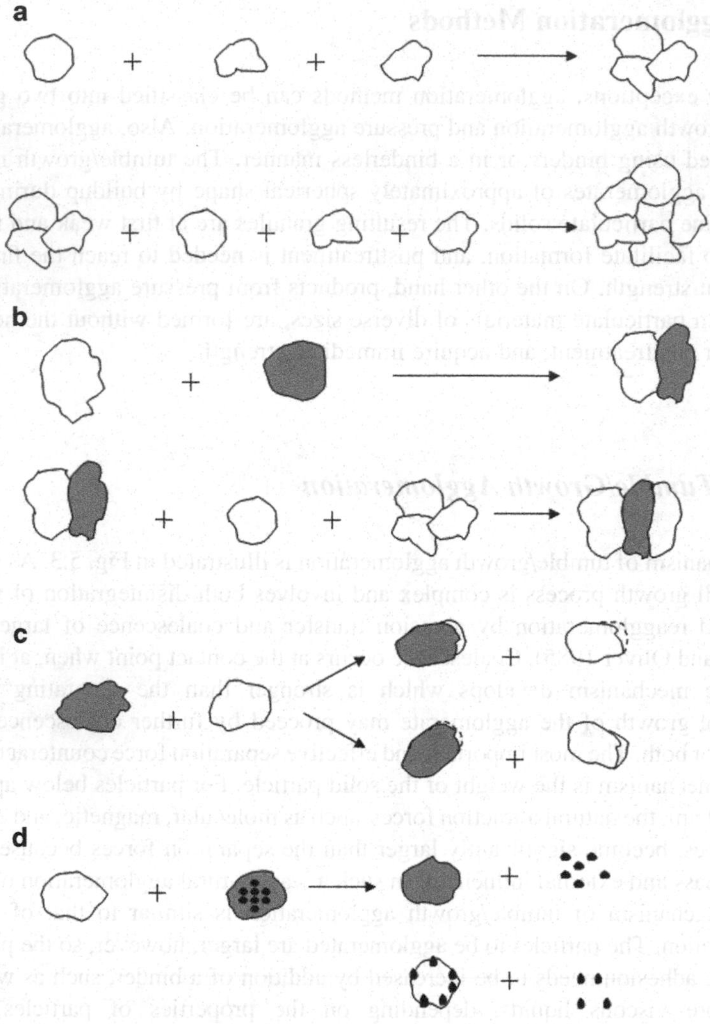

a

b

c

d

Fig. 5.3 Kinetics of tumble/growth agglomeration: (**a**) nucleation, (**b**) random coalescence, (**c**) abrasion transfer, (**d**) crushing and layering

Recirculation of undersize fines provides nuclei to which feed particles adhere more easily to form agglomerates.

In the whole process, tumble/growth agglomeration renders first weak agglomerates known as green products. These wet agglomerates are temporarily bonded by surface tension and capillary forces of the liquid binder. This is the reason why, in most cases, tumble/growth agglomeration requires some sort of posttreatment. Drying and heating, cooling, screening, adjustment of product characteristics by crushing, rescreening, conditioning, and recirculation of under-size material are some of the processes which have been used as posttreatment

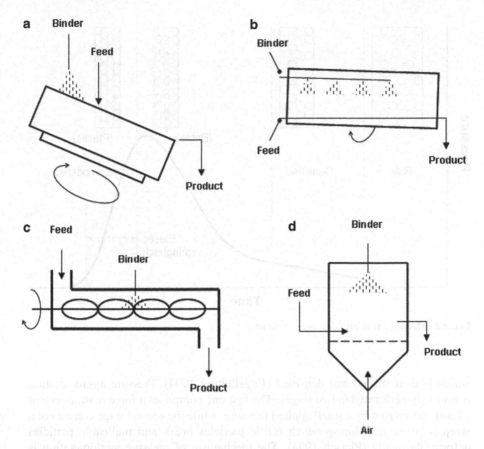

Fig. 5.4 Equipment for tumble/growth agglomeration: (**a**) inclined rotating disc, (**b**) inclined rotating drum, (**c**) ribbon powder blender, (**d**) fluidized bed

processes in tumble/growth agglomeration. Sometimes, a large percentage of recycle material must be rewetted for agglomeration and needs to be processed again, causing an economic burden to this technology (Pietsch 1983).

For larger size or mass of the particles being agglomerated by tumble/growth methods, the forces trying to separate newly created bonds during growth become significant, until further size enlargement by tumbling is not possible, even if strong binders are added. There is, therefore, a definite limitation on the coarseness of a particle size distribution, being in the range of x_{sv} between 200 and 300 μm.

5.3.2 Pressure Agglomeration

In contrast to tumble/growth agglomeration, where no external forces are applied, in pressure agglomeration pressure forces act on a confined mass of particulate solids,

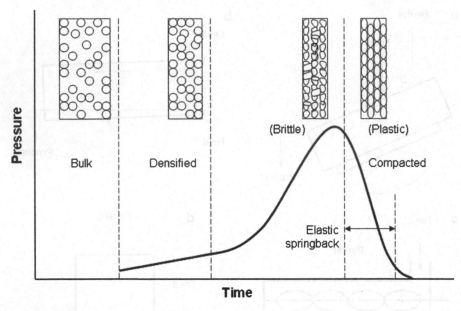

Fig. 5.5 Mechanism of pressure agglomeration

which is then shaped and densified (Engelleitner 1994). Pressure agglomeration is normally performed in two stages. The first one comprises a force rearrangement of particles caused by a small applied pressure, while the second stage consists of a steep pressure rise during which brittle particles break and malleable particles deform plastically (Pietsch 1994). The mechanism of pressure agglomeration is illustrated in Fig. 5.5. There are two important phenomena which may limit the speed of compaction and, therefore, the capacity of the equipment: compressed air in the pores and elastic springback. Both can cause cracking and weakening, which, in turn, may lead to destruction of the pressure-agglomerated products. The effect of these two phenomena can be reduced if the maximum pressure is maintained for some time, which is known as dwell time, prior to its release.

Pressure agglomeration can be performed employing a low-, medium-, or high-pressure mode. When low- or medium-pressure agglomeration is used, relatively uniform agglomerates can be obtained. Under these conditions, the porosity of the feed material is changed but no change in particle size or shape occurs. The feed mixture is often prepared with fine particles and binders, thus giving a sticky mass which may be formed by forcing it through holes in differently shaped screens or perforated dies. Agglomeration and shaping are, therefore, due to pressure forcing the material through the holes, as well as frictional forces. High-pressure agglomeration is characterized by a large degree of densification, resulting in low product porosity. Typically, the products from high-pressure agglomeration feature high strength immediately after discharge from the equipment. To increase the strength a bit further, addition of small amounts of binders or use of posttreatment methods is possible. High-pressure agglomeration is considered a versatile technique by which particulate material of any kind and size, i.e., from nanometers to centimeters, can be successfully processed under certain conditions.

Pressure agglomeration can be performed in different types of equipment. Extrusion, tabletting, and roller compaction are examples of techniques employed for agglomeration in the food industry. Extrusion involves a process of forcing a material through an orifice by means of pressure. Materials can be extruded in the solid state using a liquid, which acts as a plasticizer, or in the molten state using a heated extrusion barrel (melt extrusion). The extruded product is referred to as an extrudate. Extruders may be classified depending on their mode of operation into discontinuous (batch type) and continuous units. Extrusion agglomeration has proved to be a universally acceptable and economic method for compacting lumpy, long-fibered, powdery, and pasty materials, which have not been preground. Depending on the industry, the end product is called a pellet, a granulate, a cob, or a briquette. Generally, low- and medium-pressure agglomeration are achieved in extruders, including the screen extruder, the screw extruder, and the intermeshing-gears extruder. In tabletting or roller compaction, the particles are subjected to high pressure, leading to dense and mechanically stable agglomerates. High-pressure agglomeration is performed in presses such as the punch-and-die press, the compacting roller press, and the briquetting roller press. Low- and medium-pressure agglomeration yield relatively uniform agglomerates of elongated spaghetti-like or cylindrical shape, whereas high-pressure agglomeration produces pillow- or almond-like shapes. Figure 5.6 presents equipment used for low- and medium-pressure agglomeration, and Fig. 5.7 illustrates some common machinery for high-pressure agglomeration.

5.4   Selection Criteria for Agglomeration Methods

There are a large variety of techniques and equipment available to perform agglomeration processes in the food and processing industries. Some guidelines are given in the literature or can be provided by manufacturers. Table 5.2 summarizes some of the preliminary considerations to start the process of selecting an agglomeration method by a practicing engineer. In general terms, features of the feed, the product, and the method are the most important consider in the selection process (Pietsch 1991).

5.4.1   Feed Characteristics

The first characteristics of feed to be considered are the particle size and distribution. A limit in the range of few hundred micrometers defines the applicability

Fig. 5.6 Equipment used
for (**a**) low-pressure and
(**b**) medium-pressure
agglomeration: (*a.1*) screen
extruder, (*a.2*) basket
extruder, (*a.3*) cylindrical-die
screw extruder, (*b.1*) flat-die
extruder, (*b.2*) cylindrical-die
extruder, (*b.3*) intermeshing-
gears extruder

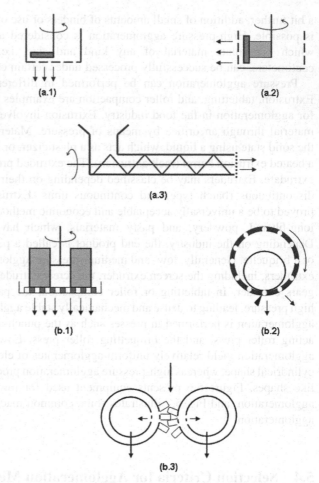

of methods using growth mechanisms based on coalescence in moving beds
of particles. Larger particles, which may also refer to seed agglomerates, can only
be incorporated if an adequate amount of binder or enough small particles are
present. Since small particles embed in larger ones, the strength of the agglomerate
is caused by the matrix of fine powder in this case. In general terms, it is difficult to
agglomerate narrow particle-size distributions or monosized particles. Adding a
binder can cause relatively large particles to agglomerate. It may be more economical, however, to crush larger particles to obtain material suitable for growth
agglomeration. This is particularly crucial when a high-porosity product is desired.
Pressure agglomeration is more suitable for larger feed size particles, e.g., sand-like
material or particles up to 20–30 mm in size. Since the external forces acting upon
the mass result in particle disintegration or deformation, the upper limit of feed
particle size is determined more by restrictions of the feeder than by the ability to

Fig. 5.7 Equipment used for high-pressure agglomeration: (**a**) compacting roller press, (**b**) briquetting roller press

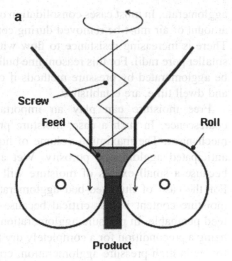

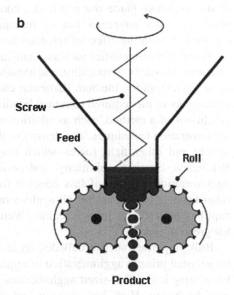

Table 5.2 Guidelines for the selection of an agglomeration process

Feed variables	Product variables	Method variables
Particle size and distribution	Agglomerate size and shape	Continuous or batch
Moisture content	Green strength	Capacity
Material characteristics	Cured strength	Wet or dry operation
Special features of material	Porosity and related features	Simultaneous processing
Bulk characteristics		Energy requirements
Binding characteristics		Costs

agglomerate. In most cases consolidation occurs in a short period, so a considerable amount of air must be removed during compaction to obtain sound agglomerates. There is increasing resistance to flow with decreasing particle size owing to the smaller pore radii. For this reason, fine bulk solids of about 150 µm or less can only be agglomerated by pressure methods if certain preconditions, such as low speed and dwell time, are established.

Free moisture can play an important role in growth agglomeration by coalescence. In such a case, moisture provides the binder or prevailing binder mechanism. The maximum volume of liquid must not be more than 95% of the anticipated agglomerate porosity. Wet agglomeration is sensitive to this limit because a small excess of moisture will cause the entire mass to turn to mud. For the case of fluidized-bed agglomerators, which may also act as driers, the moisture content is less critical because it has to be high enough to make the feed pumpable. In pressure agglomeration, moisture must be kept low, sometimes being a precondition for a completely dry feed. Owing to the extreme compression forces in high-pressure agglomeration, crushing, rearrangement, and deformation of the solid take place and result in a considerable reduction of porosity. Excess water is either squeezed out or remains in the mass as an incompressible component, with the effect of resultant low strength of the agglomerate.

Material characteristics such as chemical properties, particle density, brittleness, elasticity, plasticity, wettability, and abrasivity all play important roles in the choice of an agglomeration method. Particular chemical characteristics may be necessary to bring about the required chemical bonding, or may be incompatible with certain conditions of a method, such as addition of water or other liquids in most growth agglomeration techniques. The density of the feed particles determines the particle weight and other field forces which may counteract addition by coalescence. Brittleness, elasticity, plasticity, and abrasivity are most important for pressure agglomeration and are of less concern for growth methods. Wettability, on the other hand, is paramount for all agglomeration methods using surface tension and capillary forces in the growth regime. Wetting of particle surfaces is a requirement for green strength.

Bulk properties, such as bulk density and flowability, as well as temperature, can be adjusted prior to agglomeration to improve size enlargement. In preparation for briquetting into ration-sized agglomerates, vegetables, fruit juices, and food pulps may be frozen. High bulk density and unacceptable flowability are sometimes correlated by using two agglomeration methods in series. For example, fine feed food powders are preagglomerated to reduce the compaction stroke and improve the flow of feed into a die. This increases the speed of the turret for rotary table tabletting machines. At the same time, this technique avoids segregation of the feed mix by stabilizing the blend in a granular form.

Finally, the binding features of a given material must be considered for a possible agglomeration process. The binding characteristics should be assessed to decide whether agglomeration can be performed binderless, owing to the presence of an inherent binder in the material, or with the addition of a binder such as water and some other liquid.

5.4.2   Product Properties

Some properties such as the shape, dimensions, and particle size distribution of the agglomerated product also influence proper selection of a suitable method. Agglomerated products are normally expected to show an improvement in properties such as free-flowing characteristics and dust-free features. Granular, free-flowing, dust-free products can be manufactured using almost all methods of size enlargement. The task of narrowing down the size distribution of the discharge is done by screening out undersize and oversize components. While the fines are recirculated to the agglomerator, oversize particles are crushed and either rescreened or directly recirculated with the fines. Granular products can also be obtained by crushing and screening large agglomerates using criteria such as product porosity, product density, solubility, and reactivity. The size of the product is another important property. Spherical products are often desirable from an agglomeration operation, and such a shape can be obtained using all growth agglomeration methods. Contrastingly, spherical products cannot normally be obtained with high-pressure agglomeration, unless extremely accurate feed control can be established. By using some types of pressure agglomeration equipment, such as tabletting machines, one can obtain approximations to the spherical shape such as pillow-, lens-, and almond-shaped compacts.

Strength is a relevant property for the final product, but also plays a role during the size enlargement operation. In growth agglomeration, green agglomerates are formed fist and then must be cured to obtain permanent bonding. A weak state can exist when the binding mechanism of the green agglomerate disappears before the permanent, cured bond sets in. Unless a large amount of matrix binders is used, or agglomerates are cured at extremely high temperatures or by some chemical reaction, growth agglomeration products will normally be weaker than pressure agglomeration ones.

Different strength levels develop primarily because agglomerates growing by coalescence have higher porosity than those growing by pressure agglomeration. Materials which may disperse easily, and are only agglomerated to improve handling of the intermediate product, should just have enough strength to survive their short existence. In some other cases, a large specific surface is more important than high density and strength. Normally, an increase in external forces acting on the particulate matter during size enlargement will cause porosity and related characteristics to decrease, and density and strength to increase.

5.4.3   Alternative Methods

Agglomeration processes can be performed in a batch or a continuous manner, depending on specific requirements and applications. Batch modes generally have low capacity, but are characterized by better control than that exercised in a continuous process. Most large-volume applications operate in a continuous form, but may be accompanied by significant variations in quality. In growth agglomeration, uncontrolled buildup must be removed, whereas in pressure agglomeration, worn parts must be replaced. Most of the growth methods are wet processes using binder liquids to form green agglomerates, whereas high-pressure techniques are normally operated as dry processes. Agglomeration can sometimes be performed simultaneously with some other process. Simultaneous processing happens in mixer-granulators, granulator-driers, and even mixer-granulator-driers. Mixers are often also granulators in which both processes occur in different zones. In fluidized-bed granulators, however, agglomeration and drying can take place simultaneously.

Agglomeration has potential for diverse applications such as recycling of wastes containing valuable ingredients and disposal of particulate wastes without value in an environmentally safe and acceptable way. These applications cannot always find economic justification as typically they must be performed in compliance with legislation. Finely divided particulate material is often released to the environment in processes such as dry milling of cereals. Often, this is precipitated or removed by pollution control devices, but recontamination of the environment is an obvious concern and is normally regulated. In these cases, agglomeration methods can be employed to obtain a size-enlarged material in order to handle and dispose of it in a convenient way. Since, in most cases, one of the reasons for size enlargement is improvement of material handling, an agglomeration facility must be located near the source of the particulate solid. A suitable method must consider, therefore, the availability and cost of utilities and ancillary devices, such as binders and energy sources when wet granulation may require these. Sometimes, since disposal may represent the main aim of the process, the same task may be accomplished using roller presses for dry compaction, and granulation by crushing and screening.

5.5 Design Aspects of Agglomeration Processes

Agglomeration processes consist of varied operations within a complete system aimed at obtaining a desired product. Mixing and screening are two common operations which are part of an agglomeration process. There are some other varied posttreatment processes, such as those previously mentioned, which are used, mainly, to give strength to the agglomerate.

In tumble/growth agglomeration, if more than one feed powder is treated, the components must be metered and premixed. Homogenization may also be necessary owing to the risk of selective agglomeration as the particulates fed would involve many different sizes. During mixing, some of the liquid or dry binders can be added. It is also possible to feed all or part of the recycle material into the mixer. Aeration of the premixed material is an important factor, so particles will be loose and able to move randomly, in order to pick up the binder and agglomerate upon impact. A metered addition to the agglomerator improves and accelerates agglomerate growth by seeding the charge. This is because the recycle material, despite it representing an undersize product, consists largely of somewhat preagglomerated material. Control of the growth mechanism also requires addition of some of the liquid or dry binders to the agglomerator. Tumble/growth equipment produces green agglomerates that are better bonded by liquids. The agglomerate sizes and shapes are very varied, ranging within wide limits. Sometimes it is possible to screen the green agglomerates and feed only a narrow particle size distribution to the posttreatment stage. The moist recycle material should be sent directly to the agglomeration unit. Green agglomerates are often weak and sticky, so they tend to blind-screen quite easily. For this reason, separation of oversize or undersize material at this step may be avoided. The discharge from the posttreatment may also be screened to remove fines which may be formed by abrasion and breakage or, contrastingly, to retain oversize agglomerates which may have developed by secondary agglomeration of the still moist and sticky green agglomerates. Oversize agglomerates can then be crushed to obtain a recycle material which is normally dried, so it should be returned directly to the mixer and incorporated back into the process. In this way, tumble/growth agglomeration is an efficient process because recycling is performed continuously and losses are minimal. The problem of attrition and production of fines normally occurs in the handling and distribution of agglomerated food products after they have been released from the agglomeration process. Such a problem is severe and will be discussed in a separate chapter of this book.

With regard to pressure agglomeration, posttreatment is normally needed only in the case of low- and medium-pressure agglomeration. These methods typically require liquid binders to ensure easy formability. On the other hand, high-pressure agglomeration does not include a posttreatment and, in most applications, only dry additives may be added. In contrast to tumble/growth agglomeration, which requires fine particulates as well as dispersion and aeration features, pressure agglomeration operates well using particles of wide size distributions and without aeration. In fact, aeration of the feed prior to agglomeration must be avoided to facilitate the operation, and the maximum particle size which can be handled increases with increasing pressure. Large particles do not segregate and are easily incorporated during the forming of the agglomerate under pressure. As previously mentioned, when high forces are applied, brittle disintegration and plastic deformation occur. Also, a considerable volume reduction takes place with densification ratios as high as 1:5. Agglomerate strength increases with higher pressures during densification and forming. Knives can be used to cut extrudates and diverse types of separators may be used to break strings of briquettes into single units. Pressure agglomeration can also be used to obtain granulate products. In this case, a separator is used as a prebreaker. The product is obtained between the two decks of double-deck screens. The oversize material is crushed and rescreened, and the undersize material is recirculated. Multiple-step crushing and screening operations may be employed to improve the yield and obtain cleaner granular products.

5.6 Applications of Agglomeration

Agglomeration has many applications in food processing. In the context of instantizing, tumble/growth agglomeration is used in the food industry to improve reconstitutability of a number of products, including flours, cocoa powder, instant coffee, dried milk, sugar, sweeteners, fruit beverages powders, instant soups, and diverse spices. With regard to shaping, extrusion has been extensively used in grain process engineering to obtain an array of products from diverse cereals, principally ready-to-eat breakfast cereals.

References

Cardew PT, Oliver R (1985) Kinetics and Mechanics in Multiphase Agglomeration Systems. Notes of course on Agglomeration Fundamentals. 4th International Symposium on Agglomeration. Toronto University, Waterloo, Ontario, Canada.

Engelleitner WH (1994) Method Comparison. Notes of course on Briquetting, Pelletizing, Extrusion, and Fluid Bed/Spray Granulation. The Center for Professional Advancement. Chicago, IL.

Green DW, Perry RH (2008) Perry's Chemical Engineers' Handbook. 8th Ed. McGraw-Hill, New York.

Pietsch W (1983) Low-energy production of granular NPK fertilizers by compaction-granulation. In: Proceedings of Fertilizer'83, pp 467-479. British Sulphur Corp., London, UK.

Pietsch W (1991) Size Enlargement by Agglomeration. John Wiley & Sons, Chichester, UK.

Pietsch W (1994) Parameters to be considered during the selection, design, and operation of agglomeration systems. In: Preprints of 1st International Particle Technology Forum, Part I, pp 248-257. American Institute of Chemical Engineers, New York.

Chapter 6
Mixing and Emulsification

6.1 Introduction: Mixing of Fluids and Mixing of Solids

The unit operation in which two or more materials are interspersed in space with one another is one of the oldest and yet one of the least understood of the unit operations of process engineering. Mixing is used in many industries with the main objective of reducing differences in properties such as concentration, color, texture, and taste between different parts of a system. Food materials are multicomponent and multiphase systems, so blending of phases and components is, therefore, involved in many processes in the food industry and is one of the most commonly encountered unit operations. However, mixing mechanisms, especially those operated within food systems, are not well understood. Since the components being mixed can exist in any phase, a number of mixing possibilities arise. The mixing cases involving a fluid, e.g., liquid–liquid and solid liquid, are most frequently encountered, so they have been extensively studied.

The ultimate aim of mixing is to achieve a uniform distribution of the components by means of flow, which is normally generated in a mechanical way. The degree of uniformity attained differs widely. When miscible liquids, or soluble solids in liquids, are blended, very intimate mixing is possible and the process is governed mainly by fluid mechanics. With immiscible liquids, paste-like materials, and dry powders, the degree of uniformity obtainable is invariably less. In most of these cases, the processes cannot be simply studied using fluid mechanics principles, so use of theoretical concepts from rheology is needed. Mixing of dry solids presents the lowest probability of a uniform distribution of components. In practice, the best mix attainable is that in which there is a random distribution of ingredients. In mixing of dry solids, neither fluid mechanics principles nor rheological concepts are completely useful to describe the process.

Despite the importance of the mixing of particulate materials in many processing areas, fundamental work of real value to either designers or users of solid mixing equipment is still relatively sparse. It is through studies in very specific fields, such as powder technology and multiphase flow, that important advances in

E. Ortega-Rivas, *Non-thermal Food Engineering Operations*,
Food Engineering Series, DOI 10.1007/978-1-4614-2038-5_6,
© Springer Science+Business Media, LLC 2012

understanding the mixing of dry solids and paste-like materials have been made. A significant proportion of research effort in the food industry is directed toward the development of new and novel mixing devices for food materials.

6.2 Mixing of Liquids and Pastes

The efficiency of a mixing process depends on effective use of the energy input that will generate flow. The provision of an adequate amount of energy, the design of the mechanical device providing the energy, the configuration of the containing vessel, and the physical properties of the components are all important in the design of a mixer for food materials able to flow. The materials fed into a mixer can range from low-viscosity liquids to highly viscous pastes. There are three types of basic mixing systems that can be considered for blending of liquids and paste-like materials in the food industry:

1. A stationary vessel containing an impeller or impelling assembly mounted in a rotating shaft, which is widely used for mixing low-viscosity liquids or free-flowing solid–liquid solutions and suspensions.
2. A stationary vessel containing moving paddles, vanes, knives, plows, etc. Such mixers have been developed for mixing higher-consistency materials, such as viscous liquids, batters, and paste-like materials.
3. A moving vessel containing either moving or stationary paddles, vanes, knives, plows, screws, etc. Mixers in this category are used for the highest-consistency ingredients found in food processes, such as dough, pastes, and plastic materials.

The three types of mixers described above are used for mixing low-, intermediate-, and high-consistency mixes. The second and third types described may also handle dry powders and particulates. The purpose common to all three types of mixers is the promotion of flow.

6.2.1 Power Requirements: Dimensional Analysis

For low-viscosity-liquid mixing systems, the power introduced into a mixer by an agitator is determined by its speed of rotation, the configuration of the mixer, and the physical properties of the mixture. For intermediate- and higher-consistency mixing, some of the relations developed for low-viscosity mixing can be applied by using correction factors. Those mixes comprising some non-Newtonian fluids may be difficult to adapt to equations of power requirements, since they do not follow defined mathematical functions.

As the performance characteristics of fluid mixers involve a great number of variables, the use of dimensionless groups in this method is an obvious advantage. The basic concepts underlying dimensionless analysis and scale-up procedures have been used for a long time in classical mechanics. Perhaps the most common method of dimensional analysis is that attributed to Lord Raleigh known as the pi theorem. The basic steps of dimensional analysis can be found elsewhere (Szirtes 2007; Zlokarnik 2002), but it is known that the procedure consists in combining possible variables of the problem together into groups in order to find the least number of dimensionless groups. Then, by the use of physical arguments and experimental data, relationships between the dimensionless groups are derived.

In the particular case of mixing of fluids, if linear dimensions such as the depth of the liquid in the tank, the diameter of the tank, and the number, dimensions, and position of the baffles are all in a definite geometrical ratio with the impeller diameter, then the power input P to the agitator can be expressed as a function of the following variables:

$$P = f(\mu, g, \rho, N, D),\tag{6.1}$$

where μ is the viscosity of the liquid, g is the acceleration due to gravity, ρ is the density of the liquid, N is the rotational speed of the impeller, and D is the diameter of the impeller.

Dimensional analysis indicates that Eq. 6.1 may be reexpressed to include powers in all the variables as follows:

$$P = k(\mu^a g^b \rho^c N^d D^e),\tag{6.2}$$

where k is a constant.

If the dimensions are given in terms of fundamental units, i.e., mass (M), length (L), and time (T), then by substitution of such units of the variables, Eq. 6.2 transforms into

$$\frac{ML^2}{T^3} = k\left[\left(\frac{M}{LT}\right)^a \left(\frac{L}{T^2}\right)^b \left(\frac{M}{L^3}\right)^c \left(\frac{1}{T}\right)^d L^e\right].\tag{6.3}$$

By equating fundamental units on both sides of Eq. 6.3, and by algebraic means, one can group common powers together to give

$$P = k\left[(D^5 N^3 \rho)\left(\frac{D^2 N \rho}{\mu}\right)^a \left(\frac{DN^2}{g}\right)^b\right],\tag{6.4}$$

where k, a, and b depend on the system and its geometry. Equation 6.4 can be rewritten as follows:

$$\frac{P}{D^5 N^3 \rho} = k\left[\left(\frac{D^2 N \rho}{\mu}\right)^a \left(\frac{DN^2}{g}\right)^b\right].\tag{6.5}$$

Equation 6.5 is often referred to as the power equation. The term on the left-hand side of the equation is known as the dimensionless power number, the first term after the constant k on the right-hand side of the equation is the dimensionless Reynolds number Re, and the last term of the equation is the dimensionless Froude number Fr.

A number of workers have applied Eq. 6.5 to liquid mixing using impeller-type agitators in vertical cylindrical tanks. The Reynolds number represents the ratio of applied forces to the viscous drag forces. The Froude number represents the ratio of applied forces to gravitational forces. Vortex formation is a gravitational effect, so when it is suppressed, e.g., by using baffles in the mixing tanks, or at very low velocities represented by low values of the Reynolds number (less than 300), the contribution of the Froude number in Eq. 6.5 can be disregarded. The power equation for these cases is, therefore, simply represented as

$$\frac{P}{D^5 N^3 \rho} = k \left(\frac{D^2 N}{\mu} \right)^a. \tag{6.6}$$

Plots of the power number against the Reynolds number in log–log coordinates, so-called power curves, are available in the literature for particular mixer configurations. Power curves are independent of the vessel size and are particularly useful in the scale-up of liquid mixers from pilot plant studies. It must be considered, however, that a given curve is only applicable to the geometrical configuration for which it was developed.

6.2.2 Mixing of Liquids

The most commonly used mixer for liquids of low and moderate viscosity consists of a tank with an impeller agitator mounted. The tank commonly used is a vertical cylinder with a dished bottom in order to minimize dead spaces in corners. The filling ratio, i.e., the ratio of liquid depth to vessel diameter, is usually 0.5–1.5, with 1.0 being recommended for most purposes. The impeller is fixed to a rotating shaft, which creates currents within the liquid, and such currents should travel throughout the tank. Turbulent conditions of flow are necessary to maintain an appropriate degree of mixing. When a moving stream of liquid comes into contact with stationary or slow-moving liquid, shear occurs at the interface and low-velocity liquid is entrained in the faster-moving streams, mixing with the liquid therein. To achieve mixing in a reasonable time, the volumetric flow rate must be sufficiently fast to allow the entire volume of the mixing vessel to be swept out in proper time.

Three velocity components are created by an impeller within a mixing vessel: a radial component acting in a direction perpendicular to the shaft, a longitudinal component acting parallel to the shaft, and a rotational component acting in a

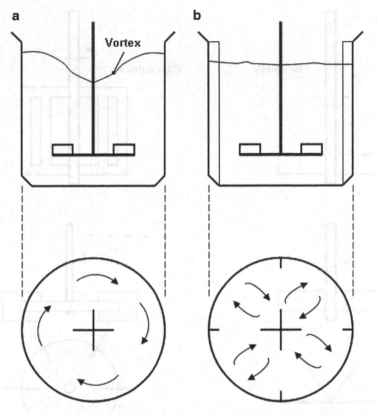

Fig. 6.1 Flow patterns in (a) unbaffled and (b) baffled vessels with an impeller agitator

direction tangential to the circle of rotation of the shaft. The tangential component generally adds little to the mixing process since, in the common design of an impeller rotating on a vertical shaft mounted centrally in the mixing vessel, this component promotes flow in a circular path around the shaft. Perhaps the most undesirable contribution of the tangential component in the design mentioned is the tendency to develop a vortex at the surface of the liquid. This vortex deepens with an increase in the rotation velocity of the impeller, and the centrifugal force developed can cause particles to separate in solid–liquid mixing, or segregation of phases in liquid–liquid mixing. Vortex formation can be avoided or greatly reduced by positioning the agitator off-center in a mixing vessel, or by the use of baffles inside the mixing tank, as shown in Fig. 6.1.

Three main types of agitators are used in mixing tanks for liquid mixing purposes: paddle agitators, turbine agitators, and propeller agitators. The most commonly used designs of these types of agitators are shown in Fig. 6.2. Paddle agitators generally rotate at speeds in the range of 20–150 rev/min with paddles generally measuring half to three quarters of the vessel diameter and the width of the blade generally being one tenth to one sixth of its length. Single-paddle agitators provide a gentle

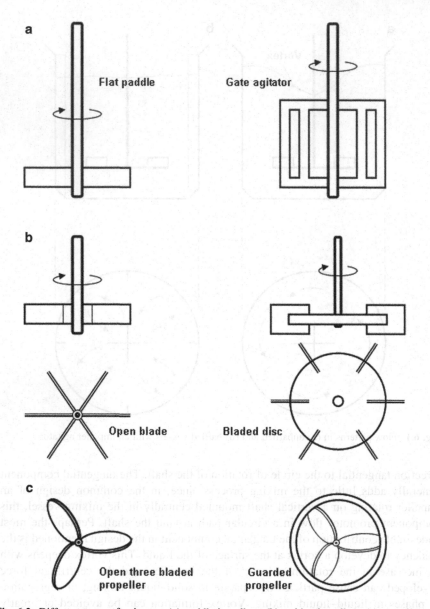

Fig. 6.2 Different types of agitators: (**a**) paddle impellers, (**b**) turbine impellers, (**c**) propellers

mixing action, which is often desirable when handling fragile, crystalline materials. Turbine agitators are formed from an impeller with more than four blades mounted on the same boss and fixed to a rotating shaft. The blades are generally smaller than paddles, measuring 30–50% of the vessel diameter, and commonly rotate at speeds in the range of 30–500 rev/min. Impeller blades may be pitched to increase vertical flow, and vane-disc impellers may be useful for dispersing gases into liquids.

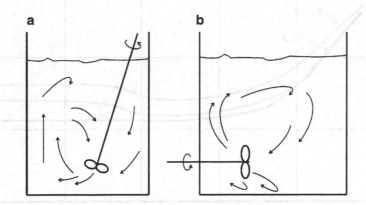

Fig. 6.3 Flow patterns in propeller-agitated systems: (**a**) off-center mounted propeller, (**b**) side-entering mounted propeller

Turbine agitators are used for mixing liquids of all sorts and are particularly effective for moderately viscous liquids. Propeller agitators consist of short-bladed impellers usually measuring less than one quarter of the vessel diameter, and rotating at higher speed in the range of 500 rev/min to several thousand revolutions per minute. Because of the predominantly longitudinal flow developed with these agitators, they are not particularly effective if mounted on vertical shafts located at the center of the vessel. They are commonly mounted off-center and with the shaft at an angle to the vertical, for example, as in Fig. 6.3a. In large tanks, propeller shafts may be mounted through the side wall of the tank in a horizontal plane, but off-center as shown in Fig. 6.3b. Propeller agitators are most effective in mixing low-viscosity liquids, as well as in dispersing solids and in emulsification operations.

A number of power curves have been derived for particular configurations of mixing vessels. For turbine agitators, curves such as the one shown in Fig. 6.4 can be used in conjunction with the diagram illustrated in Fig. 6.5. Scale-up of mixing tanks is facilitated by the use of power curves, along with diagrams of the particular dimensions of the mixing vessels.

6.2.3   Mixing of Pastes and Plastic Solids

As discussed in Chap. 3, rheology embraces the study of flow behavior in very general way, and defines two main types of flow: viscous and elastic. Whereas the first occur in fluids, the second is common in solids. An intermediate behavior of flow, known as viscoelastic flow, is also found. As stated earlier, mixing of liquids may be described by fluid mechanics principles, because the materials being mixed follow viscous flow. In the case of mixing of high-viscosity and paste-like materials, a common process in the food industry, the rheological behavior of these materials lies in the viscoelastic type of flow, so the principles involved in their operation differ greatly from those applied in liquid mixing. Furthermore, the

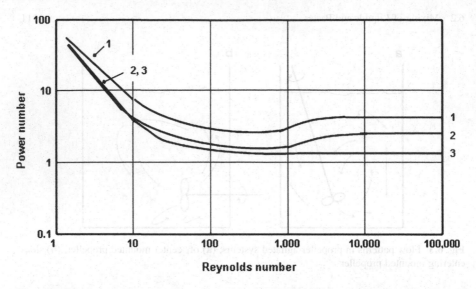

Fig. 6.4 Correlations of power versus Reynolds number for different agitators and baffles in mixing tanks according to the dimensions in Fig. 6.5. *Curve 1* six-blade bladed-disc turbine in a tank with four baffles ($D_a/W = 5, D_t/J = 12$); *curve 2* six-blade open-blade turbine in a tank with four baffles ($D_t/J = 12$); *curve 3* six-blade pitched-blade turbine (45°) in a tank with four baffles ($D_t/J = 12$) (Adapted from Geankoplis 2003)

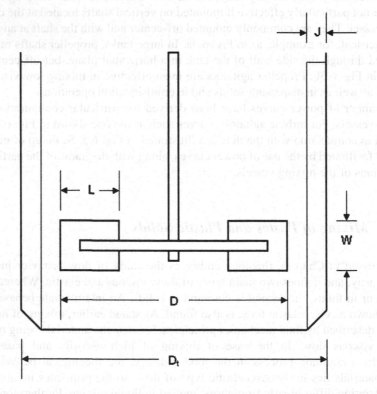

Fig. 6.5 Reference dimensions of a mixing tank and an agitator to be used in conjunction with the curves represented in Fig. 6.4 (Adapted from McCabe et al. 2005)

Fig. 6.6 Pan mixers:
(a) stationary, (b) rotating

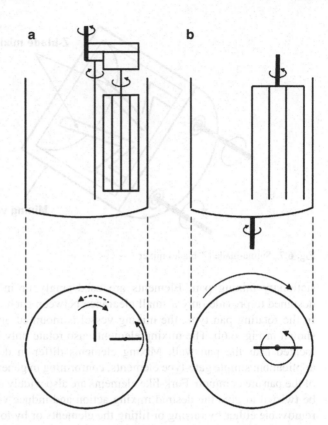

objectives of this type of mixing may not only be attaining a uniform mix, but also obtaining a product with certain desirable physical characteristics, for example, a dough for further processing. Many mixers are designed for specific duties, and may be related to specific unit operations (e.g., kneading), so few general principles apply for blending of pastes and plastic solids.

One general principle of mixers for viscoelastic-flow materials is that their performance depends on direct contact between the mixing elements and the materials of the mix, something that does not occur in liquid mixing. Thus, the material must be brought to the mixing elements or these elements must travel to all parts of the mixing vessel. The local actions responsible for mixing have been described as kneading, in which the material is pressed against adjacent material and against the vessel walls, and folding, in which fresh material is enveloped by already mixed material. The material is subjected to shear and is often stretched and torn apart by the action of the mixing elements. In general, as the consistency of the mixture becomes higher, the diameter of the impeller system is greater and the speed of rotation is slower. Mixers of reasonably versatile nature which follow these general principles and have a number of applications include pan mixers, horizontal trough mixers, and continuous paste mixers.

Pan mixers are of two general types, as shown in Fig. 6.6. In the stationary pan mixer the mixing elements move in a planetary path, moving to all parts of the

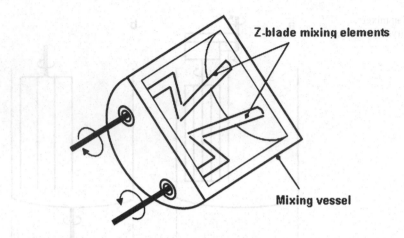

Fig. 6.7 Sigma-blade (Z-blade) mixer

stationary mixing pan. Elements are used singly or in pairs, and are usually designed to provide only a small clearance between each other and the pan walls. In the rotating pan type, the mixing vessel is mounted on a rotating turntable as shown in Fig. 6.6b. The mixing elements also rotate only in one position, and are located near the pan wall. Mixing elements differ in design depending on the application; simple gate-type elements, conforming in general shape to the contours of the pan are common. Fork-like elements are also widely used, and the blades may be twisted to give the desired mixing action and induce vertical motion. Pans are removable either by raising or tilting the elements or by lowering the pan support.

Horizontal trough mixers consist of a pair of heavy blades rotating on horizontal axes in a trough with a saddle-shaped bottom, as illustrated in Fig. 6.7. The blades rotate toward each other at the top of their cycle and may follow tangential or interlocking paths. The material is drawn over the point of the saddle and is kneaded and sheared between the blades, the container walls, and the bottom. The blades generally rotate at different speeds and may be independently driven or linked by gears. Blade elements differ in design, but a commonly found shape is the Z-blade, also known as a sigma blade. The vessel may be open or closed and usually tips up for emptying. It may also be jacketed for temperature control.

6.2.4 Applications

Mixing of liquids and paste-like materials is an important unit operation in food process engineering. Solutions of different sugars are used in an important number of processes, from the beverage to the confectionary industry. Syrups of many kinds are subjected to mixing and are used in different formulations of food products, as well as end products. As stated earlier, kneading is widely used in baking.

6.3 Mixing of Dry Solids

Mixing is more difficult to define and evaluate with powders and particulates than it is with fluids, but some quantitative measures of mixing of dry solids may aid in evaluating mixer performances. In actual practice, however, the proof of a mixer is in the properties of the mixed material it produces. These devices may be effective for many applications since they produce a mixed product with the required blending characteristics. Because of the complex properties of food systems which can themselves vary during mixing, it is extremely difficult to generalize or standardize mixing operation for wider applications of mixing devices, either novel or traditional. Developments in mathematical modeling of food-mixing processes are scarce and established procedures for process design and scale-up are lacking. As a result of this, it is virtually impossible to devise relationships between mixing and quality (Niranjan 1995), especially for blending of food powders.

Pertaining to solid foods, Niranjan (1996) mentioned as characteristic features of food mixing the fragile and differently sized nature of food products, as well as the segregating tendency of blended food systems on discharge. These characteristics along with some others, such as cohesiveness and stickiness, make food particulate mixing a complicated operation.

6.3.1 Mixing Mechanisms: Segregation

Three mechanisms have been recognized in mixing of solids: convection, diffusion, and shear. In any particular process one or more of these three basic mechanisms may be responsible for the course of the operation. In convective mixing masses or groups of particles transfer from one location to another, in diffusion mixing individual particles are distributed over a surface developed within the mixture, and in shear mixing groups of particles are mixed through the formation of slipping planes developed within the mass of the mixture. Shear mixing is sometimes considered part of a convective mechanism. Pure diffusion, when feasible, is highly effective, producing very intimate mixtures at the level of individual particles but at an exceedingly slow rate. Pure convection, on the other hand, is much more rapid but tends to be less effective, leading to a final mixture which may still exhibit poor mixing characteristics on a fine scale. These features of diffusion and convective mixing mechanisms suggest that an effective operation may be achieved by combination of both in order to take advantage of the speed of convection and the effectiveness of diffusion.

Compared with fluid mixing, in which diffusion can be normally regarded as spontaneous, particulate systems will only mix as a result of mechanical agitation provided by shaking, tumbling, vibration, or any other mechanical means. Mechanical agitation will provide conditions for the particles to change their relative positions either collectively or individually. The movement of particles during a mixing operation, however, can also result in another mechanism which may retard, or even reverse, the mixing process and which is known as segregation. When particles differing in physical properties, particularly size and/or density, are mixed, mixing is accompanied by a tendency to demix. Thus, in any mixing operation, mixing and demixing may occur concurrently and the intimacy of the resulting mix depends on the predominance of the former mechanism over the latter. Apart from the properties already mentioned, surface properties, flow characteristics, friability, moisture content, and tendency to cluster or agglomerate may also influence the tendency to segregate. The closer the ingredients are in size, shape, and density, the easier the mixing operation and the greater the intimacy of the final mix. Once the mixing and demixing mechanisms reach a state of equilibrium, the condition of the final mix is determined and further mixing will not produce a better result.

When particles having significantly different densities are mixed, it can be observed that the denser particles tend to settle to the bottom of the mixture, presumably to lower the total potential energy of the system. On the other hand, if coarse and fine particles are set in motion, the fines tend to segregate to the bottom by a possible percolation mechanism, in which the fines can pass through the interstices between larger particles. Williams (1968/1969) also showed that a single large particle placed on a bed of smaller particles will, on vibrating the bed, tend to rise toward the top of bed, even if its density is greater than that of the finer material. The explanation given for this is that the large solid particle, which will generally be denser than the loosely packed bed, causes a compaction of the bed immediately beneath it. In consequence, the freedom of movement of such a large particle, in response to vibration, will be restricted to the lateral and upward directions, and the net result will be a tendency to rise. These explanations of segregation are only applicable to specific situations. A general theory of segregation, regardless of the particular circumstances in which the operation takes place, has not yet been proposed to explain the segregation phenomena in particulate systems. In any blending operation the mixing and demixing mechanisms will be acting simultaneously. The participation of each of these two sets of mechanisms will be dictated by the environment and the tendency of each component to segregate out of the system. Since these two mechanisms will be acting against each other, an equilibrium level will be obtained as the final state of the mixture.

The importance of segregation for the degree of homogeneity achieved in mixing of solids cannot be overemphasized. Any tendency for segregation to occur must be recognized when selecting equipment for mixing solids. Segregation in a mixture of dry solids is readily detected by use of a heap test. A well-mixed sample of the solids is poured through a funnel so as to form a conical heap. Samples taken from the central core and from the outside edge of the cone should have essentially the same compositions if segregation is not to be a problem. When the two samples have significantly different compositions, it can be assumed that segregation will very likely occur unless the equipment is chosen very carefully. It is generally accepted that the efficiency of a mixing process must be related to both the flow properties of the components and the selection or design of the mixer.

6.3.2 Statistical Approach to Mixing of Solids

In the mixing of particulate solid materials, the probability of obtaining an orderly arrangement of particles, which would represent the perfect mixing, is virtually zero. In practical systems the best mixture attainable is that in which there is a random distribution of the ingredients. An ideal random distribution of two solid components in equal proportions would resemble a chess board, i.e., white and black squares in a perfect alternate pattern. In practice, however, a perfectly random mixture is commonly defined as one in which the probability of finding a particle of a constituent of the mixture is the same for all the points. Over the years, many workers have attempted to establish criteria for the completeness and degree of mixture. To accomplish this, very frequent sampling of the mix is using required and, tending to be statistical in nature, such an exercise is often of more interest to mathematicians than to process engineers. Thus, in practical mixing applications, an ideal mixture may be regarded as the one produced at minimum cost and which satisfies the product specifications at the point of use.

6.3.2.1 Sampling

Sampling is a crucial step in the mixing process because any form of control of mixing operations involves sampling procedures. The sample must be representative of the mixture and postsampling handling must not alter it. As sampling has a statistical aspect, sampling procedures following a purely mathematical approach are not completely practical in industrial situations. The confidence that can be placed in any results obtained from the sampling and analysis of a mixture is greatly influenced by several factors, including the method of sampling, the number of samples, the size of the sample, and the location in the bulk material from which the sample is taken. If sampling is not performed carefully, every mixture determination could be considered meaningless. Harnby (1992) recommended collecting samples from the outflow of a mixer, in the moving stream, instead of taking them from a static mass inside the mixer. In such a way, the possibility of bias in sample retrieval is minimized.

It can be demonstrated by statistical means that the larger the number of samples, the more reliable the results. For example, in the statistical theory of sampling, it is stated that the most representative sample will approach an infinite number of samples. In other words, the only way of including every member of a population being sampled is to take this whole population as sample. Since this would be unfeasible and unreasonable, for most practical purposes in mixing of food powders it has been established that at least 50, but not fewer than 20 samples, should be taken to obtain representative results. The size of the sample is also important. If a simple particle is drawn from the mixture, no mixing is evident. In contrast, if the whole mixture were to be analyzed, provided the ingredients were present in the correct proportions, complete homogeneity would appear to be achieved. As both of these extremes are impractical and unreliable, the recommended sample volume, often called the scale of scrutiny or characteristic sample size, falls between them and it is defined as the size of the sample which may be taken to correspond with the product usage. In animal feed manufacture, for instance, feed contains carbohydrates and proteins, balanced with added nutrients. In a particular feed, an animal must receive the correct balance of components. Provided that the required quantities of the necessary ingredients are present in the food consumed at each feed, intimate mixing is not essential. Thus, the volume of the sample which would give such a balance would be the useful one, regardless of its perfection in statistical terms.

6.3.2.2   Mixture Quality: Mixing Index and Rate

From some previously discussed aspects of the mixing process, it can be seen that food mixing is a complicated task that is not easily described by mathematical modeling. Mixture quality results from several complex mechanisms operating in parallel which are hard to follow and fit to a particular model. Dankwertz (1952) defined the scale and intensity of segregation as the quantities necessary to characterize a mixture. The scale of segregation is a description of unmixed components, whereas the intensity of segregation is a measure of the standard deviation of the composition from the mean, taken over all points in the mixture. In practice is difficult to determine these parameters, since they require concentration data from a large number of points within the system. They provide, however, a sound theoretical basis for assessing mixture quality. Taking into account the complexity of the components and interactions in mixing of food solids, one finds it rather difficult to define a unique criterion to assess mixture quality. A mixing end point or optimum mixing time can also be considered a very relative definition because of the segregating tendency of food powder mixing.

The degree of uniformity of a mixed product may be measured by analysis of a number of spot samples. Food powder mixers act on two or more separate materials to intermingle them. Once a material has been randomly distributed in another, mixing may be considered to be complete. On the basis of these concepts, the well-known statistical parameters mean and standard deviation of component concentration can be used to characterize the state of a mixture. If spot samples are taken at random from a mixture and analyzed, the standard deviation of the analyses s about the average value of the fraction of a specific powder $\bar{x}$ is estimated by the following relation:

$$s = \sqrt{\frac{\sum_{i=1}^{N}(x_i - \bar{x})^2}{N - 1}}, \qquad (6.7)$$

where x_i is every measured value of the fraction of one powder and N is the number of samples.

The value of the standard deviation on its own may be meaningless, unless it can be checked against limiting values of either complete segregation s_0 or complete randomization s_r. The minimum standard deviation attainable with any mixture is s_r and it represents the best possible mixture. Furthermore, if a mixture is stochastically ordered, s_r will be zero. On the basis of these limiting values of standard deviations, Lacey (1954) defined a mixing index M_1 as follows:

$$M_1 = \frac{s_0^2 - s^2}{s_0^2 - s_r^2}. \tag{6.8}$$

The numerator in Eq. 6.8 is an indicator of how much mixing has occurred, and the denominator shows how much mixing can occur. In practice, however, the values of s, even for a very poor mixture, lie much closer to s_r than to s_0. Poole et al. (1964) suggested an alternative mixing index, i.e.,

$$M_2 = \frac{s}{s_r}. \tag{6.9}$$

Equation 6.9 clearly indicates that for efficient mixing or increasing randomization M_2 will approach unity. The values of s_0 and s can be determined theoretically. These values are dependent on the number of components and their size distributions. Simple expressions can be derived for two-component systems. For a binary multisized particulate mixture Poole et al. (1964) demonstrated that

$$s_r^2 = \frac{pq}{\left(\frac{w}{q\left(\sum f_a w_a\right)_p + p\left(\sum f_a w_a\right)_q}\right)}, \tag{6.10}$$

where p and q are the proportions by weight of the components within a total sample weight w and f_a is the size fraction of one component of average weight w_a in a particle size range. For a given component in a multicomponent and multisized particulate system, Stange (1963) presented an expression for s_r, as follows:

$$s_r^2 = \frac{p^2}{w}\left[\left(\frac{1-p}{p}\right)^2 p\left(\sum f_a w_a\right)_p + q\left(\sum f_a w_a\right)_q + r\left(\sum f_a w_a\right)_r + \cdots\right]. \tag{6.11}$$

Equations 6.8 and 6.9 can be used to calculate mixing indices defined by Eq. 6.7. Another suggestion for the characterization of the degree of homogeneity in mixing of powders was given by Boss (1986), with the degree of mixing M_3 defined as

$$M_3 = 1 - \frac{s}{s_0}. \tag{6.12}$$

Some other mixing indices have been reviewed by Fan and Wang (1975).

McCabe et al. (2005) presented the following relationship to evaluate the mixing time t for blending of solids:

$$t = \frac{1}{k} \ln \frac{1 - 1/\sqrt{n}}{1 - 1/M_2},$$ (6.13)

where k is a constant and n is the number of particles in a spot sample. Equation 6.13 can be used to calculate the time required for any required degree of mixing, provided k is known and segregating forces are not active.

Mixing times should not be very long because of the, unavoidable, segregation nature of most mixtures of food solids. Instead of improving efficiency, long mixing times often result in poor blending characteristics. A graph of the degree of mixing versus time is recommended to select the correct mixing time quantitatively. Most cases of mixing of powders will attain a maximum degree of homogeneity in less than 15 min, when the correct equipment and working capacity have been chosen.

6.3.3 Powder Mixers

In general terms, mixers for dry solids have nothing to do with mixers involving a liquid phase. According to the mixing mechanisms previously discussed, solids mixers can be classified into two groups: segregating mixers and non-segregating mixers. The former operate mainly by a diffusive mechanism, whereas the latter practically involve a convective mechanism. Segregating mixers are normally non-impeller-type units, such as tumbler mixers, whereas non-segregating mixers may include screws, blades, and plows in their designs, and examples of them include horizontal trough mixers and vertical screw mixers.

Food powders can also be mixed by aeration using a fluidized bed. The resulting turbulence of passing air through a bed of particulate material causes the material to blend. The mixing times required in fluidized beds are significantly lower than those required in conventional powder mixers. Van Deemter (1985) discussed different mixing mechanisms prevailing in fluidized beds.

6.3.3.1 Tumbler Mixers

Tumbler mixers operate by tumbling the mass of solids inside a revolving vessel. These vessels take various forms, such as those illustrated in Fig. 6.8, and may be fitted with baffles or stays to improve their performance. The shells rotate at speeds up to 100 rev/min and have working capacities around 50–60% of the total. They are manufactured using a wide variety of materials, including stainless steel. This type of equipment is best suitable for gentle blending of powders with similar physical characteristics. Segregation can be a problem if particles differ particularly in size and shape.

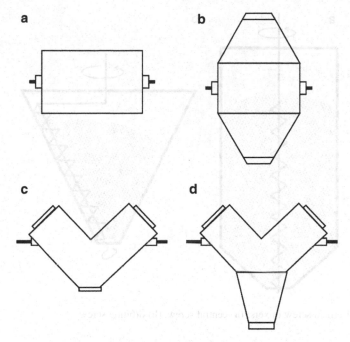

Fig. 6.8 Tumbler mixers used in food powder blending: (**a**) horizontal cylinder, (**b**) double cone, (**c**) V-cone, (**d**) Y-cone

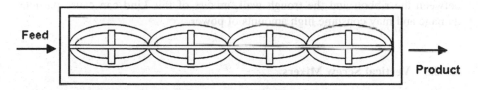

Fig. 6.9 Plain view of an open ribbon mixer

6.3.3.2   Horizontal Trough Mixers

Horizontal trough mixers consist of a semicylindrical horizontal vessel in which one or more rotating devices are located. For simple operations, single or twin screw conveyors are appropriate, and one passage through such a system may be good enough. For more demanding duties, a ribbon mixer, like the one shown in Fig. 6.9, may be used. A typical design of a ribbon mixer consists of two counteracting ribbons mounted on the same shaft. One moves the solids slowly in one direction and the other moves the solids quickly in the opposite direction. There is a resultant movement of solids in one direction, so the equipment can be used as a continuous mixer. Some other types of ribbon mixers operate on a batch basis.

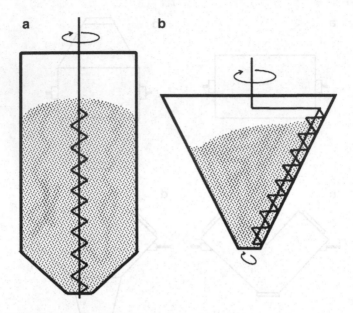

Fig. 6.10   Vertical screw mixers: (**a**) central screw, (**b**) orbiting screw

In these designs troughs may be closed, so as to minimize the hazard from dust, or
may be jacketed to allow temperature control. Because of the small clearance
between the ribbon and the trough wall, mixers of this kind can cause particle
damage and may consume high amounts of power.

6.3.3.3   Vertical Screw Mixers

In vertical screw mixers a rotating vertical screw is located in a cylindrical or cone-
shaped vessel. The screw may be mounted centrally in the vessel or may rotate or
orbit around the central axis of the vessel near the wall. Such mixers are shown
schematically in Fig. 6.10a and b, respectively. The latter arrangement is more
effective, and stagnant layers near the wall are eliminated. Vertical screw mixers
are quick, efficient, and particularly useful for mixing small quantities of additives
into large masses of material.

6.3.4   Selection and Design Criteria

Before selecting equipment for mixing solids, a careful study ought to be made of
several performance characteristics. As previously stated, mixing of solid foods is a
complex operation and mathematical modeling can be hardly used. Many factors

affect the operation of blending of solids, so process features such as mixing homogeneity and time, loading and discharging arrangements, power consumption, and equipment wear need to be analyzed and properly weighed up in order to make the most appropriate decision.

As already mentioned, blending uniformity and mixing rate are best evaluated using design graphs, bearing in mind the tendency to segregate and considering that long mixing times tend to worsen, rather than improve, efficiency. Mixing of solids can be a batch or a continuous operation. In batch mode, appropriate mixing design will produce the desired blend in a few minutes. Determination of the residence time in continuous operation is a more difficult task but, considering the main properties of blending of solids, such times also tend to be short, on the order of a few minutes or even seconds. The ribbon-type mixer is often used for continuous mixing, although it is also employed for batch mixing. Continuous mixing should be considered an option only if a single formulation can be run for an extended period, or when the fluctuations of the outgoing product are within the process requirements. When any of these factors are compromised, the batch mode of operation is preferred so as to ensure the most attainable mixing uniformity.

Loading and discharge are also important aspects, being more critical in continuous operation mode. The total handling system must be considered in order to obtain optimum charging and discharging conditions. This includes the efficient use of weight hoppers and surge bins, minor-ingredient premixing, location of discharge gates, and any other ancillary device used to aid the continuity of the process.

Power requirements are not a major concern when choosing solid mixers since other considerations usually predominate. Sufficient power must be provided in order to handle the maximum needs, as well as to prevent changes during the mixing operation. When the materials and operating conditions are subject to variation, enough power should be made available for the heaviest bulk-density materials and for the extreme conditions of operation. If the loaded equipment is to be started from rest, there should be sufficient power for this. When speed variations are desirable, this should be taken into account when planning the power requirements.

The ease, frequency, and thoroughness of cleaning are crucial when batches of different nature are to be mixed alternately in the same equipment. Plain tumbling vessels are easy to clean, provided that adequate openings are available. Areas which could be difficult to clean are seals or stuffing boxes, crevices at the baffle support, any corners, and discharge arrangements. If cleaning between different batches is time-consuming, several small mixers should be considered instead of a large single unit.

Dust formation should be avoided for safety reasons and when loss of dust may significantly affect the composition of the batch. Minimization of dust formation can be achieved by using less dusty but equally satisfactory batch ingredients, by employing palletized forms of extremely dusty materials, by proper venting so as to enable filtering or displaced air rather than unregulated loss of dust-laden air, or by addition of liquids if this is tolerable. Addition of water in small quantities is effective in minimizing dust upon discharge from the mixer, and it will also render the batch less dusty in subsequent handling stages. Water or any other liquid should be directed into the batch material instead of onto the bare surface of the mixer, since this could cause buildup. Spraying by using a nozzle is considered the most convenient way of incorporating moisture into the mixing batch. The nozzle spray pressure should be sufficient for the liquid to penetrate the batch, but not so high as to cause heavy splashing. The liquid should be added to the well-mixed batch, particularly when premature addition of liquid could impair the adequacy of blending. Also, both the time of addition and the time of application are important and should be carefully considered.

Equipment wear can be a crucial issue especially with abrasive materials such as grinding-wheel grains. An abrasion-resistant coating, such as a rubber coating, special alloys, or platings, needs to be considered in these cases. Any internal agitator device may wear even when operating at slow speed. Particularly when highly abrasive materials are to be mixed, the benefits of an agglomerated-breaking device must be weighed up against potential contamination and maintenance costs.

The capacity is an important factor when the mixed batches differ considerably in size from time to time. There are some features of the mixing operation which are not flexible in terms of capacity. For example, certain agitation devices in tumbler mixers do not function properly unless a given capacity is maintained. In general, the effect of the percentage of the mixer volume occupied by the batch on the adequacy of mixing should be considered when changes from recommended operating volumes are planned.

Food powder mixers should be selected or designed for a particular operation, firstly by analyzing and careful examining the applicable areas discussed above. Mixer selection should also involve consideration of the place of the mixer within the overall process. Possible consolidation of multistep processing of food solids deserves scrutiny at this time. If there is no machinery available which fulfills all the necessary requirements, to modifications should be considered in order to obtain the most desirable combination of features.

Pilot tests are relevant in the final decision for selection or design of a specific mixing process. In general, as the pilot unit becomes larger, the prediction of large-scale performance is more reliable. Published scale-up data for mixing of solids are very scarce, especially in food applications. With geometrically similar tumblers, if the speeds are adjusted to give comparable motion and the mixer volume fraction occupied by the load is the same, scale-up of results will be straightforward. The presence of internal rotating devices leads to difficulties in scaling up clearances, blade area to mixture volume, and sizes as well as speeds of the rotating devices. The actual materials to be processed in the industrial operation should be used if possible in the scale-up procedure. If substitute materials need to be used, they should have the same mixing characteristics. Differences in construction materials for the pilot and the production unit should be considered, since these may have a bearing on caking, abrasion, and some other adverse effects.

6.3.5   Applications

The applications of powder mixing in food systems are diverse and varied and include blending of grains prior to milling, blending of flours and incorporation of additives in flours, preparation of custard powders and cake mixes, blending of soup mixes, blending of spice mixes, incorporation of additives in dried products, and preparation of baby formula.

6.4   Emulsification

The operation in which two normally immiscible liquids are intimately mixed is defined as emulsification. One of the immiscible liquids, defined as the discontinuous or internal phase, is dispersed in the form of small droplets or globules in the other liquid, known as the continuous or external phase. The discontinuous phase is also referred to as the dispersed phase, and the continuous phase is also named the dispersing phase. In most emulsions the two liquids involved are oil and water. Pure oil and pure water are seldom involved, however. The oil phase may consist of oils, hydrocarbons, waxes, resins, and other substances, which generally behave as hydrophobic materials. The water phase may consist of solutions of salts, sugars, and other organic and colloidal materials, which are normally hydrophilic materials. To form a stable emulsion, a third substance, known as an emulsifying agent, needs to be included.

6.4.1   Introduction and Theory

When water and oil are mixed, two types of emulsion are possible. The oil may become the dispersed phase, giving an oil-in-water emulsion, or the water may become the dispersed phase, producing a water-in-oil emulsion. The emulsion formed tends to exhibit most of the properties of the liquid that forms the external phase. An oil-in water emulsion may be diluted with water and colored with water-soluble dyes, and conducts electrical current efficiently. On the other hand, a water-in-oil emulsion can only be successfully diluted with oil and colored with oil-soluble dyes, and conducts electrical current poorly. Two emulsions of similar composition may, therefore, show different properties depending on whether the oil or the water is the external phase. Factors influencing the type of emulsion obtained when mixing oil and water include the emulsifying agent, the relative proportion of the phases, and the method used to prepare the emulsion.

Emulsifying agents perform two functions in emulsification processes: they reduce the interfacial tension between the liquids to be emulsified and they protect the emulsion formed by preventing the droplets of the internal phase from coalescing. Numerous substances are used as emulsifying agents and include naturally

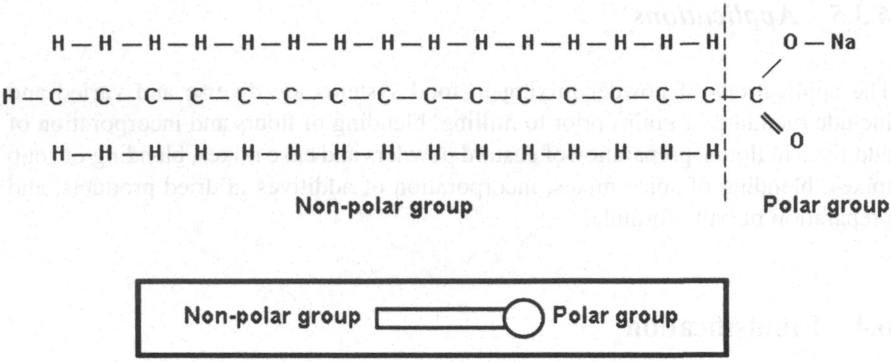

H — H — H — H — H — H — H — H — H — H — H — H — H — H — H   O — Na

H — C — C — C — C — C — C — C — C — C — C — C — C — C — C ⊢ C

H — H — H — H — H — H — H — H — H — H — H — H — H — H — H   O

Non-polar group   **Polar group**

Non-polar group [⬭────⬭] O Polar group

Fig. 6.11 Sodium palmitate molecule

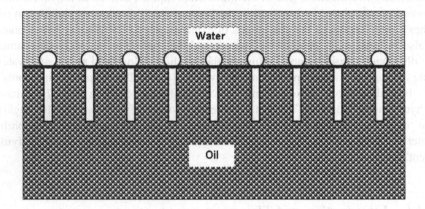

Fig. 6.12 Orientation of emulsifying agent molecules at the oil–water interface

occurring materials, such as proteins, phospholipids, and sterols, as well as wide range of synthetic materials, such as esters of glycerol, propylene glycol, sorbitan esters of fatty acids, cellulose esters, and carboxymethylcellulose. Most emulsifying agents consist of chemical compounds with molecules containing polar and non-polar groups: a simple example is soap (sodium palmitate) with the structural formula given in Fig. 6.11. In this case the hydrocarbon portion of the molecule is the non-polar group, whereas the –COONa portion is the polar group. As shown in Fig. 6.11, for convenience the non-polar portion of an emulsifying agent can be represented by a rectangle and the polar portion can be represented by a circle. In an emulsion, as represented in Fig. 6.12, the molecules of the emulsifying agent will orientate themselves at the interface so that the non-polar groups point toward the oil phase, and the polar groups point toward the aqueous phase. Thus, a layer of film of emulsifying agent is formed at the interface, which acts as a protective coating on the droplets of the internal phase, preventing them from coalescing under the influence of the interfacial tension.

The phase in which the emulsifying agent is most soluble, as a general rule, tends to become the external phase. This occurs when the polar or non-polar groups in the emulsifying agent are slightly out of balance. When the groups in the molecules of the emulsifying agent are perfectly balanced, then no definite trend to promote the formation of one type of emulsion over the other will be observed. On the other hand, if the polar and non-polar groups are definitely unbalanced, then the substance will be highly soluble in one of the phases, and will not remain at the interface. Therefore, it will not act as an emulsifying agent.

An emulsifying agent should be, as far as possible, specific in the type of emulsion it promotes. For food emulsions, the agent should be non-toxic, relatively odorless, colorless, and tasteless, and should be physically and chemically stable under the prevailing conditions of processing, handling, and storage.

6.4.2   Emulsification by Mixing

To form an emulsion, work must be done on the system to overcome the resistance to the creation of new interfaces arising from interfacial tension. Theoretically, the work of emulsification is equivalent to the product of the newly created surface and the interfacial tension. In addition, energy must be supplied to keep the liquids in motion and overcome frictional resistances. As a general principle, the work can be done in the liquids by subjecting them to energetic agitation. A type of agitation well suited to emulsification is one causing the droplets of the internal phase to be subjected to shear. By such an action, the droplets will be deformed and broken into smaller, more finely dispersed, ones. Provided the conditions are suitable, the protective film of emulsifying agent is absorbed at the interface and a stable emulsion is formed.

The time required for emulsification varies with the emulsion formulation and the technique employed and should be determined, preferably, by experimentation. For each case there is an optimum time below which a relatively unstable emulsion is formed. If agitation is continued beyond this optimum time, there is a risk of damaging the protective film and breaking the emulsion. The two phases are best prepared separately and the emulsifying agent is generally added to the external phase, but certain hydrophilic gums and colloids are best dispersed in the oil phase to minimize swelling and formation of lumps. Where premixing of the phases is practiced, the internal phase is usually added gradually to the external phase while the latter is agitated.

Slow-speed paddle agitators only find limited application for premixing or emulsification operations because of their relatively mild mixing action. Pan and Z-blade mixers find more applications in emulsification as a result of the shear developed in the mass of viscous materials. Other types of slow-speed agitators used in emulsification operations consist of rotating vessels in which the contents are tumbled. High-speed mixers of the turbine and propeller type are much more effective than slow-speed mixers as emulsion premixes and emulsifies, particularly for low-viscosity systems.

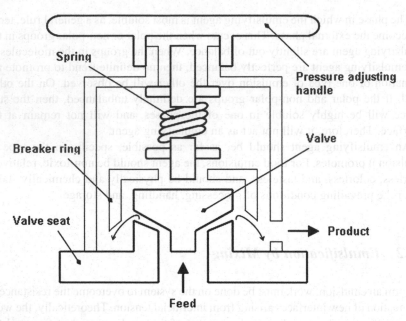

Spring

Pressure adjusting
handle

Breaker ring

Valve

Valve seat

Product

Feed

Fig. 6.13 Principle of the simple poppet-type homogenization valve

6.4.3   Homogenization

The term "homogenization" is used to describe the process in which the desired reduction in the size of droplets of the internal phase of a crude emulsion is brought about by forcing this emulsion through a narrow opening at high velocity. Equipment for homogenization includes pressure homogenizers, colloid mills, and ultrasonic emulsification devices.

A pressure homogenizer consists essentially of a homogenizing valve and a high-pressure pump. The valve provides an adjustable gap, on the order of few thousandths of a centimeter, through which the crude emulsion is pumped at pressures up to 69 MN/m^2. When they enter the gap, the liquids are greatly accelerated and the droplets of the internal phase shear against each other, are then deformed, and finally disrupted. In many valves, as the liquids leave the gap, they impinge on a hard surface set normal to the direction of flow, and this promotes disruption of the unstable droplets of the internal phase. The sudden drop in pressure as the liquids leave the gap and the collapse of bubbles due to cavitation may also contribute to the reduction in droplet size.

A common type of valve design known as the poppet valve is shown in Fig. 6.13. As can be observed, the liquids travel between the valve and its seat, causing it to lift against a strong spring, and leave the annular gap to impinge on a breaker ring. The homogenization pressure can be varied by adjusting the tension of the spring.

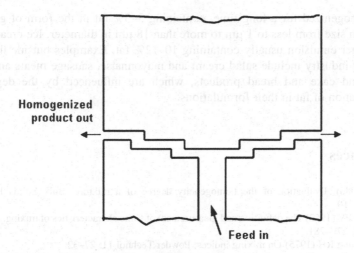

Fig. 6.14 A homogenizer valve with a stepped valve and seat

Since the clearances between the valves and seats are critical, the valves and seats must be carefully ground in and made of very hard materials such as Stellite, Monel metal, or stainless steel. An alternative design for a valve is illustrated in Fig. 6.14; this valve can sometimes rotate backward and forward in its seat.

Positive-displacement pumps are necessary to supply the feed to the homogenizer valve. For efficient results a steady feed rate is desirable and the most common system used is a multiple-cylinder plunger pump. An example of this type of pump is the triplex pump, which has three cylinders with pistons working in sequence. In such a system the output varies up and down by about 20%. This can be further reduced by the use of high-speed short-stroke plungers. The feed is normally introduced to the homogenizer as a crude premixed emulsion with droplet sizes on the order of 0.1–0.2 μm. Homogenization in two stages may be necessary to obtain satisfactory dispersion in some products. When proteins act as emulsifying agents, the small droplets formed after one passage through a valve at high pressure tend to cluster and clump together. This appears to arise from a poor distribution of the emulsifying agent over the newly created surfaces and the fat globules becoming entangled within solid films of the agent. To overcome this difficulty, such materials may be passed through a second homogenizing valve at a lower pressure (e.g., 2.8–3.4 MN/m^2), so the clusters are broken up.

6.4.4 Applications

Homogenization has typical applications in the dairy industry. Butter making is a unique example of emulsion technology since the raw material is an oil-in-water emulsion (milk) and the final product is an water-in-oil emulsion. Milk itself has been homogenized for a long time, containing 3–5% fat in the form of globules ranging in size from less to 1 μm to more than 18 μm in diameter. Ice cream is an oil-in-water emulsion usually containing 10–12% fat. Examples outside the milk and dairy industry include salad cream and mayonnaise, sausage meats and meat pastes, and cake and bread products, which are influenced by the degree of emulsification of fat in their formulations.

References

Boss J (1986). Evaluation of the homogeneity degree of a mixture. Bulk Solids Handl 6: 1207–1210.

Dankwertz PV (1952) The definition and measurement of some characteristics of mixing. Appl Sci Res 3A: 279–281.

Fan LT, Wang RH (1975) On mixing indices. Powder Technol 11: 27–32.

Geankoplis CJ (2003) Transport Processes and Separation Process Principles. Pearson Education Inc, Upper Saddle River, NJ.

Harnby N (1992) The selection of powder mixers. In: Harnby N, Edwards MF, Nienow AW (eds) Mixing in the Process Industries, pp 42–61. Butterworth-Heinemann, Oxford.

Lacey PMC (1954) Developments on the theory of particle mixing. J Appl Chem 4: 257–268.

McCabe WL, Smith JC, Harriot P (2005) Unit Operations in Chemical Engineering. 7th Ed. McGraw-Hill, New York.

Niranjan K (1995). An appraisal of the characteristics of food mixing. In: Singh RK (ed) Food Process Design and Evaluation, pp 47–68. Technomics, Lancaster, PA.

Niranjan K (1996) Mixing in the food industry. Int J Multiphase Flow 22: 98–98(1).

Poole KR, Taylor RF, Wall GP (1964). Mixing powders to fine scale homogeneity: studies of batch mixing. Trans Inst Chem Eng 42: T305-T315.

Stange K (1963) Die Mischgute einer Zufallsmischung aus drei und mehr Komponenten. Chem Ing Tech 35: 580–582.

Szirtes T (2007) Applied Dimensional Analysis and Modeling, 2nd Ed. Butterworth-Heinemann. Oxford.

Van Deemter JJ (1985). Mixing. In: Davidson JF, Clift R, Harrison D (eds) Fluidization, 2nd Ed, pp 331–354. Academic Press, London.

Williams JC (1968/1969) The mixing of dry powders. Powder Technol 2: 13–20.

Zlokarnik M (2002) Scale-up in Chemical Engineering. Wiley-VCH, Weinheim.

Chapter 7
Separation Techniques for Solids and Suspensions

7.1 Classification of Separation Techniques

Separation techniques are involved in a great number of processing industries and represent, in many cases, the everyday problem of a practicing engineer. In spite of this, the topic is normally not covered efficiently and sufficiently in higher-education curricula of some engineering programs, mainly because its theoretical principles deal with a number of subjects ranging from physics principles to applied fluid mechanics (Svarovsky 2000). In recent years, separation techniques involving solids have been considered under the general topic of powder and particle technology, as many of these separations involve removal of discrete particles or droplets from a fluid stream.

Separation techniques are defined as those operations which isolate specific ingredients of a mixture without a chemical reaction taking place. Several criteria have been used to classify or categorize separation techniques. One of these consists in grouping them according to the phases involved, i.e., solid with liquid, solid with solid, liquid with liquid, etc. A classification based on this criterion is shown in Table 7.1. Dry separation techniques constitute all those cases in which the particle to be isolated or segregated from a mixture is not wet, and include particular examples of the solid mixture and gas–solid mixture cases listed in Table 7.1. The most important dry separation techniques in processing industries have been reviewed by Beddow (1981). In food processing, there are important applications of dry separation techniques, such as the removal of particles from dust-laden air in milling operations, the recovery of the dried product in spray dehydration, and the cleaning of grains prior to processing. Solid–liquid separation is an important industrial process used for recovery of solids from suspensions and/or purification of liquids. Most of the process industries in which particulate slurries are handled make use of some form of solid–liquid separation technique. In the food industry, solid–liquid separations are widely used in a number of tasks, such as concentration and clarification of fruit juices, reduction of microorganisms in fermentation products, separation of coffee and tea slurries, desludging of fish oils, recovery of sugar crystals, and treatment of wastewater.

E. Ortega-Rivas, *Non-thermal Food Engineering Operations*,
Food Engineering Series, DOI 10.1007/978-1-4614-2038-5_7,
© Springer Science+Business Media, LLC 2012

Table 7.1 Classification
of separation techniques
according to the phases
involved

Type of mixture	Technique
Liquid–liquid	Distillation
	Extraction
	Decantation
	Dialysis and electrodialysis
	Parametric pumping
Solid–solid	Screening
	Leaching
	Flotation
	Air classification
Solid–gas	Cycloning
	Air filtration
	Scrubbing
	Electrostatic precipitation
Solid–liquid	Sedimentation
	Centrifugation
	Filtration
	Membrane separations

This chapter presents an overview of the main separation techniques used in food processing operations aimed at removing discrete particles suspended within a fluid or forming part of a mixture of two or more different solid food materials. The overview of such separation techniques will focus not only on the food manufacturing aspect, but also on the food preservation characteristics. Some separation techniques, such as membrane separations, can be used to remove microorganisms and pasteurize, or even sterilize, fluid foods without a temperature increase. By using physical removal of microorganisms, one can virtually guarantee safety, and sensory attributes are efficiently preserved. Many functional and quality aspects of processed foods can be improved by appropriate use of separation techniques.

7.2 Solid–Solid Separations

In the food industry the need to separate solid pieces, solid granules, or solid powders can be found in several stages of processing lines, from cleaning or sorting to specific processes such as flour fractionation. Screening can be considered the most important solid–solid separation technique, and can be used in any of the applications just mentioned, including cleaning and grading as discussed in Chap. 2. Screens are used for food sorting, including sorting based on size and sorting based on shape. Screening is also widely used in grain processing. Some other solid–solid separation techniques include classification using a fluid, such as air or water, and separation of a component on an intimate solid mixture using a solvent in the operation known as leaching.

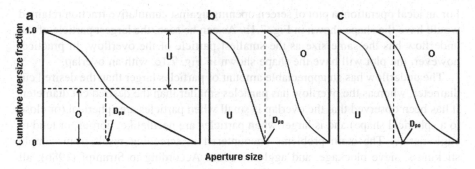

Fig. 7.1 Screening diagrams: (**a**) feedstock, (**b**) perfect separation, (**c**) actual screening

7.2.1 Screening

Screening is a technique for separating a mixture of solids particles of various sizes into several fractions, based on size difference. It consists in forcing the mixture through a screen with an aperture of specific size. Low-capacity plane screens are often called sieves. As a result of vibration or oscillation of a screen, particles smaller than a given aperture size pass through, thus being separated from the remaining mixture. Screens are made from metal bars, perforated or slotted metal plates, woven wire cloth, or fabric, such as silk bolting cloth. The metals used include steel, stainless steel, bronze, copper, nickel, and Monel. The screen surface may be planar (horizontal or inclined) or it may be cylindrical. The sizes of apertures of screens range from about 0.1 to 250 mm. In exceptional cases, the aperture may be as large as approximately 460 mm. The material passing through a given screen is termed undersize material, fines, or minus (−) material, and the material retained by screen of a given size is called oversize material, tails, or plus (+) material. Either stream may be the desired (product) stream or the undesired (reject) stream. Screening has two main applications: as a laboratory technique for particle size analysis, and as an industrial operation for fractionation and classification of particulate solids. Although the screen aperture, defined as the space between the individual wires of a wire mesh screen, is the preferred terminology for screening operations, the former designation of mesh number, defined as the number of wires per lineal inch, is still widely adapted.

7.2.1.1 Screening Fundamentals

The objective of a screening operation is to separate a feed stream into two fractions, an underflow that passes through the screen and an overflow that is rejected by the screen. An ideal screen would sharply separate the feed in such a way that the smallest particle in the overflow would be just larger than the largest particle in the underflow. Such an ideal separation would define a cut diameter D_{pc}, which would represent the point of separation between the fractions (Fig. 7.1).

For an ideal operation, a plot of screen opening against cumulative fraction retained would have the shape shown in Fig. 7.1b. As can be seen, the largest particle of the underflow has the same size as the smallest particle of the overflow. In practice, however, the plot will have the shape shown in Fig. 7.1c, with an overlap.

The underflow has an appreciable amount of particles larger than the desired cut diameter, whereas the overflow has particles smaller than the desired cut diameter. It has been observed that the overlap is small when particles are spherical (or close to a spherical shape) and is larger when particles are needlelike, fibrous, or tend to agglomerate. The main problems encountered in screening result from sample stickiness, sieve blockage, and agglomeration. According to Strumpf (1986), all these problems increase exponentially as the size of the screen aperture decreases.

7.2.1.2 Mass Balances in Screening

The efficiency of a screening operation may be evaluated by simple mass balances. Let F be the mass flow rate of the feed, O the mass flow rate of the tails, and U the mass flow rate of the fines; also, let X_F be the mass fraction of the tails in the feed, X_O the mass fraction of the tails in the overflow, and X_U the mass fraction of the tails in the underflow. Furthermore, the fractions of fines in the feed, overflow, and underflow will be $1-X_F$, $1-X_O$, and $1-X_U$, respectively. Since the total of the material fed to the screen must leave either as overflow or as underflow,

$$F = O + U. \tag{7.1}$$

The tails in the feed must also leave in the two streams, so

$$FX_F = OX_O + UX_U. \tag{7.2}$$

Elimination of U from Eqs. 7.1 and 7.2 gives

$$\frac{O}{F} = \frac{X_F - X_U}{X_O - X_U}. \tag{7.3}$$

Similarly, elimination of O gives

$$\frac{U}{F} = \frac{X_O - X_F}{X_O - X_U}. \tag{7.4}$$

The effectiveness of a screen is a measure of how well it performs the separation of tails and fines. If the screen functioned perfectly, all of material O would be in the overflow and all of material U would be in the underflow. A way of determining screen efficiency involves calculating the ratio of oversize material O that is actually in the overflow to the amount of material O entering with the feed, i.e.,

$$E_O = \frac{OX_O}{FX_F}. \tag{7.5}$$

Similarly, considering the fine material,

$$E_U = \frac{U(1 - X_U)}{F(1 - X_F)}. \tag{7.6}$$

The overall combined efficiency may be defined as the product of Eqs. 7.5 and 7.6, and it may be denoted simply as E:

$$E = \frac{OUX_O(1 - X_U)}{F^2 X_F(1 - X_F)}. \tag{7.7}$$

Substituting Eqs. 7.3 and 7.4 into Eq. 7.7 gives

$$E = \frac{(X_F - X_U)(X_O - X_F)X_O(1 - X_U)}{(X_O - X_U)^2(1 - X_F)X_F}. \tag{7.8}$$

Equation 7.8 is an alternative expression to evaluate screen efficiency without involving the streams and only using the fractions.

7.2.1.3 Capacity and Efficiency

The efficiency of separation, along with capacity, is the most important variable involved in industrial screening. Capacity and efficiency are opposing factors as maximum efficiency is related to low capacity, whereas high capacity is only attainable at the expense of efficiency. A reasonable balance between capacity and efficiency is desired in practice. Although accurate relationships are not available for estimating operating characteristics in screen operation, certain fundamentals apply and may be used as guidelines when running and designing a screening process.

The capacity of a screen is measured by the mass of material which can be fed per unit time to a unit area of screen, and can be simply controlled by varying the feed rate to the equipment. The efficiency obtained for a given capacity is dependent upon the specific nature of the screening operation. The chance of an undersize particle passing through the screen is a function of the number of times the particle strikes the screen surface, as well as its probability of passage in a single contact. If a screen is overloaded, the number of contacts is small and the chance of passing on contact is reduced by particle interference. The improvement of efficiency obtained at the expense of reduced capacity is a result of more contacts per particle and better chances of passing through the screen aperture on each contact.

A particle would have the ideal opportunity for passage when striking the surface perpendicularly, which would only be possible if it were oriented with its minimum dimensions parallel to the screen surface. Additional conditions would be no interference by other particles, as well as not sticking to, or wedging into, the screen surface. None of these conditions apply to actual screening, but this ideal situation can be used as a basis for estimating the effect of mesh size and wire dimensions on screen performance. If the width of a screen is negligible in comparison with the size of the openings, the wires will not interfere with passage of a particle and, practically, the entire screen surface will be active. In such a case, the probability of passage of a striking particle would approach unity. In actual screening, the diameter of the wire, or the fraction of the surface not constituting openings, is significant and the solid meshes strongly affect screen performance, especially by retarding the passage of particles nearly as large as the screen openings.

7.2.1.4   Factors Affecting Efficiency

The probability of passage of a particle through a given screen mainly depends on the fraction of the total surface represented by openings. Other factors are the ratio of the diameter of the particle to the width of an opening in the screen, and the number of contacts between the particle and the screen surface. If all these factors were constant, the average number of particles passing through a single screen opening in unit time would be constant and independent of the size of the screen opening. The capacity of a screen in mass per unit time divided by the mesh size would, therefore, be constant for any specified conditions of operation. In practice, however, a number of complicating factors appear and cannot be treated theoretically. Some of these disturbing factors are the interference of the bed of particles with their particular motion, the cohesion of particles to each other, the adhesion of particles to the screen surface, and the oblique direction of approach of the particles to the surface. When large and small particles are present, the large ones tend to segregate in a layer next to the screen, preventing the smaller particles from reaching the screen surface. All these factors tend to reduce capacity and lower efficiency.

Pertaining moisture, both dry particles and particles moving in a stream of water pass more easily through a screen opening than damp particles, which are prone to stick to the screen surface and to each other. As the particle size is reduced, screening becomes progressively more difficult, and capacity and efficiency tend to decrease. Blinding or clogging of the openings is particularly likely to occur when particles have sizes very similar to size of the screen aperture. In general terms, there are three possibilities for a given particle facing a screen aperture: (1) the particle is too large in relation to the aperture, so it will be easily retained, (2) the particle is too small in relation to the aperture, so it will go through easily, (3) the particle has a critical dimension, so it will be trapped and promote blinding or clogging of the screen surface.

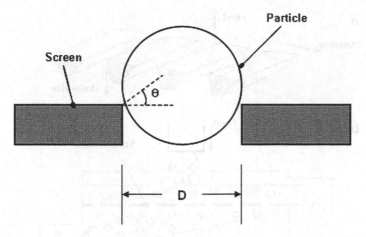

Fig. 7.2 Relationship of the particle size to sieve opening to cause blinding in screening

The above-mentioned critical dimension has been defined when the particle has an approximate size of $1.1D$ (see Fig. 7.2), which happens if the angle θ between the particle and the screen aperture is less than $\tan^{-1}\mu$, μ being the coefficient of friction between the particle and the screen material. The extreme case of blinding produces complete clogging of the screen with consequent damage which would impair separation and operation efficiency. Damaged screens should, therefore, be repaired or replaced immediately. It has been demonstrated that clogging is affected by the size of the screen aperture and the particle shape (Beddow 1980). It has also been reported that particle shape has a significant effect on efficiency for circular and rectangular screen apertures, but only a minor effect for square screen apertures (Nakayima et al. 1978).

7.2.1.5 Equipment Used for Screening

Screening as a unit operation may be performed with different types of equipment. Three types are very common: grizzlies (bar screens), screens, and trommels. The basic designs of each of these types of equipment are shown in Fig. 7.3. Grizzlies are used for screening larger particles (pieces greater than 25 mm). They consist of a set of parallel bars, spaced to the desired separation. The bars are often wedge-shaped to minimize clogging. They may be used horizontally or inclined at angles up to 60°. Vibrating grizzlies are available, the feed material passing over the screening surface in a series of jerks.

Screens are of many types: sifter, vibrating, shaking, centrifugal, and revolving, to name only a few. Sifter screens can be conveniently divided into circular-motion, gyratory-motion, and circular-vibrator types. They may be mounted in several decks and the rate of throughput can be increased by inclining the screen surface.

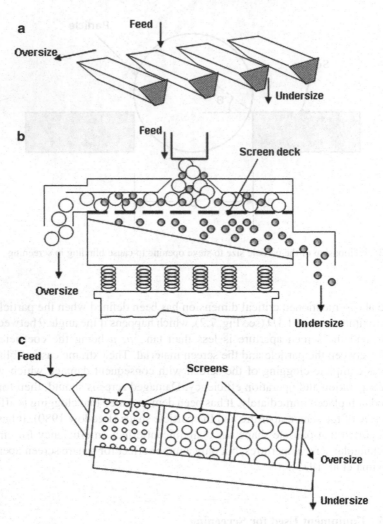

Fig. 7.3 Different types of industrial screens: (a) grizzlies (*parallel bars*), (b) high-capacity sifter with steep inclined plane, (c) revolving trommel

In centrifugal screens the surface consists of a vertical cylinder rotating at a constant speed with a gyratory motion. Gravity moves the oversize material down the length of the cylinder as fines are forced through the openings. These screens are normally inclined to the horizontal and may be multideck units, a series of screens being mounted beneath each other, permitting separation of a given feed stock into several size ranges.

Reels or trommels are revolving cylindrical screens mounted almost horizontally. Again, the screening surface may consist of wire mesh or perforated sheet. Hexagonal cross sections are also used since these lead to agitation, which aids the separation of fine material. The capacity of a trommel increases with increasing speed of rotation until a critical speed is achieved. At speeds greater than this, the material does not cascade over the surface but is carried round by centrifugal force and separation is seriously impaired. The critical speed of a trommel is given by

$$N = \frac{42.3}{D^{1/2}},$$
(7.9)

where N is the number of revolutions of the trommel per minute and D is the diameter of the trommel in meters.

7.2.1.6   Selection and Design

From the information given so far, a processing engineer should be able to evaluate the capacity and efficiency for a particular industrial screening operation. To avoid problems and come up with the most suitable choice for a process involving screening, apart from the characteristics discussed above, other equally important details should be given proper consideration.

Structural supports will be subjected to varying conditions, so attention must be given to adequately sizing them, not only for the loads to be carried but also for the deflection and vibratory conditions that could prevail in operation. Feed and product chutes also deserve careful consideration. The feed to the screen must be delivered so as to cause minimal amount of abrasion or disturbance of the bed of material on the screen. The trajectory of material being discharged from feed conveyors or other units of equipment must be considered, as must be the force of the falling material directed against an abrasive-resistant wear plate, or a dead bed of the material itself.

The screening operation should be considered within the context of the flowchart of the whole process. A description of the unit operations immediately preceding and following screening is quite relevant. Description of the equipment used adjacent to the screening step must be carefully considered. For example, there is little point in dry screening a dusty material if the succeeding stages are wet. The screening operation involved should also describe the methods used for controlling the feed rate, product collection, required screen efficiency, number and size of products, etc.

Several relevant properties of the material being separated must be known or determined to properly select or design adequate screening equipment. Some of these include the particle size distribution, particle shape, bulk density, moisture content, abrasiveness, and corrosiveness. The particle size distribution is essential to correctly size the screening unit, as well as to specify the type of screen to be installed. As previously mentioned, particle shape is determinant in promoting, or avoiding, blinding or clogging of screen units. Long or splinter-like, round or oval, or cubic particles will have slightly different screening characteristics, and may

Table 7.2 Screening operations in flour processing related to equipment

Operation name and details	Type of screen
Scalping: removal of a small amount of oversize material from a feed predominantly with fines	Grizzly
Coarse separation: separation of fractions larger than 4 mesh (4.76 mm)	Vibrating screen
Fine separation: separation of fractions between 4 mesh and 48 mesh (4.76 and 0.297 mm)	Vibrating screen, high speed, low amplitude
Ultrafine separation: separation of fractions smaller than 48 mesh (0.297 mm)	Vibrating screen, high speed, low amplitude plus sifter screens, static sieves, or centrifugal screens

have a great influence on the choice of cloth opening. Bulk density permits the determination of the volume of flow, and is a measure of the load to be carried by the screen. Moisture content, as stated earlier, may cause difficulties due to stickiness. Information on moisture content, along with data on the process following screening, will make selection of dry, damp, or wet screening possible. In some dry screening applications, when the moisture content is low, the choice of an appropriate screen medium will eliminate problems. Predrying of materials using heated screen cloths, or making them wet by adding water sprays, is equally effective in damp screening. The abrasive characteristics of the material have a great influence on the choice of construction materials, as well as on the selection of methods for loading, collecting, and transporting products. Corrosion properties will also influence the choice of construction materials used in screen frames, media, chutes, feedboxes, and other elements of the system. This property is relevant when choosing between dry and wet screening.

Worn or damaged screens will allow oversize particles to pass through the damaged area and the efficiency of the separation will be impaired. Damaged screens should, therefore, be detected and changed immediately in order to maintain the efficiency of the separation. Fine screens are very fragile and should be treated with extreme care.

7.2.1.7 Applications

One of the main applications of screening is in the flour industry, to separate the different fractions of flour. Specific terms relate to the fractions being removed as a function of the equipment used, as presented in Table 7.2. In this important application, the term "scalping" is often simply used to refer to removal of large particles, whereas "dedusting" would be employed when referring to separating small particles (Brennan et al. 1990). Cleaning and sorting of food pieces such as fruits, vegetables, or grains and pulses is another important application of screening, and was discussed in Chap. 2. Cleaning may be performed in trommels or flat-bed screens. The operation may be arranged so as to retain oversize material from the main stream while discharging a cleaned product, or else the screen may be used to retain the cleaned material as oversize material while discharging undesired material. Sorting may be performed using flat-bed screens, trommels, or drum screens.

7.2.2 Other Solid–Solid Separation Techniques

Some other solid–solid separation techniques include extraction from a solid piece or granulated solids using a specific solvent, and fractionation or classification of solids using a fluid, for example, air. A common example of solvent extraction (leaching) is the production of sugar from sugar beet. Sugar is extracted from sliced beets using hot water as a solvent. Multistage, countercurrent static bed systems are often used for this purpose. The beets are sliced to provide an increased surface area for extraction, while limiting also the amount of cell damage. Excessive cell damage can result in undesirable non-sugar compounds being released into the solution. Temperature control is also critical, as too high temperatures can cause peptization of the beet cells, with consequent contamination of the solution with non-sugar compounds. The final extracted solution contains about 15% dissolved solids. The solution is purified by settling and filtration and concentrated by vacuum evaporation. Sugar crystals are separated from the syrup by centrifugation and finally dried using hot-air dehydrators. Another example of extraction of solids is the manufacture of instant coffee. The ground roasted beans are extracted with hot water to produce a solution containing approximately 30% dissolved solids. Extraction is usually performed in a countercurrent, multistage static bed system, consisting of up to eight units. At any given time, one cell is isolated from the circuit to discharge the spent solids and replace them with fresh ground beans. The rate and degree of extraction are influenced by several factors, but the most important are water temperature and ground bean size. Too high a water temperature may increase the yield of dissolved solids, but can also impart undesirable flavors owing to excessive hydrolysis. If the grain is too fine, movement of the solution in the system is impaired, affecting efficiency. Final recovery of the coffee powder can be performed using centrifuges or hydrocyclones, followed by a final dehydration operation.

Air classification is a method of separating powdery, granular, or fibrous materials in accordance with the settling velocity and combined with the influence of particle size, particle density, and particle shape. The procedure of winnowing or aspiration is a traditional way to separate chaff from grain after threshing and is one of the simplest forms of air classification, as described in Chap. 2. Classifying and fractioning food powders as a function of size also have relevant applications. Ideally, the separation effect of an air classifier should be such that all particles which exceed a certain cut size are transported into a coarse fraction, and particles smaller than that size are transported to a fines fraction. In this sense, air classification basically consists in dividing particle size distributions of given powders and, as such, is a technique commonly used in combination with size-reduction equipment, normally to eliminate fines, which may affect properties such as wettability and dispersibility. The major interest in air classification is that it provides a means for separating small particles, in a dry manner, which cannot be readily achieved by sieving, i.e., below 50 μm. Ortega-Rivas and Svarovsky (2000) reported a successful sharp split of the particle size distribution of calcium carbonate into a fine fraction with mean particle size as fine as 6 μm.

The mode of operation of a typical air classifier is as follows. The inlet air is mixed with the material being separated. The feed particles are subjected to a centrifugal force from a revolving rotor and a drag force produced by the air current, which moves in a spiral direction toward the central shaft. As previously mentioned, the separation is based on differential mass, density, and shape. The larger and denser particles are influenced by the mass-dependent centrifugal forces and move toward the outside of the chamber, where they are removed by a screw conveyor or some alternative means. The smaller, less dense particles are more subjected to the frictional forces of the air current moving, therefore, with the air stream and leave from the center of the classifying chamber to be separated from the air stream by a cyclone. The relative magnitude of these two forces can be changed by altering the rotational speed of the disc and the air velocity. Varying either of these will result in an effect on the cut size. A common design of an air classifier consists of a rotating wheel with zigzag channels over its surface, each of them comprising six components. A diagram of this type of classifier is shown in Fig. 7.4.

Classifiers with the facility to change rotational speed and air velocity independently will be very flexible in operating terms. Although the separation mainly takes place within the classifying chamber, some preliminary removal of the coarsest particles may be achieved outside the chamber. The disc or turbine can be mounted on a horizontal or vertical axis. The latter produces a centrifugal force in the horizontal plane, favoring high throughputs but low precision in cut size. Air classifiers are categorized by reference to a number of factors, such as the presence or absence of a rotor, the drag force on the air, the relative velocity and direction of the air and particles, the use of directional devices such as vanes and cones, and the location of the fan and fines collection devices. Other important features include the capacity of the classifier and the energy utilization. A comprehensive treatise on classifier types has been presented by Klumpar et al. (1986).

Air classification is used in the food industry in important applications such as wheat flour fractionation to separate the coarse, low-protein fraction, from the fine, high-protein fraction. Other applications include classification of confectionery products, soy flour, potato granules, rice flour, lactose, and oleaginous fruits, removal of shells or hulls from disintegrated peanuts, cottonseed, rapeseed or cocoa beans, preparation of oat bran, and separation of gossypol from cottonseed protein.

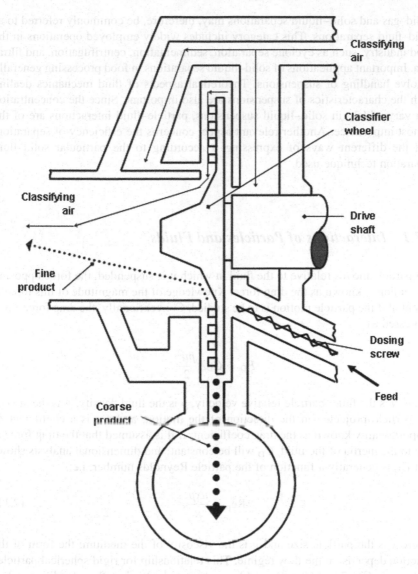

Fig. 7.4 A rotating, zigzag wheel air classifier

7.3 Properties of Fluids and Solids Relevant for Separation Techniques

As previously stated, many relevant applications in the food industry have to do with different types of phase separations. Two-phase systems in which one of the phases is a fluid and the other is a finely divided powder are particularly important.

Solid–gas and solid–liquid separations may, therefore, be commonly referred to as solid–fluid separations. This category includes widely employed operations in the food industry, such as cyclone separation, sedimentation, centrifugation, and filtration. Important applications of solid–liquid separations in food processing generally involve handling of suspensions. Theoretical aspects of fluid mechanics dealing with the characteristics of suspensions are also important. Since the concentration can vary widely in solid–liquid suspensions, particle–fluid interactions are of the utmost importance. Another relevant aspect concerns the efficiency of separation, and the different ways of expressing it according to the particular solid–fluid separation technique used.

7.3.1 Interactions of Particles and Fluids

If a particle moves relative to the fluid in which it is suspended, the force opposing the motion is known as the drag force. Knowledge of the magnitude of this force is essential if the particle motion is to be studied. Conventionally, the drag force F_D is expressed as

$$F_D = C_D A \frac{\rho u^2}{2}.$$ (7.10)

where u is the fluid–particle relative velocity, ρ is the fluid density, A is the area of the particle projected in the direction of the motion, and C_D is a coefficient of proportionality known as the drag coefficient. If it is assumed that the drag force is due to the inertia of the fluid, C_D will be constant, and dimensional analysis shows that C_D is generally a function of the particle Reynolds number, i.e.,

$$\mathrm{Re_p} = \frac{u x \rho}{\mu},$$ (7.11)

where x is the particle size and μ is the viscosity of the medium; the form of the function depends on the flow regime. This relationship for rigid spherical particles is shown in Fig. 7.5. At low Reynolds numbers under laminar flow conditions when viscous forces prevail, C_D can be determined theoretically from Navier–Stokes equations and the solution is known as Stokes's law and is represented by

$$F_D = 3\pi \mu u x.$$ (7.12)

This is an approximation that gives the best results for $\mathrm{Re_p} \to 0$; the upper limit of its validity depends on the error that can be accepted. The usually quoted limit for the Stokes region of $\mathrm{Re_p} = 0.2$ is based on an error of about 2% in the terminal

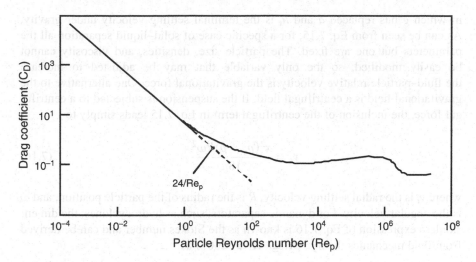

Fig. 7.5 Drag coefficient versus particle Reynolds number for spherical particles

settling velocity. Equations 7.10, 7.11, and 7.12 combined give another form of Stokes's law as follows:

$$C_D = \frac{24}{Re_p} \quad (Re_p < 0.2). \tag{7.13}$$

Equation 7.13 is shown in Fig. 7.5 as a straight line. For $Re_p > 1,000$, the flow is fully turbulent, with inertial forces prevailing, and C_D becomes constant and equal to 0.44 (the Newton region). The region between $Re_p = 0.2$ and $Re_p = 1,000$ is known as the transition region, and C_D is either described by a graph or by one or more empirical relations. For a particle of mass m under the influence of a field of acceleration a, the equation of motion is

$$m\left(\frac{du}{dt}\right) = ma - ma\left(\frac{\rho}{\rho_s}\right) - F_D, \tag{7.14}$$

where ρ_s is the solids density and t is the time.

Many solid–liquid separations of industrial interest are concerned with fine particles, which are the most difficult to separate, so the Reynolds numbers are low, often less than 0.2, owing to the low values of x and u. Therefore, it is reasonable to consider only the Stokes region. If this applies, considering also that the time necessary for the particle velocity to approach terminal settling velocity is short under the influence of gravity force, one can solve Eq. 7.14 to give

$$u_t = \frac{x^2(\rho_s - \rho)g}{18\mu}, \tag{7.15}$$

in which g has replaced a and u_t is the terminal settling velocity under gravity. As can be seen from Eq. 7.15, for a specific case of solid–liquid separation all the parameters but one are fixed. The particle size, densities, and viscosity cannot be easily modified, so the only variable that may be adjusted to increase the fluid–particle relative velocity is the gravitational force. One alternative to the gravitational field is a centrifugal field. If the suspension is subjected to a centrifugal force, the inclusion of the centrifugal term in Eq. 7.15 leads simply to

$$v_r = \frac{x^2(\rho_s - \rho)R\omega^2}{18\mu},$$ (7.16)

where v_r is the radial settling velocity, R is the radius of the particle position, and ω is the angular velocity. For dynamic separators, such as hydrocyclones, the dimensionless expression of Eq. 7.16 is known as the Stokes number and can be derived from fluid mechanics theory (Ortega-Rivas 2007) as

$$\mathrm{Stk}_{50} = \frac{x_{50}^2(\rho_s - \rho)v}{18\mu D_c},$$ (7.17)

where x_{50} is the cut size (i.e., the limiting size for a particle to be separated from the stream treated in a separating device), ρ_s is the solids density, v is the superficial characteristic velocity in the separator, and D_c is the characteristic dimension of the separating device.

As the concentration of the suspension increases, particles become closer together and interfere with each other. If the particles are not disturbed uniformly, the overall effect is a net increase in the settling velocity since the return flow caused by volume displacement predominates in particle-sparse regions. This is the well-known effect of cluster formation which is significant only in nearly monosized dispersions. With most practical widely dispersed suspensions, clusters do not survive long enough to affect the settling behavior and, as the return flow is more uniformly distributed, the settling rate steadily declines with increasing concentration. This phenomenon is referred to as hindered settling and can be theoretically approached in three different ways: as a Stokes's law correction by introduction of a multiplying factor; by adopting effective fluid properties for the suspension different from those of the pure fluid; and by determination of bed expansion with a modified version of the well known Carman–Kozeny equation. All the approaches can be shown to yield essentially identical results. Svarovsky (1984) reviewed some important correlations accounting for the hindered settling effect and demonstrated that their differences are minimal. According to this, the simple Richardson and Zakii equation is an obvious choice in practice. Such a relation can be expressed as

$$\frac{u}{u_t} = (1 - C)^{4.65},$$ (7.18)

where u is the settling velocity at concentration C and u_s is the settling velocity of a single particle.

7.3.2 Rheology of Suspensions

The dispersions and suspensions resulting from dissolving solids in liquids normally have non-Newtonian characteristics. The general classification of fluids and the definition of non-Newtonian fluid characteristics and the models were described and discussed in Chap. 3. For the case of non-Newtonian slurries and suspensions, modified forms of the Reynolds number and the Stokes number are necessary to describe the flow of the system within pipes and equipment, as well as the settling of particles being transported in the non-Newtonian flow stream.

The rheological properties of suspensions have been studied since the beginning of the twentieth century. The first research was done by Einstein (1906, 1911), in his classical study of the viscosity of dilute suspension of rigid spheres. His approach was purely hydrodynamic, and his model consists of an isolated sphere situated in a simple shear flow field in an infinite fluid. A number of workers have extended Einstein's analysis, and three approaches have been used: the theoretical basis of viscosity computation, the effect of particle texture and shape, and the effect of concentration. Several relations have been developed for suspensions in with nearest-neighbor interactions cannot be disregarded, such as the Guth and Simha (1936) equation for the first-order effect of spheres interacting with one another:

$$\frac{\mu}{\mu_o} = 1 + 2.5\varphi + 14.1\varphi^2,$$
(7.19)

where μ is the viscosity of the suspension, μ_o is the viscosity of the pure solvent, and ϕ is the volume fraction of the spheres in the suspension. A later correction gave a value of 12.6 for the last constant (Simha 1952).

For more concentrated systems, other approaches have been used. Several authors have made experimental studies of the effects of concentration on the viscosity of a suspension of spheres. The data they obtained were considerably scattered. Thomas (1963) tried to find the sources of scatter in the data. He found that scatter is caused in large part by variations in particle size. For small particle sizes (diameters less than 1–10 μm), colloidal forces become important and the viscosity begins to increase as the particle size decreases. The viscosity is also shear-rate-dependent in this case. For particles larger than 1–10 μm, the particle Reynolds number becomes significant. The inertial effects result in an increase in the relative viscosity with increasing particle size. Thomas developed a unique curve, by eliminating the diameter and shear rate effects. For low-concentration data, such a graph can be represented by

$$\frac{\mu}{\mu_o} = 1 + 2.5\varphi + 11.4\varphi^2.$$
(7.20)

Equations 7.19 and 7.20 can be used for correcting the viscosity of concentrated suspensions following Newtonian behavior.

Many concentrated suspensions follow non-Newtonian behavior and because the viscosity is no longer constant, such fluids should be characterized properly to use their parameters of characterization instead of the viscosity in calculations using some dimensionless groups. Expressions for the Reynolds number to account for the non-Newtonian flow behavior, particularly for power-law fluids, were described and discussed in Chap. 3. For the case of settling of non-Newtonian suspensions, it has been reported (Ortega-Rivas 2007) that the Stokes number Stk_{50} can also be expressed in terms of the parameters of characterization of non-Newtonian suspensions, using a procedure similar to that described for the Reynolds number. A generalized Stokes number $(Stk^*)_{50}$ for settling of power-law suspensions can be expressed as follows

$$Stk^*_{50}(r) = \frac{x_{50}^{n'+1}(r)(\rho_s - \rho)v^{2-n'}}{18K(3)^{n-1}\left(\frac{2n+1}{3n}\right)^n D_c},$$ (7.21)

where $x_{50}(r)$ represents the "reduced" cut size including the dead flux effect of particles following partition flow by inertia, ρ_s is the solids density, ρ is the liquid density, v is the superficial characteristic velocity of the separator, and D_c is the characteristic linear dimension of the separating equipment.

7.3.3 Efficiency of Separation Processes

As previously described when referring to primary properties of particulate solids in Chap. 3, the solid phase of suspensions to be treated in solid–liquid separation equipment generally consists of an immense number of particles of diverse sizes and shapes. All this population of particles needs to be identified or characterized. The frequency of occurrence of particles of every size present, arranged and presented in a statistical manner, is known as the particle size distribution. The most common way of presenting such distributions as well as the types of distributions important to particle technology can be found in specialized literature (Barbosa-Canovas et al. 2005). A variety of quantities have been used to represent the particle size in a particle size distribution, and are described in the literature (Ortega-Rivas 2005). The most common ones are equivalent sphere diameters, equivalent circle diameter, and statistical diameters. The equivalent sphere is the diameter of a sphere which has the same property as the particle itself. Such a property could be the settling velocity. The diameter derived from the settling velocity is known as the Stokes diameter and is a very useful quantity for solid–liquid separations, especially for those techniques in which the particle motion relative to the fluid is the governing mechanism. The amount of particle matter which belongs to specified size classes on the particle axis may be represented in several ways. The number of particles and the mass of particles are the most commons ones, but surface area and volume are used as well.

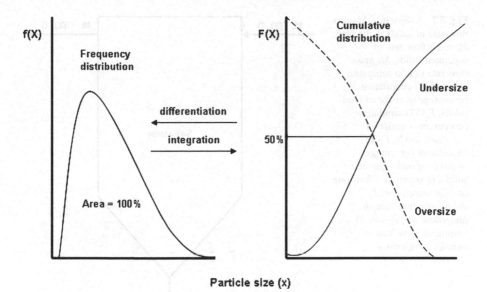

Fig. 7.6 Relationship between frequency and cumulative distributions

For solid–liquid separation techniques, the most convenient form of expressing the amount of particle matter is by mass, because the balances necessary to define the performance are normally mass balances.

In general, the particle size distribution can be presented as frequencies $f(X)$ or cumulative frequencies $F(X)$, which are related to each other by the following equation:

$$f(x) = \frac{\mathrm{d}F(X)}{\mathrm{d}X}. \tag{7.22}$$

The graphical representation of a particle size distribution is usually plotted in a cumulative form. In a typical cumulative plot, points are entered showing the amount of particulate material contributed by particles below or above a specified size. Hence, the curve has a continuously rising or deceasing character. These oversize and undersize distributions, as illustrated in Fig. 7.6, are simply related by

$$F(X)_{\text{oversize}} = 1 - F(X), \tag{7.23}$$

where $F(X)$ is the cumulative fraction undersize.

A cumulative plot will, therefore, include a broad range of particle sizes. It is often convenient, however, to refer to a single characteristic size for the system. Many characteristic sizes have been proposed, most of them involving a mathematical formula. An important one, which can be read off any cumulative plot of the particle size data, is the median particle size. It is defined as that particle size for

Fig. 7.7 A separator. M mass flow rate of solids in the feed, M_c mass flow rate of separated solids, M_f mass flow rate of non-separated solids, $F(X)$ cumulative percentage oversize of feed solids, $F_c(X)$ cumulative percentage oversize of separated solids, $F_f(X)$ cumulative percentage oversize of non-separated solids, Q volumetric flow rate of feed suspension, U volumetric flow rate of underflow suspension, O volumetric flow rate of overflow suspension

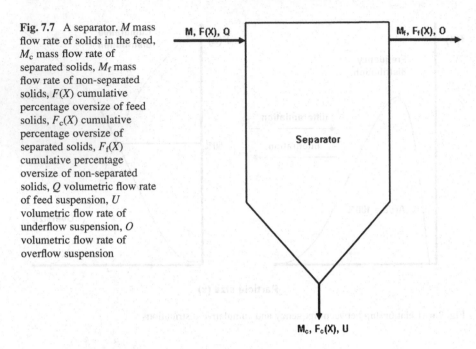

which the particle amount equals 50% of the total. If the particle size is represented by a number, such a point is called the number median size. If mass is used as the measure of particle amount, this parameter is known as the mass median size.

7.3.3.1 Total Gravimetric Efficiency

To evaluate the efficiency of separation it is necessary to take into account that solid–liquid separation is basically an imperfect process. Whereas the underflow is always wet slurry, the overflow can be considered as a turbid liquid. A single efficiency number can never be capable of fully describing the result of separation, except when it is ideal. The imperfection of solid–liquid separation has resulted in the need to express efficiency by different means. Svarovsky (2000) presented a good review of these expressions for the efficiency and reported that for most solid–fluid separation applications, common relationships for efficiency can be established. The first and most obvious definition of separation efficiency is simply the overall mass recovery as a fraction of the feed flow rate. According to Fig. 7.7,

$$E_t = \frac{M_c}{M},$$ (7.24)

where all the components are as defined in Fig. 7.7.

If there is no accumulation of solids in the separator, then

$$M = M_c + M_f,$$ (7.25)

and there is a choice of three possible combinations of the material streams for the total efficiency testing. It can be shown (Trawinski 1977) that if all the operating conditions are equal, the most accurate estimation of the local efficiency comes from the two leaving streams.

7.3.3.2 Partial Gravimetric Efficiency

The total efficiency defined by Eq. 7.24 includes all particle sizes present in the feed solids. If only a narrow range of particle sizes is of interest, another efficiency of separation particular to that range can be defined. The mathematical expression of such partial efficiency is

$$E_p = \left(\frac{M_c}{M}\right)_{x_1/x_2},$$ (7.26)

where x_1 and x_2 represent the particle size limits of a definite range.

7.3.3.3 Grade Efficiency and Cut Point

If the particle size range in Eq. 7.26 becomes infinitesimal, the efficiency obtained corresponds to a single particle size x and it is known as the grade efficiency, defined by

$$G(X) = \left(\frac{M_c}{M}\right)_x.$$ (7.27)

The grade efficiency has become a very useful definition, since most industrial powders consist of an infinite number of differently size particles. Thus, a single particle size really corresponds to a range of particles having almost similar sizes. Therefore, the grade efficiency of most separation equipment is a continuous function of the particle size. This function is seldom expressed analytically, and is usually expressed graphically. An S-shaped curve is usually obtained for separators in which inertial or gravitational body forces perform the separation.

As the value of the grade efficiency has the character of probability, plotting the probability for any given size fraction against particle size will give a curve as shown in Fig. 7.8. This sort of curve is normally called a grade efficiency curve.

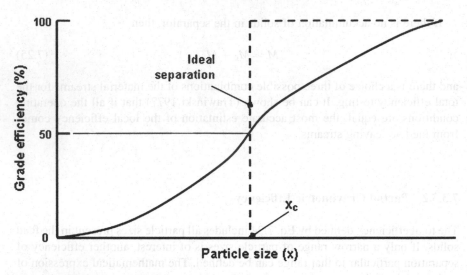

Fig. 7.8 Grade efficiency curve for dynamic separators

7.4   Solid–Gas Separations

7.4.1   Introduction

In many processes of the food and related industries, separating solids from a gas stream is very important. The typical example is the risk of dust explosion in the dry milling industry. It has been found that not only in this industry but also in many others the atmosphere may become dust-laden with particles from different sources, representing a health risk. In other cases the suspension of particles in a gas stream has been promoted, as in pneumatic conveying or spray drying, but at the end of the process there is a need to separate the phases. Separation of solids from a gas is accomplished using many different devices. Perhaps the devices most commonly used to separate particles from gas streams are cyclones and bag or gas filters.

7.4.2   Cyclone Separation

Cyclones are by far the most common type of gas–solids separation devices used in diverse industrial processes. They have no moving parts, are inexpensive compared with other separation devices, can be used at high temperatures, produce a dry product, have low energy consumption, and are extremely reliable. Their primary disadvantage is that they have a relatively low collection efficiency for particles below about 15 μm. As illustrated in Fig. 7.9, a cyclone consists of a vertical cylinder with a conical bottom, a tangential inlet near the top, and outlets at the top

Fig. 7.9 A cyclone

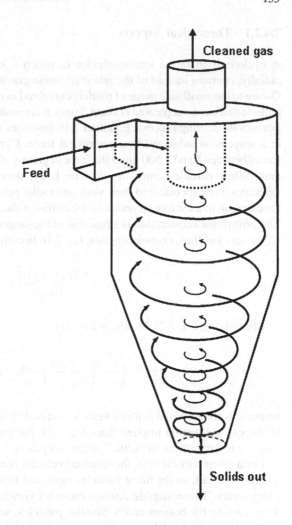

and the bottom, respectively. The top outlet pipe protrudes into the conical part of the cyclone in order to produce a vortex when a dust-laden gas (normally air) is pumped tangentially into the cyclone body. Such a vortex develops centrifugal force and, because the particles are much denser than the gas, they are projected outward to the wall, flowing downward in a thin layer along it in a helical path. They are eventually collected at the bottom of the cyclone and separated. The inlet gas stream flows downward in an annular vortex, reverses itself as it finds a reduction in the rotation space due to the conical shape, creates an upward inner vortex in the center of the cyclone, and then exits through the top of the cyclone. In an ideal operation, in the upward flow there is only gas, whereas in the downward flow there are all the particles fed with the stream. Cyclone diameters range in size from less than 0.05 to 10 m, feed concentrations cover values from 0.1 to about 50 kg/m^3, and gas inlet velocities may be on the order of 15–35 m/s.

7.4.2.1 Theoretical Aspects

A cyclone is in fact a settling device in which a strong centrifugal force, acting radially, operates instead of the relatively weak gravitational force, acting vertically. Owing to the small size range of particles involved in cyclone separation (the smallest particle that can be separated is about 5 μm), it is considered that Stokes's law primary governs the settling process. Equation 7.16 describes the settling velocity of particles in a suspension subjected to a centrifugal force. Cyclones can generate centrifugal forces between 5 and 2,500 times the force of gravity, depending on the diameter of the unit. When particles enter the cyclone body, they quickly reach their terminal velocities, corresponding to their sizes and radial position in the cyclone. The radial acceleration in a cyclone depends on the radius of the path being followed by the gas. The centrifugal acceleration is a function of the tangential component of the velocity $v_{\text{tan}} = \omega r$, and thus, considering this, Eq. 7.16 becomes

$$v_r = \frac{x^2 (\rho_s - \rho_g) v_{\text{tan}}^2}{18 \mu_g r}. \tag{7.28}$$

Multiplying Eq. 7.28 by g/g, we obtain

$$v_t = \left(\frac{x^2 (\rho_s - \rho_g) g}{18 \mu_g} \right) \frac{v_{\text{tan}}^2}{gr} = u_t \frac{v_{\text{tan}}^2}{gr}, \tag{7.29}$$

where v_t is the terminal settling velocity defined by Eq. 7.16 and μ_g is the viscosity of the gas. As can be implied from Eq. 7.29, the higher the terminal velocity, the easier it is for particle to "settle" within a cyclone.

For a given particle size, the terminal velocity is a maximum in the inner vortex, where r is small, so the finest particles separated from the gas are eliminated in the inner vortex. These migrate through the outer vortex to the wall of the cyclone and drop, passing the bottom outlet. Smaller particles, which do not have time to reach the wall, are retained by the air and are carried to the top outlet. Although the chance of a particle being separated decreases with the square of the particle diameter, the fate of a particle also depends on its position in the cross section of the entering stream and on its trajectory in the cyclone. Thus, the separation according to size is not sharp. A specific diameter, called the "cut diameter" or "cut size," can be defined as that diameter for which half of the inlet particles, by mass, are separated and the other half are retained by the gas. The cut size is a very useful variable to determine the separation efficiency of a cyclone. Since a given powder to be separated in a cyclone will have an extremely fine half of its distribution, such a half may not be easily separated using conventional pressure drops. Therefore, it is advisable to make the cut size coincide with the mean size of a powder particle size distribution to guarantee separation of the coarse part of such a distribution, as the fine part may be unattainable because of the small range involved.

Fig. 7.10 Dimensions of a Stairmand standard cyclone

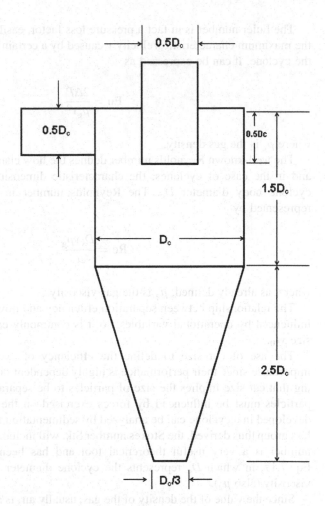

0.5D$_c$

0.5D$_c$

0.5Dc

1.5D$_c$

D$_c$

2.5D$_c$

D$_c$/3

7.4.2.2 Dimensionless Approach

Experience and theory have shown that there are certain relationships among cyclone dimensions that should be observed for efficient cyclone performance (Geldart 1986), and which are generally related to the cyclone diameter. There are several different standard cyclone "designs" and a very common one is called the Stairmand design, whose dimensions are shown in Fig. 7.10. With use of standard geometries of cyclones it is much easier to predict the effects of variable changes and scale-up calculations are greatly reduced. Such calculations may be performed by means of dimensionless relationships. Selection and operation of cyclones can be described by the relationship between the pressure drop and the flow rate, and the relationship between the separation efficiency and the flow rate (Svarovsky 1981). The pressure drop versus volumetric flow rate relationship is usually expressed as Eu = f(Re), where Eu is the Euler number and Re is the Reynolds number.

The Euler number is in fact a pressure loss factor, easily defined as the limit on the maximum characteristic velocity v caused by a certain pressure drop ΔP across the cyclone. It can be expressed as

$$Eu = \frac{2\Delta P}{\rho_g v^2},$$ (7.30)

where ρ_g is the gas density.

The well-known Reynolds number defines the flow characteristics of the system and in the case of cyclones, the characteristic dimension may be taken as the cyclone body diameter D_c. The Reynolds number in this case is, therefore, represented by

$$Re = \frac{D_c v \rho_g}{\mu_g},$$ (7.31)

where, as already defined, μ_g is the gas viscosity.

The relationship between separation efficiency and flow rate is not significantly influenced by operational variables, so it is commonly expressed in terms of cut size x_{50}.

The use of cut size to define the efficiency of cyclones is of the utmost importance since their performance is highly dependent on particle size. Considering that cut size implies the size of particles to be separated, it follows that such particles must be influenced by forces exercised on the suspension. The forces developed in a cyclone can be analyzed by sedimentation theory, and a dimensionless group thus derived, the Stokes number Stk, will include the cut size. The Stokes number is a very useful theoretical tool and has been previously defined by Eq. 7.17, in which D_c represents the cyclone diameter and μ refers to the gas viscosity (also μ_g).

Since the value of the density of the gas, usually air, is negligible in comparison with the solids density, Eq. 7.17 can also take the following form:

$$Stk = \frac{x^2 \rho_s v}{18 \mu_g D_c}.$$ (7.32)

Furthermore, if the dimension x is replaced by the specific cut size x_{50},

$$Stk_{50} = \frac{x_{50}^2 \rho_s v}{18 \mu_g D_c}.$$ (7.33)

Equations 7.30, 7.31, and 7.33, defining Euler Eu, Reynolds Re, and Stokes Stk_{50} numbers, respectively, are related by specific functions, which can be plotted as

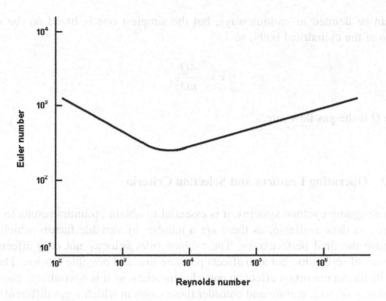

Fig. 7.11 A typical plot of Eu versus Re for cyclones

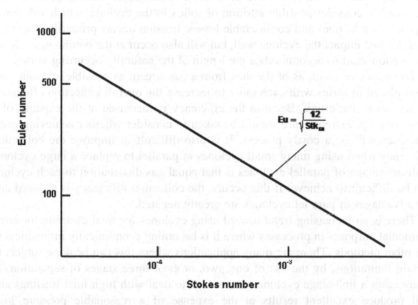

$$Eu = \sqrt{\frac{12}{Stk_{50}}}$$

Fig. 7.12 A typical plot of Eu versus Stk_{50} for cyclones

shown in Figs. 7.11 and 7.12, for a given cyclone geometry. The cyclone inner
diameter D_c is described in Fig. 7.10 and, as previously mentioned, all geometrical
proportions are related to it. In the case of scale-up procedures, proportions
must be maintained. The cyclone body velocity v is the characteristic velocity

and can be defined in various ways, but the simplest one is based on the cross section of the cylindrical body, so

$$v = \frac{4Q}{\pi D_c^2}, \tag{7.34}$$

where Q is the gas flow rate.

7.4.2.3   Operating Features and Selection Criteria

When designing cyclone systems, it is essential to obtain optimum results to have full process data available, as there are a number of variable factors which will determine the final performance. The cyclone inlet velocity not only affects the efficiency of separation but also affects pressure loss and possible erosion. The gas viscosity has an important effect on particle efficiency, so it is advisable to check its dependency on temperature and consider those cases in which a gas different from air is involved in the process. Smaller cyclone diameters increase the overall efficiency, but will also promote erosion. In addition to this, it is sometimes necessary to consider possible attrition of solids in the cyclone, which will result in production of fines and considerable losses. Erosion occurs primarily where the particles first impact the cyclone wall, but will also occur at the bottom of cyclones that are too short to accommodate the length of the naturally occurring vortex.

To remove as much as of the dust from a gas stream as possible, cyclones are often placed in series with each other to increase the overall collection efficiency relative to a single unit. Because the efficiency is increased at the expense of a pressure drop, extreme care should be taken to consider whether achieving great efficiency will be a costly process. It is also difficult to improve the collection efficiency when using many small cyclones in parallel to replace a large cyclone. A disadvantage of parallel cyclones is that equal gas distribution to each cyclone can be difficult to achieve. If this occurs, the collection efficiency is reduced and the advantages of parallel cyclones are greatly negated.

There is an increasing trend toward using cyclones for final cleaning for environmental purposes in processes where it is becoming economically impractical to use other methods. There are many applications where this can be done subject to certain limitations, by the use of one, two, or even three stages of separation. In many cases a first-stage cyclone can be used to deal with high inlet loadings and will produce excellent results at the expense of a reasonable pressure loss. The addition of a second stage to deal with the first-stage losses can then often achieve the required results, but where further cleaning is necessary a third-stage cyclone may provide the answer. The use of a third-stage cyclone invariably means that a high degree of cleanup is necessary and, therefore, the third stage should give the best possible efficiency and be capable of maintaining this efficiency for a long time. The use of several stages to try to improve efficiency or remove very fine particles may become impractical when employing only cyclones. This difficulty can be alleviated by combining methods, so for environmental purposes when certain particles need to be removed regardless of their fineness, bag filters may be coupled with cyclones.

7.4.2.4 Applications

As mentioned before, cyclones are extensively used in the food industry to reduce the particle load to safe levels in dry milling, as well as in classification of particles in closed-circuit grinding operations. They are also employed in recovering fines from spray drying and fluidized-bed drying processes. Another important application is in pneumatic conveying of diverse food products, such as grains and flours.

7.4.3 Gas Filtration

Gas–solids separations can be performed using filtration means. Gas filters are used for final particulate removal in many processes in the food industry. These filters can capture particles much smaller in diameter than a cyclone, so they are commonly placed downstream of a cyclone in diverse applications. A gas filter generally consists of a porous fabric, which can be woven to conform to the shape of a cylinder, or may be supported in a frame. The former is called a bag filter and the latter is known as an envelope filter. The main difference between both designs is in the way solids are accumulated. In bag filters dust may accumulate inside, whereas in envelope filters it would form a cake outside. The filtering arrangements are placed in a matrix so that their total area will result in a low gas velocity through the bags and, therefore, a low pressure drop through the filter. Gas velocities through the filter media are on the order of 0.005–0.02 m/s. Particulate loadings to these filters generally lie in the range of 0.2–250 g/cm^3. A diagram of the two main types of gas filters is given in Fig. 7.13. Filters used in gas–solids separations may be woven or felted fabrics of natural or synthetic fibers. There are tables listing the properties of filter media to determine whether they are suitable for applications in diverse conditions (Green and Perry 2008). Filtrations may also be made with granular solids in the form of stationary or moving beds. Many other types of materials that are porous in nature or are capable of providing a screening effect after weaving or fabrication may be suitable for certain filtration applications.

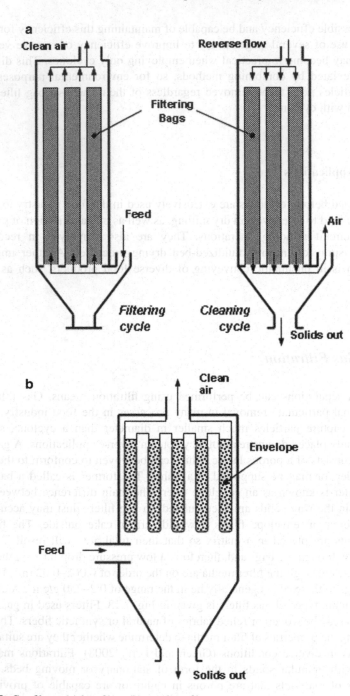

Fig. 7.13 Gas filters: (a) bag filter, (b) envelope filter

7.4.3.1 Filtering Fundamentals

In gas filtration, the collected particles build up on the surface of the filter medium and cause a gradual increase in the pressure drop through the filter. After a certain limiting pressure drop is reached, the bags are cleaned by pulsing gas back through the filter to remove the buildup of solids on the bag. As described in the previous section, standard commercial types of filters use a bag or an envelope, the latter actually being a retainer for the filter cloth. The bag filters are, in fact, elongated cylinders that may be opened at both ends to allow a cleaning cycle by using air jets blowing into the top. On the other hand, envelope filters, with a frame for support and with large flat surfaces exposed, do not have the cleaning capabilities of the bag filters, which can be shaken more or can be collapsed without risk of damage. When air is blown in the reverse direction to the filtration flow for cleaning purposes, the cleaning is more effective. The envelope filters are, therefore, best used on dusts that are easily shaken or removed from the cloth surface. The dust is removed in the envelope type by beating the screen supporting the filter and by shaking or rocking the frame by mechanical means.

The use of a blow ring in some bag- or cylinder-type mechanical filters allows continuous operation while performing the cleaning. Dust is collected on the inside while the blow ring travels up and down along the outer surface of the bag. The blow ring has an slot inside that is used to blow gas against the wall of the bag. The ring is tight enough to partially collapse the bag in order to break the dust cake, and to provide a close seal so the gas blown is fully delivered through the filter when blowing back. Dust may be collected on the outside of filter tubes or bags if a support is provided inside to prevent the filter collapsing. Sometimes, it is possible to remove the cake from the outer wall by periodically using a jet of compressed air from inside the filter in order to produce a shock to break the cake from the outer wall, whence it can be settled into the bin. The envelope- or frame-type filter collects the dust on the outer wall of the filter, as expected, since the outer wall is easier to get to in this arrangement.

7.4.3.2 Operation Characteristics

The operating variables of gas filtration are the resistance to flow, the permeability of air to the filter medium, and the resistance due to particle accumulation. With regard to the resistance to flow, the pressure drop across the filter medium ΔP_f can be represented by

$$\Delta P_f = K_c \mu_g V_f,$$ (7.35)

where K_c is a constant depending on the filter medium and V_f is the superficial velocity of the gas through the filter medium.

The resistance to the layer of particles accumulated during the filtration cycle can be calculated by determining a variable known as the cake resistance factor K_1:

$$K_1 = \frac{\Delta P_c}{V_f w},$$ (7.36)

where ΔP_c is the pressure drop through the powder layer and w is the powder mass flow rate approaching the filter.

7.4.3.3 Applications

As previously mentioned, bag filters have, practically, the same applications as cyclones, normally being coupled with these in order to remove the finest tails of particle size distributions of diverse food powders. A promising application of gas filtration in food processes is the use of ceramic candle filters, composed of multiple porous ceramic cylinders, because they can withstand high temperatures.

7.4.4 Other Solid–Gas Separation Techniques

7.4.4.1 Scrubbers

Solid particles may often be scrubbed from a gas stream by spraying a jet of liquid, usually water, into such a stream. The particles are intercepted by the droplets of the water spray and are removed by the scrubber in the form of slurry. Although scrubbers are normally more efficient than cyclones, they have the disadvantage of collecting the solids wet instead of dry. If this feature is unacceptable or impractical for processing reasons, the solids must be separated from the liquid. In such a way, the gas–solids collection problem will be replaced by a solid–liquid separation difficulty. Gas scrubbers have the same sort of applications as the previously described solid–gas separators. They are particularly useful in the separation of fumes from combustion reactions and to eliminate mists in the production of different products in many industrial processes.

7.4.4.2 Electrostatic Precipitators

Electrostatic precipitators separate solids, or liquids, from a gas stream by passing them through a strong, high-voltage field produced between two electrodes of opposite polarity. The field imposes a charge on the particles so that they migrate toward the collecting electrode. The particles are usually removed from the electrode by periodic rapping. The advantage of an electrostatic precipitator is that it can collect solids of very small size in a dry form. Electrostatic precipitators are generally large units of equipment, because the collection efficiency is proportional to the area of the collecting electrodes. For this reason, the capital costs are very high, although the operating costs are low and may justify the use of a precipitator instead of a filter for particular applications (Svarovsky 1981).

7.5 Solid–Liquid Separations

7.5.1 Introduction: Solid–Liquid Separations Used in Food Processes

Separations techniques in the food industry have been employed for a long time and examples of some applications have been mentioned before. Although solid–liquid separations in food processing have typically focused on quality aspects, such as clarification, some technologies have the capability of separating or retaining microorganisms, such as ultracentrifugation and microfiltration, so they have also been useful in pasteurizing and sterilizing liquid foods. Sedimentation may be considered an operation focused mainly on effluent treatment, but another application is separation of crystals from mother liquors. Centrifugation and filtration applications are numerous, and some have been described already. The last part of this chapter will review sedimentation, centrifugation, and filtration. Membrane separations have found relevant applications in the food industry since their appearance as an alternative to energy-demanding operations, such as evaporation. These important separation techniques will be the subject of the following chapter.

7.5.2 Sedimentation

7.5.2.1 Introductory Remarks

Sedimentation can be defined as a unit operation to separate a suspension into a supernatant clear fluid and a dense slurry containing a higher concentration of solids. In a more convenient manner, it should be established that the settling of solids of a suspension in sedimentation is due to the gravitational force, and industrial sedimentation can be described as the gravitational settling of solids suspended in liquids. The uses of sedimentation in industry fall into the categories of solid–liquid separations, solid–solid separations as in particle classification, and other operations such as mass transfer and washing. In solid–liquid separations the solids are removed from the liquid because either or both of the phases are more valuable separated or because they have to be separated before disposal. When the primary purpose is to produce the solids as a highly concentrated slurry, the process is called thickening, whereas if the purpose is to clarify the liquid, the process is referred to as clarification. Usually, the feed concentration for a thickener is higher than that for a clarifier. Some types of equipment can accomplish both thickening and clarification in a single stage, provided they are correctly designed and operated.

Sedimentation can be used for size separation, i.e., classification of solids, being one of the simplest ways to remove coarse or dense solids from a feed suspension. Successive decantation in a batch system produces closely controlled size fractions of the product. Classification by sedimentation, however, does not produce a sharp separation.

Apart from the applications described above, sedimentation is also used for other purposes. Relative motion of the particles and liquid increases the mass-transfer coefficient. This motion is particularly useful for solvent extraction in immiscible liquid–liquid systems. Another important commercial application of sedimentation is in continuous countercurrent washing, where a series of continuous thickeners are used in a countercurrent mode in conjunction with return of slurry, with the purpose of removing mother liquor or washing soluble substances from the solids.

As settling of solids in sedimentation is caused by the effect of gravity, a difference in density between the solids and the suspended liquid is a necessary prerequisite. Predicting the settling behavior of particles within a liquid in real sedimentation system is, however, rather complicated because many factors are involved in the process. Among these factors, the distance from the boundaries of the container and from other particles are critical in affecting settling. Both of these factors are directly related to the concentrations of the solids of the feed stream led to a given sedimentation unit. It has been suggested that if the relation between the diameter of the particle and the diameter of the sedimentation vessel is over 1:200 or if the solids concentration by volume is less than 0.2%, any given particle is at a sufficient distance from the boundaries of the vessel or from other particles that its fall is not affected by them. In this case the process is called free settling. Contrastingly, when the motion of the particle is impeded by other particles, which will happen if the particles are near each other even if they are not actually colliding, the process is called hindered settling. The drag coefficient in hindered settling is greater than that in free settling.

7.5.2.2   Free Settling and Hindered Settling

The approach discussed previously when deriving the Stokes's law equation (Eq. 7.15, Sect. 7.3.1) for evaluating terminal sedimentation velocity in the Stokes region (Fig. 7.5) tends to be theoretical, since complete non-interference of particles settling within a liquid is only possible for the case of these particles being solitary. There are several cases in which such an approximation is possible, for example, in measurement of the particle size distribution by sedimentation or in particle classification, but experience has shown that in many real applications of industrial sedimentation, correction factors, mainly for the suspension

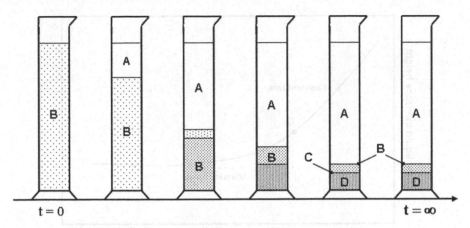

Fig. 7.14 Progressive settling in a measuring cylinder: *A* clear liquid, *B* sludge at the initial concentration, *C* transition zone, *D* thick sludge at the compression zone

concentration, are needed to minimize calculation errors. The first, ideal case of particles settling without interference is known as free settling, whereas the second, real case of particles settling but subjected to different sorts of interference is referred to as hindered settling. The main correction factor used in practice to account for interference in settling processes is the correction factor due to the concentration of the solids in the suspension described by the Richardson and Zakii equation, i.e., Eq. 7.18.

7.5.2.3 Sedimentation Rate Stages

To analyze sedimentation in greater detail, the events occurring in a small-scale experiment conducted batchwise as shown in Fig. 7.14 can be observed. Particles in a narrow range will settle with about the same velocity. When this occurs, a demarcation line is observed between the clear supernatant liquid (zone A) and the slurry (zone B) as the process continues. The velocity at which this demarcation line descends through the column indicates the progress of the sedimentation process. The particles near the bottom of the cylinder pile up, forming a concentrated sludge (zone D) whose weight increases as the particles settle from zone B. As the upper interface approaches the sludge buildup on the bottom of the container, the slurry appears more uniform as a heavy sludge (zone D), the settling zone B disappears, and the process from then on consists only of the continuation of the slow compaction of the solids in zone D.

By measuring the interface height and the solids concentrations in the dilute and concentrated suspensions, one can prepare a graphic representation of the sedimentation process as shown in Fig. 7.15. The plot shows the difference in interface height plotted against time, which is proportional to the rate of settling as well as to concentration.

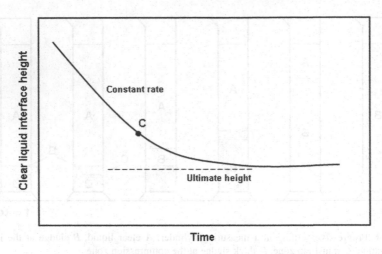

Fig. 7.15 Sedimentation rate graph

7.5.2.4 Sedimentation Units: Thickeners and Clarifiers

Sedimentation equipment can be divided into batch-operated settling tanks and continuously operated thickeners or clarifiers. The operation of the former is very simple and their use has recently diminished. They are still used, however, when small quantities of liquids are to be treated. Most sedimentation processes operate in continuous units.

The conventional thickeners are constructed from steel in sizes less than 25 m in diameter, or from concrete in larger tanks that usually range up to 100 m in diameter. The floor is often sloped toward the underflow discharge in the center. The tanks usually include a raking mechanism, which turns slowly around the central column in order to promote consolidation of solids in the compression zone while aiding the underflow discharge. The rake arms are driven by fixed connections or are dragged by cables or chains suspended from a drive arm that is rigidly connected to the drive mechanism. The rake arms are also connected to the bottom of the central column by a special arm hinge that allows both horizontal and vertical movements. This arrangement lifts the rakes automatically if the torque becomes excessive. The drive arm can be attached below the suspension level or, if scaling is a problem, above the basin. Figure 7.16 shows a diagram of a conventional circular-section thickener.

To increase the capacity, some modifications of thickeners and clarifiers have been attempted. Some examples of alternative designs of thickeners are shown in Fig. 7.17.

In the circular-basin continuous thickener, illustrated in Fig. 7.17a, after treatment with a flocculent the feed stream enters the central feed well, which dissipates the stream's kinetic energy, and disperses gently into the thickener.

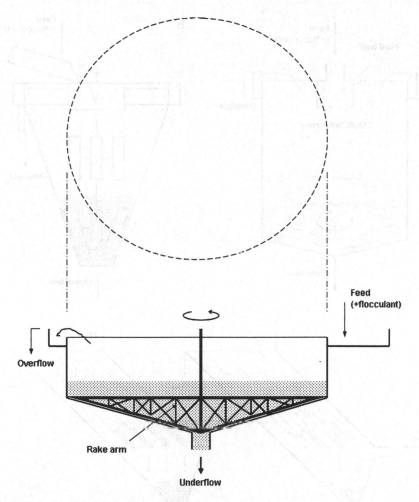

Fig. 7.16 A circular-section gravity thickener

The feed finds its height in the basin where its density matches the density of the inner suspension and spreads out at that level. The solids concentration increases downward in an operating thickener, giving stability to the process. The settling solids and some liquid move downward. The amount of the latter depends on the underflow withdrawal rate. Most of the liquid moves upward into the overflow, and is collected in a trough around the periphery of the basin. A typical thickener has three operating layers: clarification, zone settling, and compression. Frequently, the feed is contained in the zone settling layer, which theoretically eliminates the need for the clarification zone because the particles will not escape through the interface. However, in practice, the clarification zone provides a buffer for fluctuations in the feed and the sludge level. The most important dimensions of a thickener are pool area and depth. The pool area is chosen to be the largest of the three layer

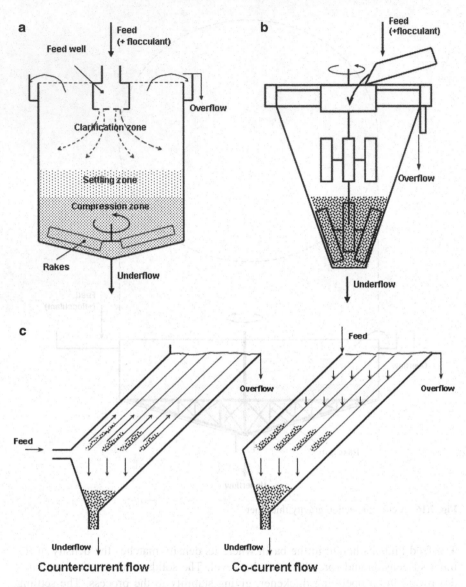

Fig. 7.17 Alternative designs for gravity thickeners: (a) circular-basin continuous thickener, (b) deep-cone thickener, (c) lamella thickener

requirements. In most cases, only the zone settling and compression layer requirements need to be considered. However, if the clarity of the overflow is critical, the clarification zone may need the largest area. As for the pool depth, only the compression layer has a depth requirement because the concentration of the solids in the underflow is largely determined by the detention time and, sometimes, by the static pressure. The thickness of the other two layers is governed

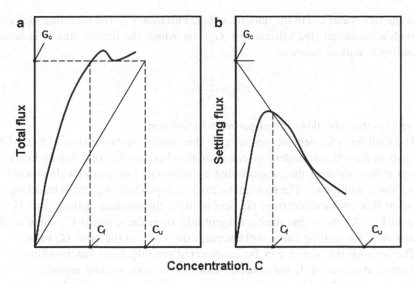

Fig. 7.18 Design curves for thickeners: (**a**) total flux versus concentration, (**b**) settling flux versus concentration

only by practical considerations. The deep-cone thickener (Fig. 7.17b), based on the deep-cone vessel used for the processing of coal and metallurgical ores, is equipped with a slow-turning stirring mechanism, which enhances flocculation in the upper part and acts as a rake in the lower section. The unit is used for densification of froth-flotation tailings at overflow rates from 6.5 to 10 m/h to a final discharge containing 25–35% moisture by weight. Other commercial deep-cone thickeners are of particular advantage where the final underflow density is increased by the large static head above the discharge point, e.g., with flocculated clays. Another development is the Swedish lamella thickener shown in Fig. 7.17c. It consists of a number of inclined plates stacked closely together. In the countercurrent design, the flocculated feed enters the stack from a side feed box. The flow moves upward between the plates and the solids settle onto the plate surfaces and slide down into the sludge hopper underneath, where they are further consolidated by vibration or raking. In theory, the effective settling area is the sum of the horizontal projected areas of all plates. In practice, however, only about 50% of the area is utilized. When sticky sludges are treated, the whole lamella pack can be vibrated continuously to assist the sliding motion of the solids down the plates. In some instances the plates are corrugated instead of flat, or they are replaced by tube bundles, becoming what is known as a tube settler.

Thickeners are designed by the traditional Coe and Clevenger, or Talmage and Fitch procedures (Fitch 1975a, b, c), which use batch-settling data to evaluate the whole concentration range right up to the underflow concentration, even though a batch-settling test cannot possibly simulate the continuous-thickener process. The design procedures are based on plotting the total flux versus concentration, as shown in Fig. 7.18a. The concentration of the solids continuously increases from the feed value C_f, to the underflow concentration C_u. The total flux plot goes through a minimum (the critical flux G_c), on which the design area A is based, according to a mass balance:

$$A = \frac{QC_f}{G_c},$$

(7.37)

where Q is the mass flow rate of particles by volume.

The total flux plot depends not only on the settling characteristics of the solids, but also on the selected underflow concentration because the total flux includes the transport flux, or dead flux, contributed by movement of particles downward as underflow is withdrawn. The plot can be made independent of C_u by subtracting the transport flux contribution from G_c, and plotting the resultant settling flux G_s, as done in Fig. 7.18b. A line drawn tangentially from the selected C_u value to the minimum of the settling curve will intercept the y-axis at the same G_c value.

The settling flux curve can be constructed directly from batch-settling tests performed at several different concentrations in the zone settling regime (Coe and Clevenger method) or from just one test curve (Talmage and Fitch method) by converting the $u = f(C)$ data into settling flux by the formula

$$G_s = uC.$$

(7.38)

The Coe and Clevenger test overestimates the critical flux, leading to underdesign of the thickener area, whereas the Talmage and Fitch procedure underestimates it leading to overdesign. The latter procedure is less laborious because it only requires one settling test.

As with clarifiers, the conventional one-pass unit employs horizontal flow in circular or rectangular vessels. Figure 7.19 shows a schematic diagram of a rectangular basin with feed at one end and overflow at the other. The feed can be preflocculated in an orthokinetic flocculator which may form part of the unit. Settled solids are pushed to a discharge trench by paddles or blades on a chain mechanism or suspended from a traveling bridge. Circular-basin clarifiers are most commonly fed through a centrally located feed well; the overflow is led into a trough around the periphery of the basin. The bottom gently slopes to the center and the settled solids are pushed down the slope by a number of motor-driven scraper blades that revolve slowly around a vertical central shaft. Like thickeners, circular clarifiers can be stacked in multitray arrangements to save space.

Clarifier performance depends on area, which is determined by the flocculation nature of the feed suspension. When the overflow clarity is independent of the overflow rate and depends only on the detention time, the required time is determined by simple laboratory testing of residual solids concentrations in the supernatant versus detention time under the conditions of mild shear. This determination is sometimes called the second-order test procedure because the flocculation process follows a second-order reaction rate. In most cases, the clarifying performance

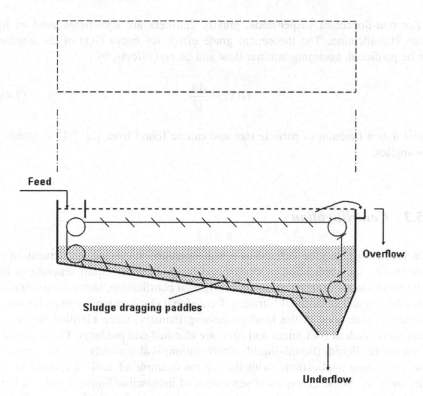

Fig. 7.19 A rectangular-basin clarifier

depends on the detention time and the overflow rate. Tests are conducted in a vertical tube that is as long as the expected depth of the clarifier, under the ideal assumption that a vertical element of a suspension, which has been clarified, maintains its shape as it moves across the tank.

If the suspension is non-flocculant, or if flocculation takes place prior to settling, the overflow clarity is independent of the detention time and depends only on the overflow rate Q according to the relation

$$Q_{\max} = A u_t. \tag{7.39}$$

In this case the settling velocity u_t is expressed by

$$u_t = \frac{H}{t}, \tag{7.40}$$

where u_t is the maximum particle settling velocity that yields a satisfactory clarity in a simple laboratory sedimentation test, H is the height of the laboratory container, and t is the time at which the supernatant liquid becomes clear.

For non-flocculant suspensions, gravity clarifiers are sometimes used as for solids classification. The theoretical grade efficiency curve $G(x)$ of the clarifier can be predicted, assuming laminar flow and no end effects, by

$$G(x) = \frac{Au_t}{Q}, \tag{7.41}$$

where u_t is a function of particle size and can be found from Eq. 7.15 if Stokes's law applies.

7.5.3   Centrifugation

This technology may be defined as a unit operation involving the separation of materials by the application of centrifugal force. In solid–liquid separations the different modes of centrifugation are centrifugal clarification, slurry centrifugation (desludging), and centrifugal filtration. Centrifugal clarification accounts for many practical applications in the food processing industry, since clarified liquors of many sorts, such as fruit juices and beer, are clarified end products. The separation of immiscible liquids (liquid–liquid centrifugation) is also widely used in important food processing applications, with the typical example of milk skimming in the dairy industry. Some examples of separation of immiscible liquids involve a light liquid phase to be separated from a dense liquid phase, with the former present as droplets in the mixture, e.g., edible oil containing a small amount of moisture. These droplets will behave as spherical particles, which will float rather than settle because of their density, but will follow the theoretical principles covered by Stokes's law and the equations derived from it.

7.5.3.1   Centrifugal Clarification: Theoretical Principles

"Centrifugal clarification" is the term used to describe the removal of small quantities of insoluble solids from a fluid by centrifugal means. If a dilute suspension containing solids with a density greater than that of the liquid is fed to a rotating cylindrical bowl, the solids will move toward the wall of the bowl. If an outlet is provided for the liquid near the center of rotation, then those solid particles which reach the wall of the bowl will remain in the bowl. Those particles which do not reach the wall of the bowl will be carried out in the liquid. The fraction remaining in the bowl and the fraction passing out in the liquid will be controlled by the feed rate, i.e., the dwell time, in the bowl. If a solid particle of diameter x moves radially in a liquid within a rotating bowl, at its terminal velocity under laminar flow conditions the radial velocity of the particle will be represented by Eq. 7.16. The time required for a particle to travel an elemental radial distance, dR, is

$$dt = \frac{dR}{v_r} = \frac{18\mu}{\omega^2(\rho_s - \rho)x^2} \frac{dR}{R}. \tag{7.42}$$

Assuming that half of all those particles present in the feed with a particular diameter, x_c, are removed during their transit through the bowl, those particles with diameters greater than x_c will be mostly removed from the liquid, whereas those particles with diameters smaller than x_c will likely remain in the liquid. In this context, x_c as defined here is known as the "cut-point" or "critical" diameter.

If clarification is taking place in a simple cylindrical centrifuge with cross section as shown in Fig. 7.20, all particles of diameter x_c contained in the outer half of the cross-sectional area of the ring of liquid will reach the wall of the bowl and will be removed from the liquid. The maximum distance that a particle in this zone has to travel to reach the wall of the bowl is $R_2 - [(R_1{}^2 + R_2{}^2)/2]^{1/2}$, as indicated in Fig. 7.20. The time required for a particle of diameter x_c to travel this distance is

$$t = \frac{18\mu}{\omega^2(\rho_s - \rho)X_{50}^2} \int_{[(R_1^2+R_2^2)/2]^{1/2}}^{R_2} \frac{dR}{R}. \tag{7.43}$$

Integrating Eq. 7.43 and substituting limits, we obtain the following relation:

$$t = \frac{18\mu \ln\left(\dfrac{R_2}{\left[(R_1^2+R_2^2)/2\right]^{1/2}}\right)}{\omega^2(\rho_s - \rho)X_{50}^2}. \tag{7.44}$$

The minimum residence time for a particle in the bowl is V/Q, where V is the volume of liquid held in the bowl at any time and Q is the volumetric flow rate of liquid through the bowl. Thus, for a particle of diameter x_c to be separated out,

$$\frac{V}{Q} = \frac{18\mu \ln\left(\dfrac{R_2}{\left[(R_1^2+R_2^2)/2\right]^{1/2}}\right)}{\omega^2(\rho_s - \rho)X_{50}^2}. \tag{7.45}$$

Equation 7.45 may be written in the form

$$Q = 2\left(\frac{g(\rho_s - \rho)X_{50}^2}{18\mu}\right)\left(\frac{\omega^2 V}{2g \ln\left(\dfrac{R_2}{\left[(R_1^2+R_2^2)/2\right]^{1/2}}\right)}\right). \tag{7.46}$$

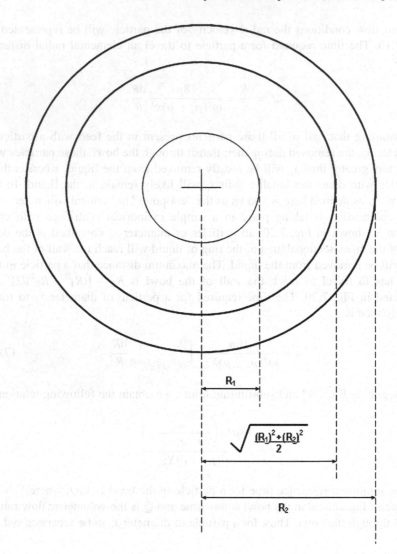

Fig. 7.20 Cross-section diagram of a tubular centrifuge

The first term in Eq. 7.46 includes the previously discussed expression for Stokes's law, i.e., Eq. 7.15. Thus, another way of expressing Eq. 7.46 is as follows:

$$Q = 2u_g\Sigma, \tag{7.47}$$

where u_g is the terminal settling velocity of a particle of diameter x_c in a gravitational field and Σ is a characteristic parameter of any given centrifuge, equivalent to the area of a gravity settling tank with settling characteristics similar to those of the centrifuge, i.e., one which will remove half of all particles of diameter x_c.

Different values of Σ are given in the literature (McCabe et al. 2005). For a simple cylindrical-bowl centrifuge,

$$\Sigma \approx \frac{\pi\omega^2 b(3R_2^2 + R_1^2)}{2g},\tag{7.48}$$

where b is the height of the bowl. Also, for a disc-bowl centrifuge,

$$\Sigma = \frac{2\pi\omega^2(S-1)(R_x^3 - R_y^3)}{3g\tan\Omega},\tag{7.49}$$

where S is the number of discs in stack, R_x and R_y are the outer and inner radii of the stack respectively, and Ω is the conical half angle of the discs (Trowbridge 1962).

7.5.3.2 Desludging, Separation of Liquids, and Centrifugal Filtration

Desludging refers to the removal of solids from a liquid by centrifugal means when the suspension is concentrated rather than dilute, which is the range handled by clarifiers. When the solids contents of a suspension exceed 5–6% by weight, centrifugal clarifiers will operate inefficiently and a slurry centrifuge will produce better separation. Slurry centrifuges, or decanter centrifuges, are provided with devices to allow continuous removal of solids from the bowl during operation. The general theory discussed in Sect. 7.5.3.1 is still applicable, but errors may occur owing to the solids movement system altering the conditions on which the formulae are based, so more complex expressions are needed to describe the process. It is critical, for example, to use correction factors for the concentration, wall effects, etc.

Immiscible liquids can be separated by centrifugation. If two liquids with significantly different densities are placed in a cylindrical rotating bowl, the denser liquid will tend to move toward the wall of the bowl and the less dense liquid will be displaced toward the center of rotation. If provision is made for introducing the liquid feed continuously into the bowl and for tapping off from the two liquid layers separately, separation of the liquids can be attained. The liquid mixture is usually fed in at the bottom of the bowl by a centrally located pipe, and the liquids are removed from each layer by a weir system. There is a cylindrical interface with radius R_i known as the neutral zone, which can be described by the following equation (Brennan et al. 1990):

$$R_i = \left(\frac{R_A^2 - (\rho_B/\rho_A)R_B^2}{1 - (\rho_B/\rho_A)}\right)^{1/2},\tag{7.50}$$

where R_A is the weir radius for the denser phase, R_B is the weir radius for the less dense phase, ρ_A is the density of the denser phase, and ρ_B is the density of

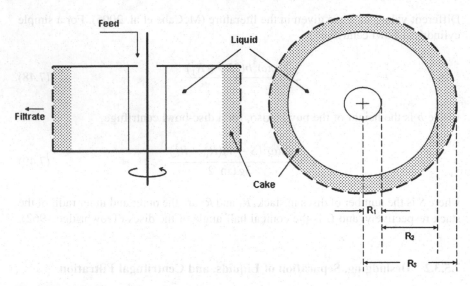

Fig. 7.21 A simple centrifugal filter

the less dense phase. When the difference between the densities of the liquids is less than about 3%, the neutral zone becomes unstable and, consequently, the separation becomes inefficient.

Centrifugal filtration describes the separation of solids from a liquid by filtration for the case when the filtrate flow is induced by centrifugal means. The general principle of centrifugal filtration is shown in Fig. 7.21. As can be observed, a suspension is fed into a rotating bowl with a perforated wall, which is lined with a suitable filter medium. The solids are forced onto the wall of the bowl, giving rise to a filter cake through which the filtrate passes, under the influence of centrifugal force. The filtrate passes through the filter medium and also the perforated wall and, finally, into an outer casing and out through a discharge port. Wash liquid may be sprayed through the solids to remove soluble material, and then the cake is spun as dry as possible by rotating the bowl at a higher speed than during the operating run. The motor is shut off and, with the bowl slowly turning, the solids are discharged by scraping them out with a scraper knife, which peels the cake from the filter medium and drops it through an opening in the floor of the bowl. The filter medium is rinsed clean, or substituted if needed, the motor turned on, and the cycle repeated. Direct comparison of pressure filtration and centrifugal filtration reveals certain differences. For example, in the latter case, both the centrifugal force and the filtering area increase with an increase in radius. Centrifugal force acts on the filtrate passing through the cake and on the cake itself, aiding the hydraulic pressure head. For a relatively simple arrangement of a centrifugal filter, assuming an incompressible cake and disregarding kinetic energy changes in the filtrate, one may express the rate of flow of filtrate through by

$$Q = \frac{\rho \omega^2 (R_3^2 - R_1^2)}{2\mu \left(\frac{\alpha M_c}{A_a A_l} + \frac{R_m}{A_m} \right)}, \tag{7.51}$$

where ρ is the density of the liquid, ω is the rotation speed, R_3 is the radius of the inner surface of the bowl, R_1 is the radius of the inner surface of the liquid ring, μ is the viscosity of the filtrate, α is the specific cake resistance, M_c is the mass of solid cake in the bowl, A_a is the arithmetic mean of the cake area, A_l is the logarithmic mean of the cake area, R_m is the resistance of the filter medium, and A_m is the area of the filter medium.

Equation 7.51 is applicable only for cakes of uniform thickness. In many practical cases cakes are thicker near the bottom than at the top of the bowl. Such systems will give a filtration rate 5–20% greater than a cake of constant thickness of similar volume and permeability (Green and Perry 2008).

7.5.3.3 Equipment for Centrifugation

Tubular-bowl and disc-bowl centrifuges are designs normally used for clarification. Figure 7.22 illustrates a tubular-bowl centrifuge consisting of a long and narrow cylindrical bowl rotating at high speed in an outer stationary casing. The feed is introduced through a stationary pipe at the bottom of the bowl, and is quickly accelerated to the speed of the bowl by means of vanes or baffles. The two liquids are removed from the annular layers formed through a circular weir system, as shown in Fig. 7.22. The control over the neutral zone radius is exercised by fitting rings with different internal diameters to the dense phase outlet. The solids capacity of a tubular-bowl machine is seldom more than 2–5 kg and for economical operation the solids content of the feed should normally not exceed about 1% by weight. Disc-bowl centrifuges (Fig. 7.23) comprise a relatively shallow, wide cylindrical bowl, rotating at moderate speed in a stationary casing. The bowl contains closely spaced metal cones, or discs, which rotate with the bowl and are placed one above the other with a fixed clearance between them. The discs have one or more sets of matching holes in order to form channels through which the feed material flows. Under the influence of centrifugal force, the denser phase, travelling toward the wall of the bowl, streams down the undersides of the discs whereas the less dense phase, displaced toward the center, flows over the upper faces of the discs. The liquids are thus divided into thin layers and the distance any drop of the liquid must travel to be caught up in the appropriate outgoing stream is very small. The separated liquids are removed by a weir system and, like in the case of tubular centrifuges, different gravity discs on the dense phase outlet need to be used to control the neutral zone position. Disc-bowl centrifuges have solids capacities in the range of 2–20 kg and are usually only suited to clarifying feeds with less than a few percent by weight of solids. Both, tubular-bowl and disc-bowl centrifuges are useful for removing small traces of solids to produce clear liquids, especially if the solids are gelatinous in nature and unsuited to filtration.

Nozzle-discharge (self-cleaning) and valve-discharge (self-opening) centrifuges are sophisticated pieces of equipment designed to give continuity to the solid–liquid centrifugation process. A diagram of a nozzle-discharge centrifuge is given in Fig. 7.24. As can be seen, this centrifuge has a biconical bowl with a number of

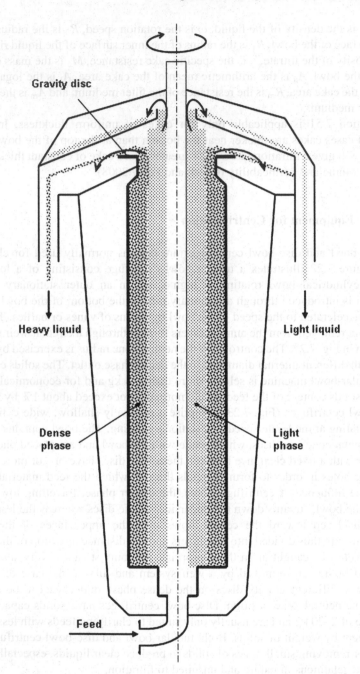

Fig. 7.22 A tubular-bowl centrifuge

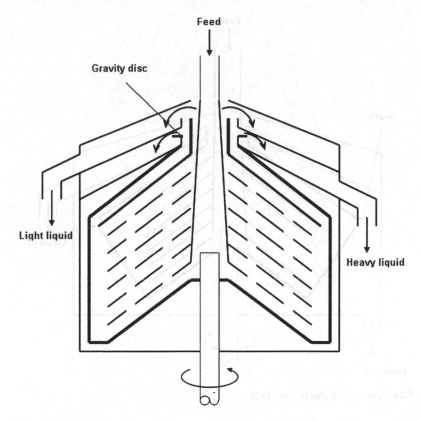

Fig. 7.23 A disc-bowl centrifuge

holes (approximately 0.3 cm in diameter) spaced around it at its largest diameter. The solids removed from the liquid are continuously discharged, in the form of thick slurry, into an outer casing. Feeds containing up to 25% solids can be handled in clarifiers of this type. The valve-discharge centrifuge has valves fitted in the solids discharge ports in the bowl which can be opened at desired intervals to get rid of the solids. Such valves may be controlled by timers or, alternatively, may be automatically opened by hydrostatic control.

To handle concentrated suspensions, basket centrifuges, screening centrifuges, and decanter centrifuges, among others, have been employed. Decanter or scroll centrifuges are conventionally used for the continuous separation of solids from liquids. They consist of a solid horizontal bowl tapered at one end and enclosed within a cylinder (Fig. 7.25). Thick slurry is pumped into the bowl along the central axis, and it is projected against the cylinder wall near the tapered section. A screw, revolving at a higher speed close to the wall of the bowl, conveys the solids along the tapered surface to the point of discharge and allows the inner layer of liquid to be drained out through a second discharge channel.

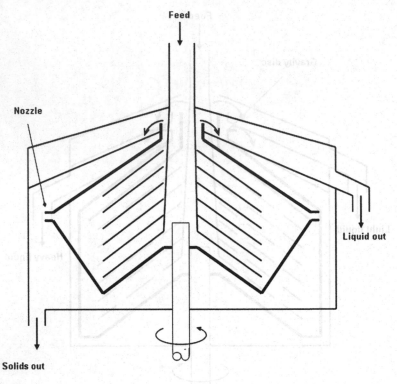

Fig. 7.24 A nozzle-discharge centrifuge

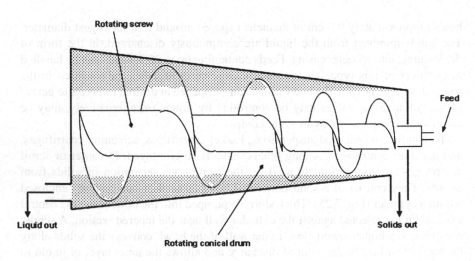

Fig. 7.25 Principle of the conveyor-bowl centrifuge

7.5.3.4 Applications

Tubular-bowl machines and disc-bowl separators have found diverse applications in the food processing industry. Some important examples include dewatering of vegetable and fish oils, clarification of sweet juices and fermenting products, separating of cream from milk, recovery of yeasts, and dewatering of different starches. In treatment of fruit juices, beer, and wines particularly, the quality of the clarified products may be directly related to the centrifugation efficiency in terms of removing protein fractions so as to avoid undesirable reactions such as browning. Centrifugation has also been used solely as the clarification step in fruit juice processing. There is a type of apple juice, for example, that is slightly clearer than the non-clarified juice but that is considerably more opaque than the filtered juice which is obtained by treating the pressed screened juice in a centrifuge only. In some other applications, for example, starch refining, the use of "sour liquid" (liquid fermentate from traditional fermentation processes) aids sedimentation and competes with centrifugation in the quality of the product obtained.

7.5.4 Hydrocyclone Separation

The use of hydrocyclones is another possibility to separate suspended solids from a liquid taking advantage of the centrifugal force. The geometrical features of hydrocyclones are similar to those of cyclones (Fig. 7.9), i.e., a hydrocyclone consists of a cono-cylindrical body, which promotes vortex formation when a suspension is pumped through it. The vortex produces a centrifugal force, which causes coarse particles to migrate against the cyclone wall and be discharged through the underflow orifice. Fine particles remain around the central axis of the cyclone and are carried out by the overflow stream. Hydrocyclones are easily manufactured and modified and have been well tested in thickening, clarification, classification, and other operations in many industries (Svarovsky 1984).

7.5.4.1 Operating Principles

Similarly to centrifuges, hydrocyclones may be evaluated in terms of separation efficiency by means of the cut size or cut point (x_c or x_{50}). The cut size, which is the only single number that in some way represents the separating capability of a hydrocyclone, is the particle size at which the grade efficiency $G(X)$ curve shows a value of 50%. A grade efficiency curve for a hydrocyclone is derived from screen analysis data for the feed, overflow, and underflow streams, and is a continuous representation of the overall mass recovery as a fraction of the mass flow rate. Since most suspended powders and fine particulate systems can be represented by a continuous size distribution, the grade efficiency curve is really derived from

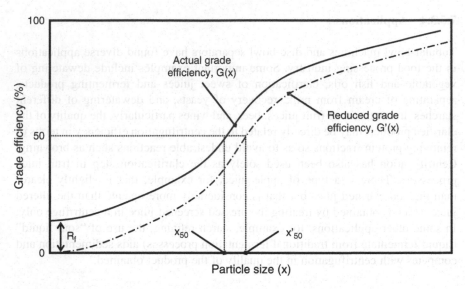

Fig. 7.26 Grade efficiency and reduced grade efficiency curve for a hydrocyclone

a stepwise calculation, drawing a line through the midpoints of size intervals (Trawinski 1977). Consequently, for hydrocyclones and dynamic separators, the grade efficiency curve is an S-shaped cumulative plot in which the 50% point represents a limit value. Thus, particles with this limit size have a 50% probability of being separated. In other words, all particles above the cut size are generally discharged in the underflow, whereas those below the cut size are normally carried away in the overflow. Ortega-Rivas (2007) reviewed some of the numerous expressions utilized for determining the cut size which have been reported in the literature.

The grade efficiency curve, as described above, gives a plot which does not pass through the origin. This can be explained bearing in mind that a hydrocyclone is a flow divider, so the underflow always contains a certain quantity of very fine particles which simply follow the flow, and are split in the same ratio as the liquid. The apparent finite efficiency for fine particles is therefore equal to the underflow-to-throughput ratio R_f, and a "corrected" or "reduced" cut size x'_{50}, will practically assess the performance of hydrocyclones when derived from the reduced grade efficiency $G'(X)$. All these definitions are given in Fig. 7.26.

7.5.4.2   Use of Relationships Between Dimensionless Groups

Since hydrocyclones do not have any rotating parts and the vortex action to produce the centrifugal force is obtained by pumping the feed suspension tangentially into the cono-cylindrical body, the literature is full of studies of the effects of the relative geometric proportions on pressure drop or capacity and separation efficiency.

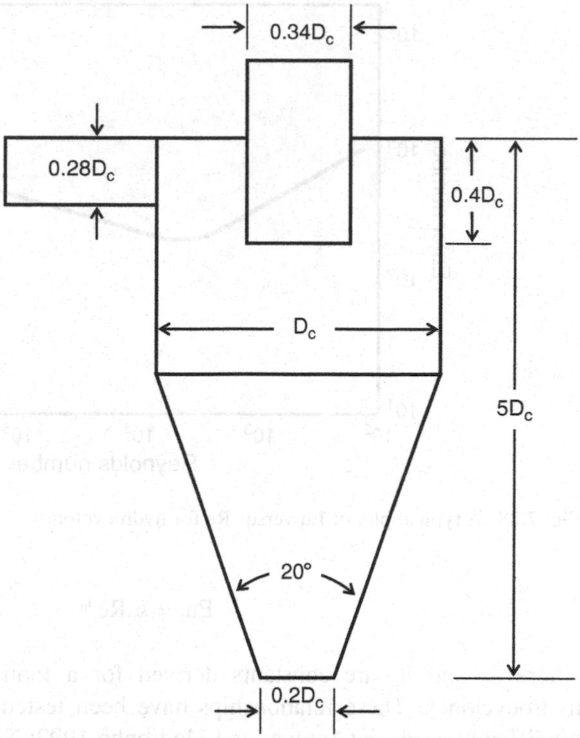

Fig. 7.27 Dimensions of
Rietema's standard
hydrocyclone

With use of this information, a hydrocyclone geometry can be selected to obtain the optimum performance in terms of cut size. In this sense, possibly the best way to predict hydrocyclone performance is the use of a dimensionless scale-up model, which has been well described elsewhere (Ortega-Rivas and Svarovsky 1993). The same three dimensionless groups that describe cyclone operation and performance are used for similar purposes in hydrocyclones, i.e., the Euler number Eu, the Reynolds number Re, and the Stokes number Stk_{50}. The only observation is that the density and viscosity in these expressions for dimensionless groups correspond to liquid (usually water) in this case, and not to gas like in cyclones. For the best application of the relationships among dimensionless groups, certain proportions must be unchanged. Such proportions are generally reported as a function of the diameter of the hydrocyclone. There are several different standard hydrocyclone designs in which the proportions remain the same regardless of size. One of the most efficient designs for separation is called the Rietema cyclone (Rietema 1961), whose proportions are illustrated in Fig. 7.27.

The dimensional analysis gives two basic relationships between the above-mentioned dimensionless groups:

$$Stk_{50} \ Eu = constant, \tag{7.52}$$

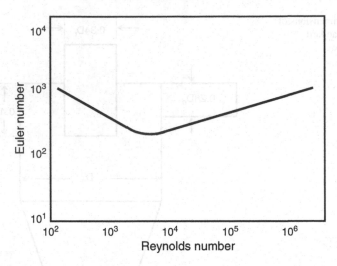

Fig. 7.28 A typical plot of Eu versus Re for hydrocyclones

$$Eu = k_p Re^{n_p}, \tag{7.53}$$

where k_p and n_p are constants derived for a family of geometrically similar hydrocyclones. These relationships have been tested over a range of conditions by different workers (Antunes and Medronho 1992; Nezhati and Thew 1988; Chen et al. 2000). Their plots are typical, as shown in Fig. 7.28 for Eq. 7.53.

At higher concentrations, the feed concentration as a fraction of the volume C has to be included as an additional dimensionless group. Svarovsky and Marasinghe (1980) reported the following expression for the effect of high solids concentrations in the feed:

$$Stk_{50}(r) = k_1(1 - R_f)e^{k_2 C}, \tag{7.54}$$

where $Stk_{50}(r)$ includes the previously described reduced cut size, which takes into account the "dead flux" effect of very fine particles simply following the flow and split in the same ratio as the liquid. R_f is the underflow-to-throughput ratio. The correlation has proved to hold well for concentrations above 8% by volume, and the values of the constants k_1 and k_2 were found to be 9.05×10^{-5} and 6.461, respectively, for limestone and an AKW® (Amber Kaolinwerke, Hirschau, Germany) hydrocyclone of 125-mm diameter.

An exhaustive study for concentrations up to 10% by volume was performed by Medronho and Svarovsky (1984) to verify the applicability of Eqs. 7.52, 7.53, and 7.54. They employed three geometrically similar hydrocyclones with Rietema's optimum geometry and obtained the following relations:

$$Stk_{50}(r)Eu = 0.047[\ln(1/R_f)]^{0.74}e^{8.96C}, \tag{7.55}$$

$$Eu = 71Re^{-0.116}(D_i/D_c)^{-1.3}e^{2.12C}, \tag{7.56}$$

$$R_f = 1218(D_u/D_c)^{-4.75}Eu^{-0.30}, \tag{7.57}$$

where D_i, D_c, and D_u are the inlet, body, and underflow diameters of the hydrocyclone, respectively.

For concentrations higher than 10% by volume, many practical slurries show non-Newtonian behavior and it can be shown (Ortega-Rivas and Svarovsky 1993) that the Reynolds and Stokes numbers can be reexpressed to consider such behavior. The correlations derived under this consideration are as follows:

$$Stk_{50}^*(r)Eu = 0.006[\ln(1/R_f)]^{2.37}e^{6.84C}, \tag{7.58}$$

$$Eu = 1686Re^{*-0.035}e^{-3.39C}, \tag{7.59}$$

$$R_f = 32.8(D_u/D_c)^{1.53}Re^{*-0.34}e^{3.70C}, \tag{7.60}$$

where $Stk_{50}^*(r)$ and Re^* are the "generalized" Stokes and Reynolds numbers, meaning that they include the parameters of characterization of non-Newtonian suspensions, i.e., the fluid consistency index K' and the flow behavior index n, instead of the medium viscosity (Ortega-Rivas and Svarovsky 1993), described by Eqs. 7.21 and 3.23 respectively.

7.5.4.3 Main Applications

In terms of biological fluids and suspensions, the standard application of small hydrocyclones is for starch refining, and they have been extensively used in corn and potato starch refining, giving good results. Hydrocyclones have been used to separate gossypol from cottonseed protein in cottonseed oil processing. They have also been used as separators in multistage mixer–separator extraction systems for soluble coffee. Hydrocyclones have been employed in thickening of wastewater sludge with promising results. Sephadex, yeasts, and blood cells have also been separated using hydrocyclones. Yeast recovery in different fermentation processes has also been performed well by hydrocyclones. Another application reported is that of kieselguhr recycling for filters in the brewing industry.

7.5.5 Filtration

Filtration may be defined as the unit operation in which the insoluble solid component of a solid–liquid suspension is separated from the liquid component by passing the suspension through a porous barrier which retains the solid particles on its upstream surface or within its structure, or both. The solid–liquid suspension is known as the feed slurry or prefilt, the liquid component that passes through the membrane is called the filtrate, and the barrier itself is referred to as the filter medium. When the solid particles are retained on the upstream surface of the filter medium, the separated solids are known as the filter cake once they form a detectable layer covering such an upstream surface. The flow of filtrate may be caused by several means. Pressure and a vacuum are two conventional ways of driving the suspension across the medium. Gravitational and centrifugal forces may also be used to enable the suspension to cross the medium. Gravitational filtration has limited use in food processes but is applied to water and sewage treatment. In general terms, filtration theory applies to cases where cake buildup occurs.

7.5.5.1 Filtering Fundamentals

Some of the more fundamental treatments of filtration theory were reviewed by Wakeman and Tarleton (1998). In the initial stages of filtration, the first solid particles to encounter the filter medium become enmeshed in it, reducing its open surface area and increasing the resistance it offers to the flow of filtrate. As filtration proceeds, a layer of solids builds up on the upstream face of the medium and this layer, or cake, increases in thickness with time. Once formed, this cake in fact becomes the primary filter medium. Filtrate passing through the filter encounters three types of resistance: a first resistance offered by channels of the filter itself, a second one because of the presence of the filter medium, and a third one due to the filter cake. The total pressure drop across the filter is equivalent to the sum of the pressure drops resulting from these three resistances. Usually the pressure drop due to the channels of the filter is disregarded in calculations. If $-\Delta P$ is the total pressure drop across the filter and $-\Delta P_c$ and $-\Delta P_m$ are the pressure drops across the cake and the medium respectively, then

$$- \Delta P = -\Delta P_c - \Delta P_m. \tag{7.61}$$

The pressure drop across the filter cake can be related to the filtrate flow by the expression (McCabe et al. 2005)

$$- \Delta P_c = \frac{\alpha \mu w V}{A^2} \left(\frac{dV}{dt} \right), \tag{7.62}$$

where α is the specific resistance of the cake, μ is the viscosity of the filtrate, w is the mass of solids deposited on the medium per unit volume of filtrate, V is the volume of the filtrate, and A is the area of the filter normal to the direction of filtrate flow.

The specific resistance of the cake, α, physically represents the pressure drop necessary to produce unit superficial velocity of a filtrate of unit viscosity passing through a cake containing unit mass of solids per unit filter area. It is related to the properties of the cake by

$$\alpha = \frac{k(1-\varepsilon)S_0^2}{\varepsilon^3 \rho_s},$$ (7.63)

where k is a constant, ε is the porosity of the cake, S_0 is the specific surface area of the solid particles in the cake, and ρ_s is the solids density.

If a cake is composed of rigid non-deformable solid particles, α is independent of $-\Delta P_c$ and does not vary throughout the depth of the cake, and is known as incompressible cake. However, if the cake contains non-rigid, deformable solid particles or agglomerates of particles, the resistance to flow will depend on the pressure drop and will vary throughout the depth of the cake. In this case the cake is called compressible and an average value of the specific resistance for the entire cake must be used in Eq. 7.62. This average specific resistance must be measured experimentally for a particular slurry.

By analogy with Eq. 7.62, the resistance of the filter medium may be defined by the following relation:

$$-\Delta P_m = \frac{R_m \mu}{A}\left(\frac{dV}{dt}\right),$$ (7.64)

where R_m is the pressure drop across the medium.

It is reasonable to assume that R_m is constant during any filtration cycle and that it includes the resistance to filtrate flow offered by the filter channels. This being the case, Eqs. 7.62, 7.63, and 7.64 can be combined to give

$$\frac{dV}{dt} = \frac{A(-\Delta P)}{\mu\left(\frac{\alpha w V}{A} + R_m\right)}.$$ (7.65)

Equation 7.65 is a general expression for the filtrate flow rate.

7.5.5.2   Constant-Pressure Filtration

When the pressure drop is maintained constant, Eq. 7.65 may be integrated thus

$$\int_0^t dt = \frac{\mu}{A(-\Delta P)}\left(\frac{\alpha w}{A}\int_0^V V dV + R_m \int_0^V dV\right)$$ (7.66)

or, if we substitute limits and transpose for time t,

$$t = \frac{\mu}{-\Delta P}\left[\frac{\alpha w}{2}\left(\frac{V}{A}\right)^2 + R_m\left(\frac{V}{A}\right)\right].$$ (7.67)

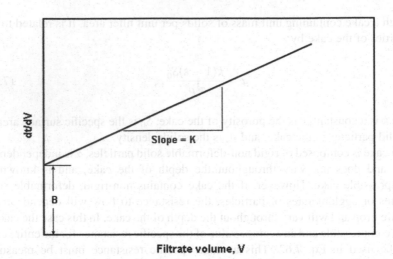

Fig. 7.29 Plot of the results from a constant-pressure-filtration run

Equation 7.67 is a general expression for the filtration time during constant-pressure filtration. To use it, values of α and R_m must be determined experimentally. This can be done by rewriting Eq. 7.65 in the following form:

$$\frac{dt}{dV} = KV + B, \tag{7.68}$$

where

$$K = \frac{\alpha w \mu}{A^2(-\Delta P)} \tag{7.69}$$

and

$$B = \frac{R_m \mu}{A(-\Delta P)}. \tag{7.70}$$

As can be seen, Eq. 7.68 represents a straight line if dt/dV is plotted against V. Therefore, if a constant-pressure filtration is performed and values of V for different values of t are recorded, a graph of dt/dV versus V can be constructed as shown in Fig. 7.29. The slope of this line is K and the intercept on the ordinate when $V = 0$ is B. Thus, by using such a graph, we can directly determine the values of values of α and R_m from Eqs. 7.69 and 7.70.

For incompressible cakes Eq. 7.67 can be used directly at different pressures. However, for compressible cakes, the relationship between α and $-\Delta P$ needs to be determined experimentally by performing filtration runs at different constant pressures. Empirical equations may be fitted to the results obtained. Two such equations have been suggested (McCabe et al. 2005):

$$\alpha = \alpha_0(-\Delta P)^s, \tag{7.71}$$

$$\alpha = \alpha_0'[1 - \beta(-\Delta P)^{s'}], \tag{7.72}$$

where α_0, α_0', s, s', and β are empirical constants. Having determined values for α_0 and s, we can then use Eq. 7.71 and hence Eq. 7.67 for constant-pressure-filtration calculations at different pressures.

7.5.5.3   Constant-Rate Filtration

If filtration is performed at constant rate, then $dV/dt = \text{constant} = V/t$. Considering this and substituting into Eq. 7.65, we can rewrite such an equation as follows:

$$-\Delta P = \left(\frac{\mu \alpha w V}{A^2 t}\right)V + \frac{\mu V R_m}{At} \tag{7.73}$$

or

$$-\Delta P = K'V + B'. \tag{7.74}$$

Once again it can be seen that Eq. 7.73 represents a straight line if $-\Delta P$ is plotted against V. The slope of the line is K' and the intercept with the $-\Delta P$ axis when $V = 0$ is B'. Thus, for incompressible cakes α and R_m can again be determined by experimental means. Equation 7.73 can then be used for cycle calculations.

For compressible cakes, the relationship between α and $-\Delta P_c$ must, again, be determined experimentally. If a relationship of the form shown in Eq. 7.71 is assumed to apply, then Eq. 7.62 may be modified to

$$(-\Delta P_c)^{1-s} = [-(\Delta P - \Delta P_m)]^{1-s} = \frac{\mu \alpha_0 w V}{A^2}\frac{V}{t}, \tag{7.75}$$

which in turn may be written as

$$[-(\Delta P - \Delta P_m)]^{1-s} = K''t, \tag{7.76}$$

where

$$K'' = \frac{\mu \alpha_0 w}{A^2}\left(\frac{V}{t}\right)^2. \tag{7.77}$$

If we assume that $-\Delta P_m$ is constant throughout a constant-rate filtration, then by plotting t versus ΔP and drawing a smooth curve through the points and extrapolating the curve to the $-\Delta P$ axis, we can obtain an approximate value for $-\Delta P_m$. If $-(\Delta P - \Delta P_m)$ is then plotted against t on log–log paper and a straight line is obtained, the slope of this line is $(1-s)$. Thus, s can be calculated and K'' and α_0 can be derived from Eqs. 7.76 and 7.77. If the first log–log plot of $-(\Delta P - \Delta P_m)$ is not a straight line, further approximations for $-\Delta P_m$ need to be made (McCabe et al. 2005).

7.5.5.4 Filtration Media and Filter Aids

Filtration equipment is adapted using consumable materials for every operations. The main consumables in diverse filtration operations are the filter medium and the filter aids. Filter media are selected, primarily, on the basis of the pore size that will be appropriate for retention of the solids in the suspension. Other important functions of the filter medium are the promotion of cake formation and the support of cake once it has formed. The medium should offer minimum resistance to flow according to the need for rapid formation of the filter cake. It must be strong enough to support the cake and retain its strength under the extreme conditions that may occur during operation. The surface characteristics of the filter medium should facilitate cake removal. It must be non-toxic and chemically compatible with the materials being filtered, and it should also be cost-effective. Filter media may be rigid or flexible. Rigid media may be loose or packed. Examples of loose rigid media are sand, gravel, diatomaceous earths, and charcoal. Packed or fixed rigid media include porous carbon, porcelain, fused alumina, perforated metal plates, and rigid wire mesh. Flexible media comprise woven fabrics of cotton, silk, wool, and jute. Flexible media made out of synthetics include nylon, polypropylene, polythene, poly(vinyl chloride), and copolymers. Glass fiber and flexible metal meshes are also used as filter media, as are non-woven materials, such as cotton fibers, wool fibers, and paper pulp. These materials may be available in preformed pads of various shapes. Table 7.3 summarizes different filter materials along with their main features.

Another way of facilitating the filtration process, mainly when the solids to be filtered are finely divided or are of a slimy highly compressible nature, is the use of filter aids. For solids of this type, the filter medium tends to block quickly, thus shortening filtration cycles considerably. To overcome this difficulty and obtain reasonable filtration cycles, materials referred to as filter aids are often used. They consist of comparatively large, inert, non-compressible solid particles of different shapes. They may be applied by mixing them with the feed slurry or they may be suspended in clear liquid, often some filtrate from a previous run, which is then passed through the filter in order to build a precoat of the filter aid on the medium. The filter aid then forms a rigid lattice structure on the medium and provides numerous channels for the filtrate to flow through, slowing the plugging of the

Table 7.3 Physical properties of fibers used in materials for filter media

Fiber	Maximum temperature (°C)	Specific gravity	Water absorbance (%)	Elongation (% to rupture)	Durability
Acetate	99	1.30	9–14	30–50	Poor
Acrylic	135–149	1.14–1.17	3–5	25–70	Good
Cotton	93	1.55	16–22	5–10	Regular
Spun glass	288–315	2.50–2.55	Up to 0.3	2–5	Poor
Nylon	107–121	1.14	6.5–8.3	30–70	Excellent
Polyester	149	1.38	0.04–0.08	10–50	Excellent
Polythene					
Low density	66–74	0.92	0.01	20–80	Good
High density	93–110	0.92	0.01	10–45	Good
Polypropylene	121	0.91	0.01–0.1	15–35	Good
Poly(vinyl chloride)	66–71	1.38			Regular
Rayon	99	1.50–1.54	20–27	6–40	Poor
Saran	71–82	1.7	0.1–1.0	15–30	Regular
Wool	82–93	1.3	16–18	25–35	Regular

medium and prolonging filtration cycles. Examples of commonly used filter aids are kieselguhr and diatomaceous earths. Paper pulp, carbon, fuller's earth, and some other materials can be used as filter aids.

7.5.5.5   Equipment Used for Filtration

The most common types of filters used in industrial processing are pressure filters such as filter presses, leaf filters, and cartridge filters, as well as vacuum filters such as the rotary drum vacuum filter and the rotary vacuum disc filter. Most filtration units produce a cake, which with long cycles may become reasonably dry at the expense of a decline in the flow rates. Maintaining a constant rate of filtration is costly as the pressure drop must be increased during the cycle.

A commonly used design for a filter press is the plate-and-frame filter press. The components of a plate-and-frame filter press are the frames and the two types of grooved plates, the filter plate and the wash plate, a shown in Fig. 7.30a. The grooved plates covered on both sides with filter medium and alternatively assembled with frames, as illustrated in Fig. 7.30b, are mounted in a rack. The whole assembly can be squeezed tightly together by a screw-driven manual mechanism or by a hydraulic or pneumatic mechanism to form a liquid-tight unit (Fig. 7.31). Both plates and frames are provided with openings at one corner so that when the press is closed these openings form a channel through which the feed is introduced. A slurry is fed into the frames and the cake builds up in the hollow center of the frames. The filtrate passes through the medium and to the grooved surfaces of the filter plates to be removed via an outlet channel in each plate. Filtration continues until the flow of filtrate drops below a practical level, or when accumulation of cake in the frames

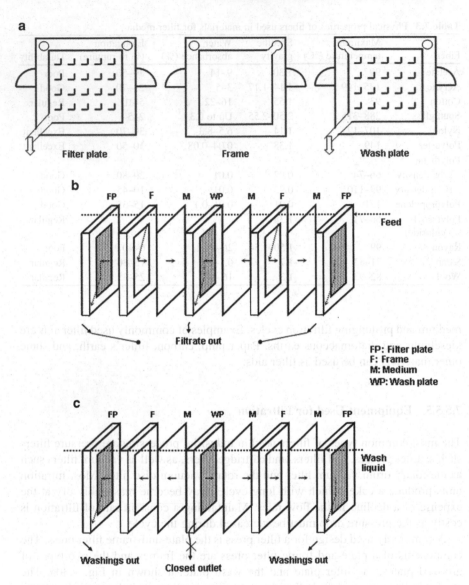

Fig. 7.30 Description and arrangements of filter press components: (**a**) detail of plates and frames, (**b**) arrangement for the filtration cycle, (**c**) arrangement for the washing cycle

increases the pressure to an extremely high level. Washing of the cake can be done by replacing the flow of feed slurry with wash liquid or by the use of special wash plates mounted in the arrangement so that every second plate is a wash plate. When a filtration cycle is run, these wash plates act as filter plates, whereas during washing their outlets are closed and the wash liquid is introduced to their surfaces through a special inlet channel. The flow path for both filtration and washing cycles using wash plates is illustrated in Fig. 7.30b, c, respectively.

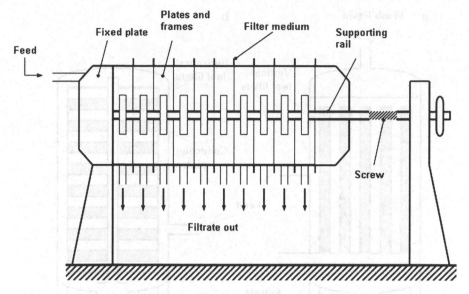

Fig. 7.31 A plate-and-frame filter press assembled and mounted

Another alternative for pressure filtration is a pressure-vessel filter, which makes use of filter leaves as the basic filter element. A filter leaf consists of a wire mesh screen or grooved drainage plate over which the filter medium is stretched. The leaf may be suspended from the top or supported from the bottom or the center. The supporting element is usually hollow and forms an outlet channel for the filtrate. Several designs of pressure-vessel filters with filter leaf elements are commercially available. Two of them are the horizontal-leaf filter and the vertical-tank, vertical-leaf filter, both shown in Fig. 7.32. Vertical-tank, vertical-leaf filters (Fig. 7.32a) are the cheapest of the pressure-vessel leaf filters and have the lowest volume-to-area ratio. Cake discharge is accomplished by sluicing, either manually or by vibration. Horizontal-leaf filters (Fig. 7.32b), in either a horizontal or a vertical tank, are advantageous when the flow is intermittent. Filtration takes place only on the upper size of the horizontal leaves, so the filtration area is limited compared with other pressure-vessel designs. Horizontal-leaf filters are most suitable when thorough cake washing is necessary. Cake discharge may be accomplished by rotation of the leaves.

In vacuum filters a subatmospheric pressure is maintained downstream of the medium and atmospheric pressure is maintained upstream. Because the pressure drop across the filter is limited to 1 atm, such filters are not suited to batch operations. Some types of leaf filters, tube filters, or edge filters can be operated under a vacuum, but continuous-vacuum filters are far most common. Since the upstream pressure is atmospheric pressure, cake discharge is facilitated, thus favoring continuous operation. The rotary drum vacuum filter is, perhaps, the most common type of vacuum equipment used in an important number of applications. As can be seen in Fig. 7.33a, a vacuum is applied to the inside of a revolving drum covered with reinforced wire mesh and filter cloth or medium. The surface of the drum consists of a number of

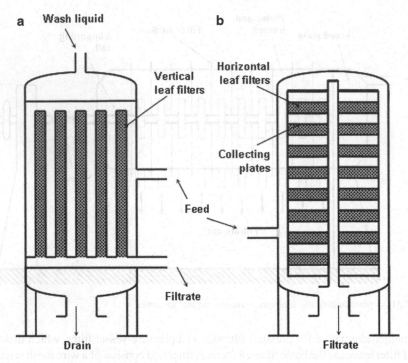

Fig. 7.32 Pressure-vessel leaf filters: (a) vertical-tank, vertical-leaf filter, (b) horizontal-leaf filter

shallow compartments formed between dividing strips running the length of the drum. Each compartment is connected by one or more pipelines to an automatic rotary valve situated centrally at one end of the drum. Filter medium covers the entire drum surface and is supported by perforated plates, grids, or wire mesh to provide drainage space between the medium and the floor of each shallow compartment. The drum dips into a slurry and rotates, a vacuum is drawn on its surface, and filtrate flows through the medium and out through the drain pipe from the compartment and is directed to the filtrate receiver by means of the valve. A layer of cake builds up on the outer surface of the medium and the cake is sucked free from filtrate as the drum turns, becoming drier near the end of the turn. Prior to the cake being submerged again in the slurry tank on completing a drum revolution, a knife removes the cake in the form of a continuous layer. Rotary vacuum precoat filters may produce filtrates which contain 0.1% or less suspended solids. Rotary vacuum disc filters, as shown in Fig. 7.33b, consist of a number of circular filter leaves mounted on a horizontal axis about which they rotate. Each disc is fitted with a cake-removal device and is divided into sectors, each of which has an individual outlet to the central shaft. These outlets form a continuous channel through which the filtrate flows from all sectors at the same angle. All the channels terminate in a rotary valve like the one used in drum filters. In operation, each disc operates as a drum filter, with the cycle being controlled by the rotary valve.

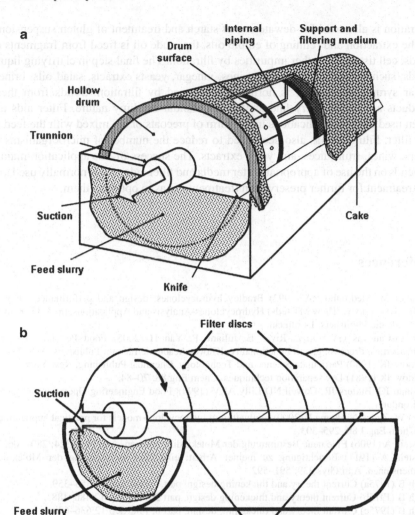

Fig. 7.33 Equipment for vacuum filtration: (**a**) continuous rotary drum, (**b**) rotary disc

7.5.5.6 Filtration Applications

Prior to the wide commercialization of membrane separations, conventional filtration was the final clarification step in many food applications, particularly for fruit juices and fermented products, such as beers and wines. Similarly to the case of centrifugation, filtration effects have focused on quality aspects, not on microbial safety. As previously mentioned, filtration applications in the food industry are numerous and varied. In the extraction of sugar from cane and sugar beet, the mill juice contains much solid impurity that is removed by sedimentation followed by filtration.

Filtration is also used for dewatering of starch and treatment of gluten suspensions. In the extraction and refining of edible oils, the crude oil is freed from fragments of seeds, cell tissue, and other impurities by filtration. The final step in clarifying liquid foods, such as fruit juices, beers, wines, vinegar, yeasts extracts, salad oils, brines, sugar syrups, and jellies, is normally performed by filtration. Solids from these products are usually very fine and/or of colloidal or slimy nature. Filter aids are often used in these applications in the form of precoats or are mixed with the feed to the filter. Filtration has also been used to reduce the number of microorganisms in beers, wines, fruit juices, and yeast extracts. The success of this application mainly depends on the use of appropriate filter media and filter aids, and is normally used as a pretreatment for further preservation treatments so as to optimize them.

References

Antunes M, Medronho RA (1992) Bradley hydrocyclones: design and performance analysis. In: Svarovsky L, Thew MT (eds) Hydrocyclones-Analysis and Applications, pp 3–13. Kluwer Academic Publishers, Dordrecht.

Barbosa-Cánovas GV, Ortega-Rivas E, Juliano P, Yan H (2005) Food Powders: Physical Properties, Processing, and Functionality, Kluwer Academic/Plenum Publishers, New York.

Beddow JK (1980) Particulate Science and Technology. Chemical Publishing, New York.

Beddow JK (1981) Dry separation techniques. Chem Eng 88: 70–84.

Brennan JG, Butters JR, Cowell ND Lilly AEV (1990). Food Engineering Operations. Elsevier, London.

Chen W, Zydek N, Parma F (2000) Evaluation of hydrocyclone models for practical applications. Chem Eng J 80: 295–303.

Einstein A (1906) Eine neue Bestimmung der Moleküldimensionen. Ann Phys 324: 289–306.

Einstein A (1911) Berichtigung zu meiner Arbeit: eine neue Bestimmung der Moleküldimensionen. Ann Phys 339: 591–592.

Fitch B (1975a) Current theory and thickening design, part 1. Filtr Sep 12:355–359.

Fitch B (1975b) Current theory and thickening design, part 2. Filtr Sep 12:480–488.

Fitch B (1975c) Current theory and thickening design, part 3. Filtr Sep 12:636–638.

Geldart D (1986) Gas Fluidization Technology. John Wiley and Sons, London.

Green DW, Perry RH (2008) Perry's Chemical Engineers' Handbook. 8th Ed. McGraw-Hill, New York.

Guth E, Simha R (1936) Untersuchungen übe die Viskositat von Suspensionen und Lösungen 3. Über die Viskositat von Kugelsuspensionen. Kolloid-Z 74: 266–275.

Klumpar IV, Currier FN, Ring, TA (1986) Air classifiers. Chem Eng 93: 77–92.

McCabe WL, Smith JC, Harriot P (2005) Unit Operations in Chemical Engineering. 7th Ed. McGraw-Hill, New York.

Medronho RA, Svarovsky L (1984) Test to verify hydrocyclone scale-up Procedure. In: Proceedings of the 2nd International Conference on Hydrocyclones, pp 1–14. BHRA The Fluid Engineering Centre, Cranfield, UK.

Nakayima YN, Whiten WJ, White MR (1978) Method for measurement of particle shape distribution by sieves. Trans Inst Min Metall 87: C194-C203.

Nezhati K, Thew MT (1988) Aspects of the performance and scaling of hydrocyclones for use with light dispersions. In: Wood P (ed) 3rd International Conference on Hydrocyclones, pp 167–180. Elsevier Applied Science Publishers, London.

Ortega-Rivas E (2005) Handling and processing of food powders and particulates. In: Onwulata C (ed) Encapsulated and Powdered Foods, pp 75–144. CRC Taylor & Francis, Boca Raton, FL.

Ortega-Rivas E (2007) Hydrocyclones. In: Ullmann's Encyclopedia of Industrial Chemistry, 28 p. Wiley-VCH, Weinheim.

Ortega-Rivas E, Svarovsky L (1993) On the completion of a dimensionless scale-up model for hydrocyclone separation. Fluid/Part Sep J 6: 104–109.

Ortega-Rivas E, Svarovsky L (2000) Centrifugal air classification as a tool for narrowing the spread particle size distributions of powders. In: Wöhlbier RH (ed) Processing Part I, pp 193–195. Trans Tech Publications, Clausthal-Zellerfeld.

Rietema K (1961) Performance and design of hydrocyclones. Chem Eng Sci 15: 198–325.

Simha R (1952) A treatment of the viscosity of concentrated suspensions. J Appl Phys 23: 1020–1024.

Strumpf DM (1986). Selected particle size determination techniques. Manuf Confect 66: 111–114.

Svarovsky L (1981) Solid-Gas Separation. Elsevier, Amsterdam.

Svarovsky L (1984) Hydrocyclones. Holt Rinehart and Winston, London.

Svarovsky L (2000) Solid–liquid Separation. Butterworths-Heinemann, London.

Svarovsky L, Marasinghe BS (1980) Performance of hydrocyclones at high feed solid concentrations. In: Proceedings of International Conference on Hydrocyclones, pp 127–142. BHRA The Fluid Engineering Centre, Cranfield.

Thomas DG (1963) Non-Newtonian suspensions. Part I: physical properties and laminar transport characteristics. Ind Eng Chem 55: 18–29.

Trawinski HF (1977) Hydrocyclones. In: Purchas DB (ed) Solid/Liquid Separation Equipment Scale-up, pp 241–287. Uplands Press Ltd, Croydon.

Trowbridge MEO'K (1962) Problems in the scaling-up of centrifugal separation equipment. Chem Eng 40: 73–86.

Wakeman RJ, Tarleton S (1998) Filtration: Equipment Selection Modelling and Process Simulation. Elsevier Science, Oxford.

Chapter 8
Membrane Separations

8.1 Introduction

Membrane separations are techniques used industrially to remove solutes and emulsified substances from solutions by application of pressure onto a very thin layer of a substance with microscopic pores, known as a membrane. Membrane separation processes include reverse osmosis, ultrafiltration, microfiltration, dialysis, electrodialysis, gas separation, and pervaporation. Reverse osmosis, ultrafiltration, microfiltration, and electrodialysis have been widely used commercially (Girard and Fukumoto 2000). A suitable way of classifying membrane separation techniques is by referring to their limiting size of retention. As shown in Table 8.1, reverse osmosis separates sizes from 0.0001 to 0.001 μm, the range of separation of ultrafiltration is from 0.001 to 0.1 μm, and microfiltration membranes have pore sizes from 0.1 to 10 μm.

Owing to the small size range of pores involved in membranes, an alternative, convenient way of referring to separating capability is by means of a molecular weight rating, expressed in terms of a rejection coefficient against a species of specific molecular weight (Chen et al. 2004). Ideally, the membranes will have a sharply defined molecular weight cutoff (MWCO). Such an ideal membrane will retain all species with molecular weight greater than the MWCO but will allow all species with lower molecular weight to pass. Membranes are available in a number of increments in MWCO ranges from 1,000 up to 100,000. For example, the MWCO range of reverse osmosis is below 300, whereas that for ultrafiltration membranes is between 300 and 300,000 (Table 8.1).

8.2 Basic Concepts of Membrane Filtration

Membrane separations originated as a consequence of the need to desalt seawater using alternatives to the thermal processes based on evaporation. Seawater was successfully desalinated by a technique known as reverse osmosis in 1958

E. Ortega-Rivas, *Non-thermal Food Engineering Operations*,
Food Engineering Series, DOI 10.1007/978-1-4614-2038-5_8,
© Springer Science+Business Media, LLC 2012

Table 8.1 Membrane retention range

Pore size	Molecular weight cutoff	Examples	Applicable technique
100 μm		Pollen	Microfiltration
10 μm		Starch	
1,000 Å		Blood cells	
		Typical bacteria	
		Smaller bacteria	
		DNA, viruses	
100Å	100,000	Albumin	Ultrafiltration
	10,000	Vitamin B$_{12}$	
	1,000		
10Å		Glucose	Reverse osmosis
1Å		Water	
		Sodium chloride	

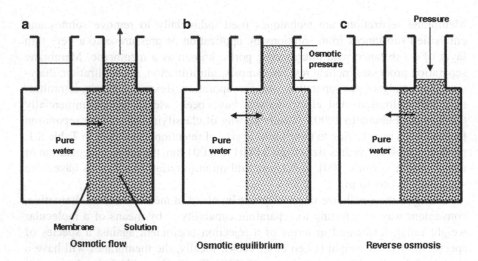

Fig. 8.1 Principle of reverse osmosis

(Sourirajan 1970). The principle of separation of solutes by reverse osmosis is illustrated in Fig. 8.1. The natural solvent transfer through a semipermeable membrane causes a pressure known as osmotic pressure. If such a natural process is inverted, the solutes concentrate even more in the solution containing them in higher proportion, and they are diluted to a minimum in the solution containing them in the lower proportion. With use of membrane separations, phases can be separated from two-phase and multiphase systems without the need for any phase state change, i.e., without requiring thermal energy. The osmotic pressure π for

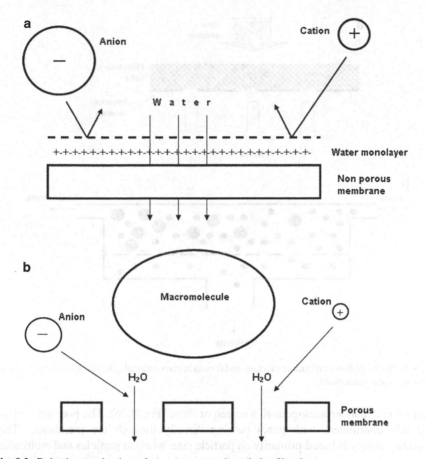

Fig. 8.2 Rejection mechanisms for reverse osmosis and ultrafiltration

dilute solutions can be established as a function of pressure and temperature, using fundamental thermodynamics, as follows:

$$\pi = MRT, \tag{8.1}$$

where M is the molar concentration of the solution, R is the universal gas constant, and T is the absolute temperature.

Ultrafiltration is often compared with reverse osmosis although the mechanism of separation is quite different. This difference is illustrated in Fig. 8.2. As can be seen, in reverse osmosis a rejection is based on electrostatic repulsion due to formation of a pure layer of water over the membrane, and the virtual charges on this layer reject charges of ionic free species of salt solutions, as shown in Fig. 8.2a. Simultaneously, by a complex mechanism of sorption, diffusion, and desorption, pure water passes through the membrane performing the separation process. Ultrafiltration membranes, on the other hand, are porous in nature, with a rigid and highly voided structure, and

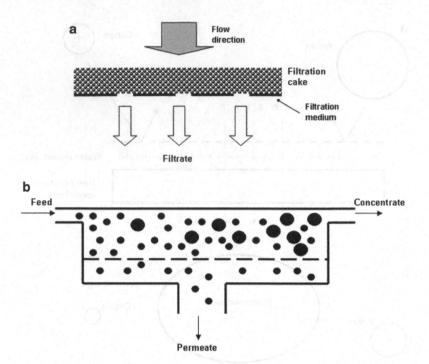

Fig. 8.3 Modes of flow in filtration: (**a**) dead-end flow as in conventional cake filtration, (**b**) cross-flow as in membrane separations

function in a manner analogous to a screen or sieve (Fig. 8.2b). The pore network is randomly distributed, with pores passing directly through the membrane. The separating ability is based primarily on particle size, wherein particles and molecules larger than the largest pore are completely retained, whereas species smaller than the smallest pore are totally permeated. In general terms, the mechanism of separation in reverse osmosis is known as salt rejection, whereas that in ultrafiltration is called organic rejection. As described, ultrafiltration can be considered as an extension of conventional filtration, with its separating ability ranging to the molecular level. The operating mode of the equipment is, however, different. As shown in Fig. 8.3, in ultrafiltration the fluid moves continuously across the membrane surface. The arrangement is known as cross-flow filtration and is used as means of sweeping the membrane surface to control the buildup of foulants and particulate matter.

Membrane separation processes operate in a continuous manner and so, unlike conventional filtration, the accumulated solids do not form a moist cake but, rather, form a dense slurry that is discharged like in centrifugation or hydrocyclone separation. This stream of concentrated solids is simply known as a concentrate, whereas the clarified stream is called a permeate. A schematic flow diagram of ultrafiltration and some other membrane separations is shown in Fig. 8.4. The flow rate Q of a membrane separation process may be represented by

$$Q = kA(\Delta P - \Delta\pi), \tag{8.2}$$

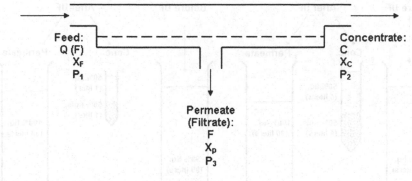

Fig. 8.4 Mass balance in a membrane separation process

where k is the membrane permeability coefficient, A is the membrane superficial area, ΔP is the pressure drop, and $\Delta \pi$ is the difference in osmotic pressure between the feed and the permeate. Equation 8.2 is more suitable for reverse osmosis applications, because of the osmotic pressure exerted by solutes in such applications. For ultrafiltration applications $\Delta \pi$ is negligible in relation to ΔP and the membrane thickness causes an effect in the process, so it should be included as a variable. Under these considerations, Eq. 8.2 can be reexpressed as

$$Q = \left(\frac{kA}{\delta}\right)\Delta P, \tag{8.3}$$

where δ is the membrane thickness.

The pressure drop ΔP in Eq. 8.3 is, really, an average pressure or pressure gradient due to the cross-flow arrangement of membrane separation processes. Referring to Fig. 8.4, a feed stream containing differently sized solids is pumped across the membrane surface at a velocity determined by the feed-side pressure gradient $(P_1 - P_2)$. This gradient, known as the hydrodynamic pressure gradient, causes the continuous movement of fluid across the membrane, which is referred to as cross-flow. As the feed stream flows across the membrane surface, smaller particles may be able to pass through the membrane and exit in the permeate stream at pressure P_3, which is usually atmospheric pressure. The rate of permeate flow is generally reported as the flux, i.e., the flow rate per unit area of membrane. The driving force for permeate flow is also a pressure gradient, but is not the hydrodynamic pressure gradient defined above but, rather, the pressure gradient that exits through the membrane from the feed side to permeate side at each point along the membrane surface. This pressure gradient is known as the transmembrane pressure (TMP) gradient or simply the TMP. Clearly, the TMP varies along the membrane surface, being a maximum at the inlet and a minimum at the outlet. The average TMP, according to the diagram in Fig. 8.4, can be defined as

$$\text{TMP} = \frac{P_1 + P_2}{2} - P_3, \tag{8.4}$$

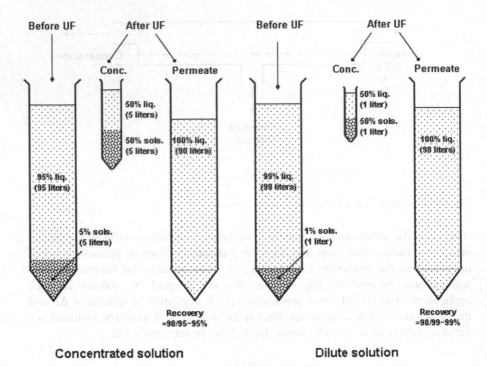

Concentrated solution   **Dilute solution**

Fig. 8.5 Recovery percentages in membrane separation processes

The permeate pressure P_3 is negligible as compared with the pressure gradient between the feed and the concentrate and, thus

$$\text{TMP} = \frac{P_1 + P_2}{2}. \qquad (8.5)$$

The recovery percentage is a way of expressing the efficiency of a membrane separation process: it is the relation of the permeate flow rate to the feed flow rate (Fig. 8.4), i.e.,

$$\text{Recovery}(\%) = \frac{F}{Q} \times 100. \qquad (8.6)$$

Recovery percentages depend on the feed concentration, being higher at lower concentrations of feed solids. A fraction of liquid approximately equal to that of the solids should remain in the concentrate to make it flow. Therefore, as illustrated in Fig. 8.5, the more dilute the feed, the higher the recovery percentage.

8.2.1 Features of Membranes

Membranes for ultrafiltration and similar processes are made out of a very thin film with microperforations of 0.1–1-μm size. Such tiny perforations on quite a thin film would be easily damaged by the applied pressures needed to perform separation in an actual operation. The ultrathin skin or film is, therefore, supported on a relatively thick (100–200-μm thickness) substructure. The retention is performed over the thin film, the pore sizes of which may vary depending on the manufacturing technique. Two common membrane structures are the plane membrane and the hollow-fiber membrane, illustrated in Fig. 8.6.

The structure of a membrane differs in terms of its chemical nature, microcrystalline structure, pore size, pore size distribution, and degree of symmetry. The permeate flux of the membrane and its solute rejection can be used as the main features needed for characterization. Since the properties of membrane materials can be influenced by environmental conditions and time, other properties such as resistance to compaction, chemical and thermal stability, and resistance to microbial attack are also important. Additional properties required for food processing are appropriate tolerance to cleaning agents and disinfecting solutions, and null toxicity of the contact materials (Cheryan 1992).

Over the years, four generations of membrane materials have appeared on the market. The first generation comprised cellulosic materials, the second included polymeric materials, the third was mainly based on inorganic materials, and the fourth is based on carbon fibers. In general terms, fabrication sources for membrane materials can be classified as organic, inorganic, and synthetic.

8.2.2 Nomenclature and Manufacture of Membranes

Conventional nomenclature of membranes includes two prefix letters (e.g., UM, PM, and XM), which refer to different polymers, and two digits which indicate the nominal MWCO. For instance, a commercial range of ultrafiltration membranes would cover from UM05 to XM50, which would mean retention of macromolecules from about 5,000 Da (UM05) to approximately 50,000 Da (XM50). Since pore sizes may differ because of the material used and the manufacturing procedure, for the same nominal size, the actual pore size may be different. For example, Table 8.2 lists some types of membranes, classified according to the nomenclature described. As can be seen, for the same MWCO of 10,000, two different membranes have different pore sizes.

Organic membranes are normally manufactured from cellulose by reaction with acetic anhydride, acetic acid, and sulfuric acid, a reaction known as acetylation. Cellulose acetate membranes are easy to manufacture and provide high flux along with high solute rejection properties. They have some disadvantages, however. For example, they do not operate over a wide temperature range, being recommended to run at a maximum temperature of 65°C. Also, the cellulose acetate is

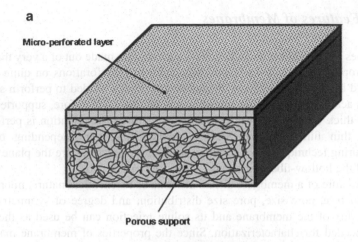

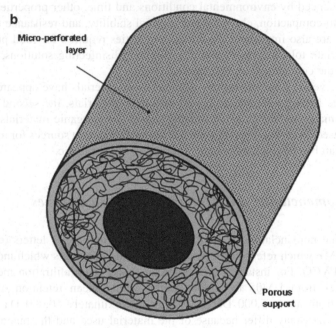

Fig. 8.6 Common structures of membranes

highly degradable under acid conditions, having a restricted pH range between 2 and 7. The lifetime of such membranes is somewhat limited, as cellulose acetate offers poor resistance to chlorine, being oxidized easily.

Inorganic mineral membranes are made from different materials, such as α-alumina, silica, stainless steel, carbon, and zirconium. They are formed by deposition of inorganic solutes onto a reusable microporous support. The mineral and ceramic membranes are extremely versatile, representing a clear advantage over

Table 8.2 Membrane nomenclature

Membrane	Molecular weight cutoff	Approximate pore size (Å)
UM05	5,000	21
UM10	10,000	30
PM10	10,000	38
PM30	30,000	47
XM50	50,000	66
XM100A	100,000	110
XM300	300,000	480

Table 8.3 Properties of membrane manufacturing materials

Material	Maximum temperature (°C)	pH range	Resistance to solvents
Cellulose acetate	30/65	2–7	Low
Polyacrylonitrile	60	2–10	High
Polyamide	60	1.5–9.5	Medium
Polysulfone	80	1.5–12	Medium
Aluminum oxide	300	0–14	High
Zirconium oxide	300	0.5–13.5	High

organic cellulosic membranes. Both the membrane and the support material provide high resistance to chemical degradation, corrosive agents, and abrasion, and operate over wide ranges of temperature and pH. The temperature operating range is up to 300°C, and the pH range covers nearly the complete scale. They can withstand pressures between 4,100 and 8,200 kPa. They are available in a wide range of pore sizes, from 0.01 to 1.4 µm, have a long life span, and are not subject to compaction.

Synthetic membranes are prepared by using synthetic materials such as nylon, poly(vinyl fluoride), polyurethane, polysulfone, and polyester. Polysulfone is preferred for ultrafiltration applications because of its high temperature operating range of up to 80°C and its wide pH operating range from 1.5 to 12. Polysulfone membranes provide good resistance to sanitation with chlorine solutions and are available in a reasonably wide range of pore sizes between 0.001 and 0.02 µm. The main limitation of polysulfone membranes is their low resistance to pressure, with a limit of about 69 kPa for flat sheet membranes and, only, 170 kPa for hollow-fiber membranes. A summary of the main characteristics of membrane manufacturing materials is given in Table 8.3.

8.3   Membrane Modules and Membrane Separation Equipment

To be incorporated into an industrial installation, membranes are configured into modules and such modules are mounted in structures comprising pipelines, valves, pressure gauges, and all the gadgetry that will constitute the membrane separation

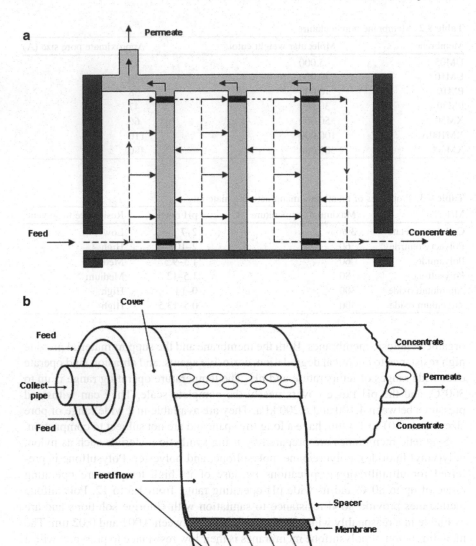

Fig. 8.7 Membrane modules for assembly into equipment: (**a**) plate-and-frame arrangement, (**b**) spiral-wound arrangement

equipment. Several arrangements or geometries of membranes are in use, including flat sheet membranes assembled similarly to a plate-and-frame filter press (Fig. 8.7a), and spiral-wound flat sheet membranes (Fig. 8.7b) consisting of two flat sheets of membrane sandwiching a porous support medium, and wrapped with a plastic spacer around a central tube.

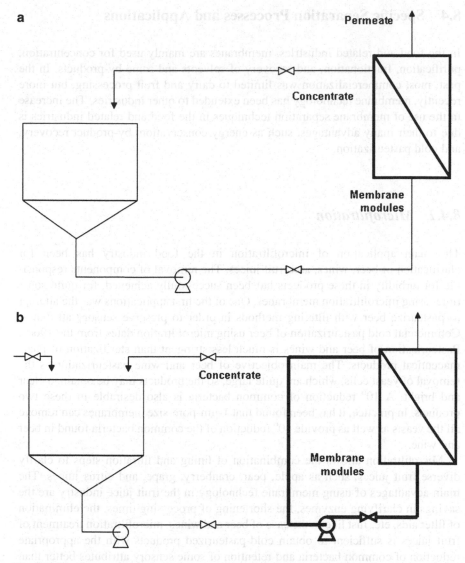

Fig. 8.8 Membrane separation operation modes: (**a**) standard batch configuration, (**b**) topped-off batch system with recirculation loop

Membrane separation equipment may be integrated in processing lines in more than one way. The simplest technique is the batch configuration illustrated in Fig. 8.8a. In this mode, an initial volume of liquid is circulated through the membrane system, and permeate is continuously removed until a final volume is achieved. The recirculating topped-off batch configuration is a variation, shown in Fig. 8.8b. This arrangement is more practical because it does not require the use of a large feed tank and allows the specification of a smaller prefilter prior to the system.

8.4   Specific Separation Processes and Applications

In the food and related industries, membranes are mainly used for concentration, purification, fractionation, and recovery of solvents and some by-products. In the past, most commercialization was limited to dairy and fruit processing, but more recently, membrane technology has been extended to other industries. The increase in the use of membrane separation techniques in the food and related industries is due to their many advantages, such as energy conservation, by-product recovery, and cold pasteurization.

8.4.1   Microfiltration

The main application of microfiltration in the food industry has been for clarification of beer, wines, and fruit juices. The removal of components responsible for turbidity in these products has been successfully achieved, for quite some time, using microfiltration membranes. One of the first applications was the attempt to pasteurize beer with filtering methods in order to preserve sensory attributes. Commercial cold pasteurization of beer using microfiltration dates from the 1960s. Pasteurization of beer and wines is much less stringent than sterilization of pharmaceutical products. The main objective of beer and wine pasteurization is the removal of yeast cells, which are quite large, so the products may be obtained clear and bright. A 10^6 reduction of common bacteria is also desirable in these two products. In practice, it has been found that 1-μm-pore-size membranes can remove all the yeasts as well as provide 10^6 reduction of the common bacteria found in beer and wine.

Microfiltration allows the combination of fining and filtration steps to clarify diverse fruit juices, such as apple, pear, cranberry, grape, and citrus juices. The main advantages of using membrane technology in the fruit juice industry are the savings in clarifying enzymes, the shortening of processing times, the elimination of filter aids, etc. Just like in the case of beer and wines, microfiltration treatment of fruit juices is sufficient to obtain cold-pasteurized products with the appropriate reduction of common bacteria and retention of some sensory attributes better than those of products treated thermally. Microfiltration and ultrafiltration can be used for the same purpose and a criterion to use either of them may be based on the sensory attributes of the pasteurized product. For example, comparison of microfiltration and ultrafiltration has been done (Wu et al. 1990) for pasteurized apple juice. It was found that the microfiltered juice contained more soluble solids and it was more turbid compared with the ultrafiltered juice.

Microfiltration has not been used as extensively as ultrafiltration and reverse osmosis in processing of milk and dairy products. Microfiltration has been used, however, to reduce the total number of lactic acid bacteria (Pafylias et al. 1996) and other microorganisms in the permeate compared with the microbiological condition of whey (Al-Akoum et al. 2002; Gésan et al. 1993). The process also resulted in defatting of the whey and was considered a gentle sterilization method (Eckner and Zottola 1991; Pearce et al. 1991). Another aspect was the use of a vibratory shear-enhanced filtration system for the separation of casein micelles from whey proteins of skim milk reconstituted from low-heat milk powder, which had a protein content similar to that of fresh milk (Al-Akoum et al. 2002). The performances of microfiltration with a 0.1-μm-pore Teflon membrane and of ultrafiltration with a polyethersulfone one with a 150,000 MWCO membrane were compared. The critical flux for stable operation was investigated in microfiltration by increasing the permeate flux in steps while monitoring the TMP.

8.4.2 Ultrafiltration

Ultrafiltration applications have provided a significant advantage in processing of dairy products since the advent of such technology. The main applications of ultrafiltration in the dairy industry include preconcentration of whey obtained as a waste stream in cheese making, production of whey protein from concentrates, and concentration of milk as an aid in cheese making. Depending on the specific application, MWCOs in the range from 25,000 to 50,000 are often used. Ostergaard (1986) defined the following main applications of these membranes: (1) membranes with a MWCO of 50,000 in acidified milk to produce quark and other specialized fresh cheeses, and (2) membranes with a MWCO of 25,000 in the production of whey protein concentrate and acidified whole milk for production of cream cheese.

Many aspects in dairy processing have been investigated over the years. The performance of ultrafiltration and nanofiltration membranes were investigated for utilization in concentration of whey protein and lactose (Atra et al. 2005). Such performances were characterized in terms of permeate flux, membrane retention, and yield, parameters which are determined by pressure, recycle flow rate, and temperature. The influence of these parameters on milk protein, whey protein, and lactose concentration was measured. The experiments were conducted using laboratory-scale ultrafiltration and nanofiltration units. The permeate flux and the protein and lactose contents in permeate and concentrate fractions were measured during the experimental runs. From a comparison of the separation behavior of the membranes it was found that the membranes investigated are suitable for concentration of the milk and whey proteins and lactose with high flux and retention. The filtration characteristics were obviously influenced by the process parameters.

Ultrafiltration can be used as a unique operation for the clarification and pasteurization of apple juice because of its operating principle (Ortega-Rivas 1995). Apple juice is normally depectinized prior to ultrafiltration treatment to increase the flux rate and reduce fouling. Some investigations have been reported regarding the quality of apple juice treated by membrane filtration. Heatherbell et al. (1977) clarified apple juice by ultrafiltration and obtained a stable clear product. Rao et al. (1987) studied retention of odor-active volatiles using different ultrafiltration membrane materials. Padilla and McLellan (1989) investigated the effect of the MWCO of ultrafiltration membranes on the quality and stability of apple juice. The use of mineral membranes for apple juice clarification has also been reported (Ben Amar et al. 1990). In general terms, the use of ultrafiltration to clarify apple juice results in a fresh appearance because the product is cold-sterilized, avoiding undesirable reactions triggered by the conventional thermal process. With regard to common pasteurization, it has been stated that the advantage of thermal processing is the inactivation of enzymes which cause browning of juices (Sapers 1991). However, since it has also been suggested (Vamos-Vigyazo 1981) that enzyme activity is associated with particulate fractions, ultrafiltration might also be used to separate such particulates from raw apple juice in order to prevent, or greatly reduce, enzymic browning.

Ultrafiltration has also been applied to some other fruit juices. Rwabahizi and Wrolstad (1988) found that strawberry juice clarified through a 10-kDa hollow-fiber membrane had an average of 55% anthocyanin loss compared with 17% loss by conventional filtration. Braddock (1982) used ultrafiltration and reverse osmosis to recover limonene from citrus-processing waste streams. The membrane flux rates declined after contact with limonene (around 0.11%). Polysulfone membranes exhibited the most severe decline, followed by cellulose acetate and Teflon membranes. Capannelli et al. (1992) tested polysulfone, poly(vinylidene difluoride), and ceramic membranes of various MWCOs with orange and lemon juices. With a given set of working conditions, the flux was largely independent of the membrane material and the MWCO. This was attributed to fibrous deposits that developed at the membrane surfaces and acted as a dynamic semipermeable barrier.

8.4.3   Reverse Osmosis

Desalination of seawater and brackish water may be considered the original application of reverse osmosis, as well as the foundation of all the other membrane separation techniques. Merten (1966) presented an early report on this subject and, since then, reverse osmosis applications have grown rapidly. Reverse osmosis has been applied in numerous food processes (Cheryan 1992), including preconcentration of egg whites, preconcentration of coffee before drying, concentration of maple sap as well as tomato juice and sugar solutions, concentration of milk, and concentration of whey.

Applications in the diary industry are important and represented a breakthrough in terms of the ability to separate specific components, without the need to change phase. Reverse osmosis has experienced, just as much as microfiltration and ultrafiltration, rapid growth in terms of fundamental research, technology transfer, and commercial distribution. In the dairy industry, investigations have focused not only on processing of milk and dairy products, but also on effluent treatment. Permeate flux and chemical oxygen demand (COD) reduction were investigated in dairy processing waters using different nanofiltration and reverse osmosis membranes (Al-Akoum et al. 2004).

Dairy process waters were simulated by ultra-high-temperature processed skim milk diluted 1:3 to obtain an initial COD of 36,000 mg O_2/l. Balannec et al. (2002) studied a new combination of a membrane-based cheese production procedure and found there was a significant increase in the cheese yield by incorporating the whey proteins. The performances of treatment of dairy process waters with membranes for recovery of milk constituents and water were analyzed in terms of COD and ion rejection. The dead-end filtration experiments permitted several nanofiltration and reverse osmosis membranes to be compared. It was found that a single membrane operation is insufficient to produce water of composition complying with the requirements for drinking water. Because of the high COD level of the dairy process waters and despite high rejection of lactose, COD, and milk ions, the concentration in the permeate remained too high even with reverse osmosis membranes. To reach the goal for target reuse of the purified water in the dairy plant, a finishing step must be added.

Another important application of reverse osmosis is in fruit juice treatment, particularly in concentration of some juices, including orange, apple, tomato, pear, grapefruit, and kiwi juices. An important difficulty with this application is that high concentrations are difficult to attain, owing to the high pressures exerted by juice concentrate. With recent advances in membrane technology, most juices can be concentrated up to 65°Brix, so reverse osmosis can be considered a pretreatment, with other technologies, such as freezing and evaporation, completing the concentration process. For example, reverse osmosis has been used commercially to concentrate tomato juice from 4.5 to 8.5°Brix, followed by an evaporation cycle to end with a 30°Brix product. Also, orange juice can be preconcentrated by reverse osmosis from 12 to 20°Brix prior to flash evaporation to obtain a high-quality concentrate. Similarly, reverse osmosis has been used for concentration of maple sap, with water removal of up to 75%.

References

Al-Akoum O, Ding LH, Jaffrin MY (2002) Microfiltration and ultrafiltration of UHT skim milk with a vibrating membrane module. Sep Purif Technol 28: 219–234.

Al-Akoum O, Jaffrin MY, Ding HL, Frappart M (2004) Treatment of dairy process waters using a vibrating filtration system, and NF and RO membranes. J Membr Sci 235: 111–122.

Atra R, Vatai G, Bekassy-Molnar E, Balint, A (2005) Investigation of ultra- and nanofiltration for utilization of whey protein and lactose. J Food Eng 67: 325–332.

Balannec B, Gésan-Guiziou G, Chaufer B, Rabiller-Baudry M, Daufin G (2002) Treatment of dairy process waters by membrane operations for water reuse milk constituents concentration. Desalination 147: 89–94.

Ben Amar R, Gupta BB, Jaffrin MY (1990) Apple juice clarification using mineral membranes: fouling control by backwashing and pulsating flow. J Food Sci 55, 1620–1625.

Braddock R J (1982) Ultrafiltration and reverse osmosis recovery of limonene from citrus processing waste streams. J Food Sci 47: 946–948.

Capannelli G, Bottino A, Munari S, et al (1992) Ultrafiltration of fresh orange and lemon juices. Food Sci Technol 25: 518–522.

Chen W, Parma F, Parkat A, Elkin A, Sen S (2004) Selecting membrane filtration systems. Chem
    Eng Prog 100: 22–25.
Cheryan M (1992) Concentration of liquid foods by reverse osmosis. In: Heldman DR, Lund DB
    (eds) Handbook of Food Engineering, pp 393–436. Marcel Dekker, New York.
Eckner KF, Zottola A (1991) Potential for the low temperature pasteurization of dairy fluids using
    membrane processing. J Food Prot 54: 793–797.
Gésan G, Merin U, Daufin G, Maugas JJ (1993) Performance of an industrial cross-flow
    microfiltration plant for clarifying rennet whey. Neth Milk Dairy J 47: 121–124.
Girard B, Fukumoto LR (2000) Membrane processing of fruit juices: a review. Crit Rev Food Sci
    Nutr 40: 91–157.
Heatherbell DA, Short JL, Strubi P (1977) Apple juice clarification by ultrafiltration. Confructa 22:
    157–169.
Merten U (1966) Desalination by Reverse Osmosis. MIT Press, Cambridge, MA.
Ortega-Rivas E (1995) Review and advances in apple juice processing. In: Singh RK (ed) Food
    Process Design and Evaluation, pp 27–46. Technomic Publishing, Lancaster PA.
Ostergaard B (1986) Applications of membrane processing in the dairy industry. In: MacCarthy D
    (ed) Concentration and Drying of Foods, pp 133–146. Elsevier Applied Science, London.
Padilla OI, McLellan MR (1989) Molecular weight cut-off of ultrafiltration membranes and the
    quality and stability of apple Juice. J Food Sci 54: 1250–1254.
Pafylias I, Chelyan M, Mehaia MA, Saglam N (1996) Microfiltration of milk with ceramic
    membranes. Food Res Int 29: 141–146.
Pearce RJ, Marshall SC, Dunkerley JA (1991) Reduction of lipids in whey protein concentrates by
    microfiltration: effect of functional properties, IDF Spec Issue 9201: 118–129.
Rao MA, Acree TE, Cooley HJ, Ennis RW (1987) Clarification of apple juice by hollow fiber
    ultrafiltration: fluxes and retention of odor-active volatiles J Food Sci 52: 375–377.
Rwabahizi S, Wrolstad RE (1988) Effects of mold contamination and ultrafiltration on the color
    stability of strawberry juice and concentrate. J Food Sci 53: 857–861.
Sapers GM (1991) Control of enzymatic browning in raw fruit juice by filtration and centrifuga-
    tion. J Food Process Preserv 15: 443–456.
Sourirajan S (1970) Reverse Osmosis. Academic Press, New York.
Vamos-Vigyazo L (1981) Polyphenol oxidase and peroxidase in fruits and vegetables. CRC Crit
    Rev Food Sci Nutr 15: 49–127.
Wu ML, Zall R, Tzeng WC (1990) Microfiltration and ultrafiltration comparison for apple juice
    clarification. J Food Sci 55: 1162–1163.

Part III
Preservation Operations

Chapter 9
Electromagnetic Radiation: Ultraviolet Energy Treatment

9.1 Introduction: The Electromagnetic Spectrum

The efficiency of thermal methods of food preservation is beyond doubt, but the existence of undesirable side effects, mainly in sensory attributes, is also a reality. This fact has provided the incentive to explore non-thermal methods of food preservation, for example, different sources of radiation. Electromagnetic radiation, commonly referred to simply as "light," comprises self-sustaining energy with electric and magnetic field components. Electromagnetic radiation has been known for a long time and its effects and applications are varied. The entire range of radiation extending in frequency from approximately 10^{23} to 0 Hz is known as the electromagnetic spectrum (Table 9.1) and includes gamma rays, X-rays, visible light and microwaves. Many forms of energy included in the electromagnetic spectrum have been explored as alternatives to conventional thermal food preservation, in order to pursue an ideal balance between microbial safety and premium sensory quality of processed food products.

Ultraviolet (UV) processing makes use of radiation from the UV region of the electromagnetic spectrum. The antimicrobial properties of UV radiation are believed to be due to damage to DNA, reducing the possibility of the microbial population reproducing in a food product exposed to UV radiation. This form of radiation has been used for diverse applications, including the decontamination of air in food processing facilities, the surface treatment of bakery products, and the treatment of drinking water. UV liquid treatment systems appear to be suitable for clear and translucent liquid foods, such as fruit juices (apple, pear, berry, kiwifruit, etc.) and vegetable juices (carrot, celery, spinach, etc.), particularly as some of these products are subject to flavor and color changes when treated thermally.

UV technology was originally used in Europe as an alternative to chlorination to ensure satisfactory disinfection of municipal drinking water supplies. It is now applied globally for the disinfection of drinking water, wastewater, process water, and industrial effluents. UV technology is particularly suited to the beverage,

E. Ortega-Rivas, *Non-thermal Food Engineering Operations*,
Food Engineering Series, DOI 10.1007/978-1-4614-2038-5_9,
© Springer Science+Business Media, LLC 2012

Table 9.1 The
electromagnetic spectrum

Frequency (Hz)	Type of energy
10^{22}	Gamma rays
10^{21}	
10^{20}	
10^{19}	X-rays
10^{18}	
10^{17}	UV light
10^{16}	Visible light
10^{15}	
10^{14}	Infrared rays
10^{13}	
10^{12}	Radar waves
10^{11}	
10^{10}	Microwaves
10^{9}	
10^{8}	
$\uparrow$ $\downarrow$	Radio waves
10^{1}	

bottled water, and food processing sectors, where extremely high standards of hygiene are expected. Contamination of the process at any point by pathogenic or spoilage microorganisms can have extremely serious consequences for manufacturers, thus making microbial disinfection of the whole process essential.

UV treatment is rapidly gaining acceptance across the whole spectrum of food and beverage industries as a highly efficient, non-chemical method of disinfection. UV treatment inactivates all known pathogenic and food spoilage microorganisms, including bacteria, viruses, yeasts, and molds. It represents a low-maintenance environmentally friendly technology, which eliminates the need for chemical treatment while ensuring very high levels of disinfection.

A typical UV disinfection system consists of a UV lamp housed in a protective quartz sleeve, which is mounted within a cylindrical stainless steel chamber. The fluid to be treated enters at one end and passes along the entire length of the chamber before exiting at the other end. Virtually, any liquid can be effectively treated with UV radiation, including spring, surface or municipal water, filtered process water, viscous sugar syrups, fruit or vegetable juices, and industrial effluents. There are two main types of UV technology, depending on the type of UV lamp used: low pressure and medium pressure. Low-pressure lamps have a monochromatic UV output (limited to a single wavelength at 254 nm), whereas medium-pressure lamps have a polychromatic UV output (with an output between 185 and 400 nm). Low-pressure systems are usually better suited for small, intermittent flow application, whereas the medium-pressure systems are better suited to higher flow rates.

9.2 Mechanisms of Microbial Inactivation

Several theories have been suggested for the antimicrobial effects of UV light. One of them states that the antimicrobial effects of UV light are due to the absorption of the energy by highly conjugated carbon double bonds in protein and nucleic acids, which causes disruption of cellular metabolism. The biological effects of UV photons derive from excitation, rather than ionization of molecules. UV light has been classified into three ranges: UVA from 315 to 400 nm (long wavelengths that cause tanning of human skin), UVB from 280 to 315 nm (may cause skin burns that may lead to skin cancer), and UVC from 200 to 280 nm (considered the germicidal range). Shorter UV wavelengths are known as the vacuum UV range, since they require a vacuum as their transmission medium. These wavelengths are absorbed by almost all substances. The effects of UV light on DNA are dependent on the total energy absorbed, and below a certain level of absorbed energy may be reversible. The absorption of UV light in matter is given by

$$I = I_0 e^{-\mu X}, \tag{9.1}$$

where I is the intensity of radiation at distance X within the absorber, I_0 is the intensity at the surface of the absorber, and μ is a coefficient of absorption depending of the photon energy and the nature of the absorber.

Another theory refers to the indirect lethal action resulting from the production of hydrogen peroxide, along with various chemical and physicochemical changes in the constituents of the cell. The production of hydrogen peroxide is not normally considered the mechanism by which UV light induces its effects, but it has been suggested that organic peroxides may induce an antimicrobial effect.

The sensitivity of target microorganisms in UV treatments is described by the parameters of the inactivation kinetics, in a similar manner as the description of microbial inactivation kinetics using thermal methods or chemical disinfectants. In such cases, the inactivation kinetics is commonly described by a first-order model. For the case of UV treatment, inactivation is defined as the reduction of the concentration of a given microorganism N due to the exposure to a UV radiation dose D during a specific contact time t. The UV dose is related to the fluence, defined as the total radiant energy passing through a unit cross-sectional area of an infinitesimally small sphere, and is the product of the fluence rate (W/m^2) multiplied by the time of exposure. The fluence is, therefore, expressed in SI units as joules per square meter. On the basis of the first-order model, the linear relationship between the logarithm of inactivation and the UV dose is described by

$$\log \frac{N_t}{N} = -kD, \tag{9.2}$$

where N_t is the microbial load after contact time t, N is the initial microbial load, and k is the inactivation constant rate.

Table 9.2 D_{UV} values for UV microbial inactivation in fruit products

	D_{UV} (min)			
Flow rate (L/min)	Saccharomyces cerevisiae[a]	Listeria innocua[a]	Escherichia coli[a]	Saccharomyces cerevisiae[b]
0.073	40.4	16.6	16.8	26.0
0.255	34.0	12.5	9.8	12.9
0.451	25.4	9.2	6.3	11.8

[a]Apple juice (Guerrero-Beltran and Barbosa-Canovas 2005)
[b]Mango nectar (Guerrero-Beltran and Barbosa-Canovas 2006)

The decimal reduction rate D_{UV} is the dose needed to inactivate 90% of microorganisms at a given flow rate and is expressed as

$$D_{UV} = -\frac{2.303}{k}. \tag{9.3}$$

Table 9.2 presents D_{UV} values for inactivation of selected microorganisms.

The first-order kinetics model is simply described by a straight line. Some microbial survival curves may have, however, a sigmoid shape with shoulders and a plateau region. These curves have to be modeled using more complex mathematical functions. The initial plateau segment indicates that the microbial load has been injured, the second segment indicates that the number of injured microorganisms has increased, and the third part or tailing region represents the remaining surviving microorganisms. The tailing phase has been attributed to diverse factors, including the presence of a resistant fraction of the microbial population or shielding by suspended solids of part of the irradiated volume (Sastry et al. 2000). A typical sigmoidal curve for microbial inactivation using UV radiation is presented in Fig. 9.1.

Some functions have been proposed to model the experimental curves obtained in actual UV treatments, like the one exemplified in Fig. 9.1. The Peleg model (Peleg 1995) is represented by

$$\log\frac{N_t}{N} = \frac{-t}{k_1 + k_2 t}, \tag{9.4}$$

where k_1 and k_2 are constants. If $k_2 > 0$, the curve has upward concavity; if $k_2 < 0$ the curve has downward concavity. A linear relationship will be observed, of course, when $k_2 = 0$.

The Weibull model (Gacula and Singh 1984) is given by the following expression:

$$\log\frac{N_t}{N} = -bt^n, \tag{9.5}$$

where b and n are constants. If $n < 1$, the curve has upward concavity, if $n > 1$, the curve has downward concavity, and if $n = 1$, the relationship is linear.

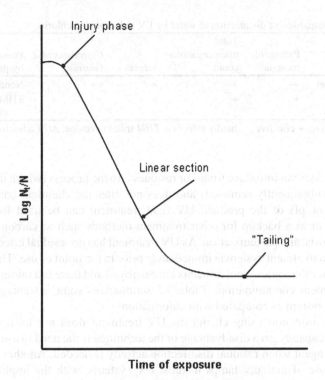

Fig. 9.1 Typical survival curve for microorganisms exposed to UV light

The Gompretz model (Gutfreund 1998) can be expressed by

$$\frac{N_t}{N} = -e^{a+kt},$$ (9.6)

where a and k are constants.

Microbial reductions of the order of four log cycles have been observed with exposures from 50 to 380 J/m^2 in drinking water (Hoyer 1998). *Mycobacterium smegmatis*, a phage, and two viruses were the most resistant, requiring doses between 200 and 380 J/m^2, whereas *Escherichia coli* ATC 23958 and *Vibrio cholerae* required only 50 J/m^2 for inactivation. However, the bacteria were made more resistant by photoreactivation, with postirradiation reactivation of the strain of *E. coli* requiring 200 J/m^2 for the same lethality effect.

9.3 Advantages and Limitations of UV Processing

UV systems can provide relatively low cost decontamination and shelf life extension of water and liquid food products, such as fresh fruit juices. UV disinfection has many advantages over other alternative methods. Unlike chemical biocides, UV

Table 9.3 Comparison of disinfection of water by UV radiation and chlorine

	Pathogenic bacteria	Total microorganism count	Viruses	*Cryptosporidia* *Giardia*	Possible harmful by-products
UV disinfection	++	+	+	++	None
Chlorination	+	+	+	−	THM, AOX, chlorite

++ very effective, + effective, − hardly effective, *THM* trihalomethane, *AOX* adsorbable organic halogens

disinfection does not introduce toxins or residues into the process (which themselves have to be subsequently removed) and does not alter the chemical composition, taste, odor, or pH of the product. UV light treatment can be used for primary disinfection or as a backup for other treatment methods such as carbon filtration, reverse osmosis, and pasteurization. As UV treatment has no residual effect, the best position for a treatment system is immediately prior to the point of use. This ensures incoming microbiological contaminants are destroyed and there is a minimal chance of posttreatment contamination. Table 9.3 summarizes some advantages of UV disinfection system as compared with chlorination.

Unlike disinfection using chemicals, UV treatment does not have a residual disinfection capacity, so a disadvantage of the technique is the need to use some sort of chemical agent when residual disinfection activity is needed. Another disadvantage is the use of mercury lamps in most UV systems, with the implied risk of contamination by mercury in the case of breakage.

For the specific case of use in food products, it has been known for a long time that even the UV light in the sunrays causes undesirable effects in some foods. For example, sunrays will quickly taint butter and milk by oxidizing their fats, and will cause potatoes to turn green as a result of chlorophyll formation. Although this effect is practically harmless, a parallel production of the toxic substance solanine often takes place. In artificial lighting, where high-intensity fluorescent tubes are used to display foods, the UV rays from such lights are sufficient to cause undesirable effects such as oxidizing fats, bleaching colors, and greening potatoes.

In the treatment of foods by UV radiation, the low penetration depth often means a high dose has to be used, which tends to result in the development of off-flavors and odors before satisfactory sterilization is achieved. When foods containing particulate solids are treated, the UV dose alone is insufficient to achieve inactivation of pathogenic microorganisms, so it has to be combined with some other hurdles, such as pH and water activity reduction. The oxidation of fat by photochemical action has been attributed to be responsible for the development of off-flavors such as rancidity, tallowiness, and fishiness (Ellickson and Hasenzahl 1958). Iwanami et al. (1997) reported the effects of UV radiation on lemon flavor composed of lemon oil, water, and ethanol. Three new aldehyde compounds were identified as photoreaction products of citral. The concentrations of limonene, terpinolene, and nonanal decreased, whereas the concentration of *p*-cymene increased after the UV treatment. The results suggest that citral is quite unstable on UV exposure and its photolysis could affect other components of lemon flavor during UV treatment.

9.4 Applications

9.4.1 Decontamination of Air

UV rays have been used intensively in disinfection of equipment, glassware, and air in different industries for many years (Fields 1978). The bactericidal effect of UV light is particularly effective in destroying airborne microorganisms. Special UV air disinfection systems are available to treat air in the ductwork of air conditioning systems serving clean rooms and other high-purity areas. Air treatment systems can also be used to disinfect displacement air for pressuring tanks or pipelines holding perishable fluids. Storage tanks are particularly susceptible to bacterial colonization and contamination by airborne spores. To prevent this, immersion UV treatment systems have been designed to fit in the tank head air space and disinfect the air present.

9.4.2 Surface Treatment

Surface disinfection systems are used to reduce microbial counts on all kinds of packaging, including glass and plastic bottles, cans, lids, and foils. By irradiating the surfaces with UV light prior to filling, one eliminates spoilage organisms, extending the shelf life of the product and reducing the risk of contamination.

Surface disinfection has also been used extensively for treatment of fruits and vegetables. UV radiation has been reported to inhibit fungal growth in grape berries (Creasy and Coffee 1988), in kumquat and orange fruits (Rodov et al. 1992), and in papaya (Moy et al. 1977). It has also been reported to be an effective and rapid method to preserve postharvest shelf life of mangoes without adversely affecting some quality attributes. Gonzalez-Aguilar et al. (2001) employed UV light exposures of 10–20 min before storing mangoes, and achieved a shelf life of 14 days at 5 or 20°C, and 7 days exclusively at 20°C. The treated fruits maintained an appropriate appearance, did not show decay symptoms, maintained firmness, and had higher levels of sugars during storage at either 5 or 20°C. The delay in senescence and other quality attributes of tomatoes were investigated by Maharaj et al. (1999). Mature green tomatoes were UV-radiated and stored at relatively high humidity and 16°C for 35 days. A dose of 3.7×10^3 J/m^2 was found to be sufficient to delay ripening and senescence, as well as to retard development of color and softening of tissue. Additionally, a delay in the climacteric response by at least 7 days was observed, and the respiration rate was reduced.

The efficacy of UV radiation in the control of soft rot and dry rot diseases of potato tubers has been reported (Ranganna et al. 1997). A dose of 15×10^3 J/m^2 was found to be effective in preventing the diseases as visual observation of potato quality showed no significant change in the firmness and color of the tubers. Lu et al. (1987) compared different alternative preservation methods, including UV

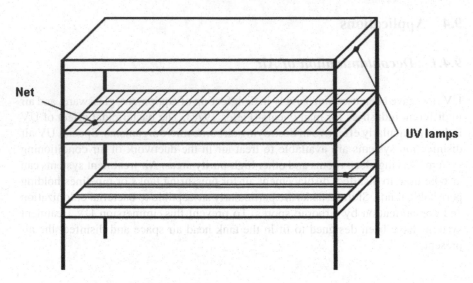

Fig. 9.2 UV unit for applying UV radiation on the surface of food products

exposure, in treatment of Walla Walla onions. UV-radiated onions contained the greatest percentage of marketable pieces after 4 weeks storage at 20–25°C. Also, UV treatment was more economical, and safer, than the other methods compared, which included gamma and electron beam irradiation.

Shortly after slaughtering, meat may become tender upon storage as a result of enzymic activity. This process is speeded up at relatively high temperatures favoring the growth of surface microorganisms. UV radiation can be used to control microbial surface growth in meat in order to allow relatively high temperature storage and minimize meat wastage. The lamps employed emit radiation, not only in the germicidal 253.7-nm range but also in the 185-nm range. Radiation in the latter range converts atmospheric oxygen into ozone, which sterilizes the irregular and shaded areas of the irradiated surface of the meat. UV light is also used in storage vats and other tanks, over conveyors, and for final treatment of both caps and stoppers (Borgström 1968).

High-intensity UVC lamps can also increase the potential of inactivating surface bacteria in fish and seafood (Colby and Flick 1993). Huang and Toledo (1982) reported on UV treatment of Spanish fresh mackerel using doses of 3×10^3 J/m^2. The microbial count on the mackerel surface was reduced by two to three log cycles. The shelf life of the mackerel that were UV-treated and stored in ice at -1°C was extended by 7 days over that of the untreated sample. UV radiation is less effective on rough surfaces. Huang and Toledo (1982) also found that fish with a rough surface, such as croaker and mullet, had little reduction in surface bacterial count when treated with UVC radiation up to 1.8×10^3 J/m^2 for up to 50 s.

UVC equipment used to apply UV radiation on surfaces of different foods consists of relatively simple units. A diagram of UVC equipment utilized to apply doses to plant materials is given in Fig. 9.2.

9.4.3 Water Purification

Although municipal water supplies are normally free from harmful or pathogenic microorganisms, this should not be assumed. In addition, water from private sources such as natural springs and wells can also be contaminated. Any water used as an ingredient in food or beverage products or coming in direct contact with the product can therefore be a source of contamination. UV radiation disinfects this water without chemicals or pasteurization. It also allows the reuse of process water, saving money and improving productivity without risking the quality of the product.

Cleaning-in-place (CIP) systems are used in many material processing industries, and are designed for automatic cleaning without major disassembly of equipment. It is essential that the CIP final rinse water used to flush out foreign matter and disinfection solutions is microbiologically safe. Fully automated UV disinfection systems can be integrated with CIP rinse cycles to ensure the final rinse water does not reintroduce microbiological contaminants. Medium-pressure lamps are ideal for this application because of their mechanical strength, meaning they are not affected by any sudden changes in the temperature of the CIP water, such as when hot (80°C) liquid is instantly followed by cold (10°C) liquid. Using UV radiation to disinfect the water used to rinse or wash process equipment and work surfaces can dramatically decrease contamination while increasing shelf life. UV radiation also reduces the amount of chlorine needed to disinfect rinse and wash water.

Stored reverse osmosis and granular activated carbon (GAC) filtrate are often used to filter water, but can be a breeding ground for bacteria. Installing UV systems after the filter is a highly effective way of disinfecting both stored reverse osmosis and GAC filtered water. GAC filters are also sometimes used to dechlorinate water following chlorine treatment. Dechlorination removes the off-flavors often associated with chlorine disinfection, meaning the final product remains untainted and free from unwanted flavors or odors. Placing UV systems ahead of GAC filters improves the performance of the filters and results in longer carbon runs, so decreasing operating costs.

To eliminate pathogens in water, a radiant exposure of at least 400 J/m^2 UV light at 254 nm is required to achieve a reduction of four log cycles. The dose must be delivered to every volume element of the water, so homogeneity of the product flow along with some other features, such as homogeneity of the radiation field and consistency of the absorption characteristics of the fluid, represent critical elements in process design criteria. A typical UV reactor for use in water disinfection is shown in Fig. 9.3.

9.4.4 Fruit and Vegetable Juices

As discussed in the previous sections, the use of UV radiation sources is well established for air and water treatment, as well as for surface decontamination.

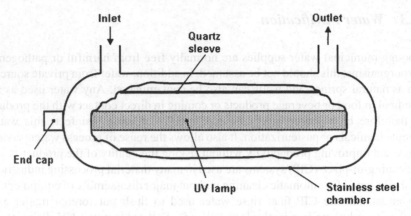

Fig. 9.3 UV reactor for water disinfection

With the approval by the US Food and Drug Administration of UV light as an alternative treatment to thermal pasteurization of fresh juice products (US Food and Drug Administration 2000), there has been renewed interest in applications of UV radiation in processing of liquid foods and beverages. Considerable research efforts have been focused on expanding UV radiation applications for treatment of liquid foods, such as fruit juices. UV light has been investigated for treatment of apple juice. Freshly pressed apple juice contains microorganisms that can cause serious deterioration within 2 days at room temperature if they are not inhibited. The microbial population of freshly squeezed juice was greatly reduced, and the shelf life prolonged, by using specially designed UV lamps (Harrington and Hills 1968). The amount by which the microbial count was reduced was affected by the clarity of the juice, the length of UV exposure, and the presence of potassium sorbate. UV radiation, using doses ranging from 4,500 to 6,500 J/m^2 in a vertical setup, was employed for treatment of model juices (Koutchma and Parisi 2004). The inactivation kinetics of *Escherichia coli* and the UV-dose distribution were studied. Linear inactivation curves were obtained, with a maximum inactivation rate of 90%. The rate of inactivation decreased as the solution absorption coefficient increased. A modified biosimetry method using the injection of bacteria as a tracer can be used to evaluate dose delivery in UV reactors.

Sugar syrups used as flavorings in the beverage, fruit juice, and bottled water industries can be a prime breeding ground for microorganisms. Although syrups with very high sugar contents do not support microbial growth, any dormant spores may become active after the syrup has been diluted. Treating the syrup and dilution water with UV radiation prior to use will ensure any dormant microorganisms are deactivated. Schneider et al. (1960) reported on the effects of reduction of counts of two types of bacteria and one type of yeast in maple syrup exposed to UV radiation at different intensities and for different times. It was observed that the two types of bacteria were equally more sensitive than the yeast.

Different designs of reactors have been devised for UV treatment of liquid foods and beverages. Continuous-flow UV reactors have been evaluated for the

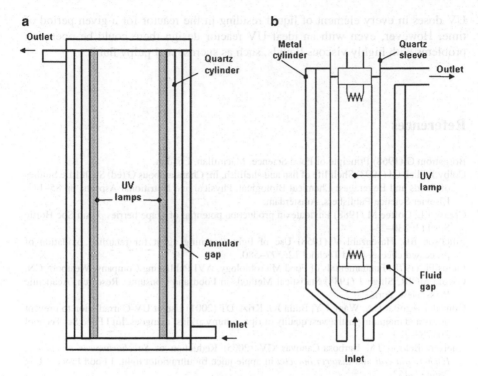

Fig. 9.4 Continuous UV reactors: (**a**) laminar thin-film reactor, (**b**) annular-type laminar reactor

pasteurization of fresh juices. A particular design consists of an extremely thin film UV reactor to decrease the path length and avoid the low penetration range. Thin-film reactors are characterized by laminar flow with a parabolic velocity profile, reaching a maximum fluid velocity in the center and being twice as fast as the average velocity of the liquid. The annular UV reactor is another design, in which liquid is pumped through the gap formed by two concentric stationary cylinders. They comprise a single large quartz sleeve as the inner reactor wall and a surrounding metal cylinder as the outer reactor wall. The length and gap size are variable to accommodate the type of liquid and different flow rates. Annular-type laminar reactors have been used for treatment of apple juice and mango nectar. Figure 9.4 illustrates a thin-film reactor and an annular reactor.

A second type of design is focused on achieving turbulence within a UV reactor to bring all material into close proximity with UV light during treatment. The higher flow rates attained in turbulent conditions will improve homogeneity of the flow in order to have each volume of the product exposed to an approximately equal UV irradiance.

Thin-film and annular-type reactors normally operate under laminar flow conditions, so their use is restricted to low-viscosity juices with no pulp. Turbulent flow reactors require larger volume of juice and large numbers of UV lamps. A desirable design for UV reactors would require a plug flow so as to guarantee equal

UV doses in every element of liquid residing in the reactor for a given period of time. However, even with an ideal UV reactor design there could be operating problems with highly viscous liquids, such as sucrose and pulpy fluids.

References

Borgström G (1968) Principle of Food Science. Macmillan, London.

Colby J, Flick CJ (1993) Shelf life of fish and shellfish. In: Charalambous G (ed) Shelf Life Studies of Foods and Beverages: Chemical Biological, Physical and Nutritional Aspects, pp 85–143. Elsevier Science Publishers, Amsterdam.

Creasy LL, Coffee M (1988) Phytoalexin production potential of grape berries. J Am Soc Hortic Sci 113: 230–234.

Ellickson BE, Hasenzahl V (1958) Use of light screening agent for retarding oxidation of processed cheese. Food Technol 12: 577–580.

Fields FL (1978) Fundamentals of Food Microbiology. AVI Publishing Company, Westport, CN.

Gacula MCJR, Singh J (1984) Statistical Methods in Food and Consumer Research. Academic Press, New York.

Gonzalez-Aguilar GA, Wang CY, Butta JG, Krizk DT (2001) Use of UV-C irradiation to prevent decay and maintain postharvest quality of ripe 'tommy atkins' mangoes. Int J Food Sci Technol 36: 767–773.

Guerrero-Beltran JA, Barbosa-Canovas GV (2005) Reduction of *Saccharomyces cerevisiae*, *Escherichia coli* and *Listeria innocua* in apple juice by ultraviolet light. J Food Process Eng 28: 437–452.

Guerrero-Beltran JA, Barbosa-Canovas GV (2006) Inactivation of *Saccharomyces cerevisiae* and polyphenoloxidase activity in mango nectar treated with ultraviolet light. J Food Prot 69: 362–368.

Gutfreund H (1998) Kinetics for the Life Sciences: Receptors, Transmitters and Catalysis. Cambridge University Press, Cambridge.

Harrington WO, Hills C (1968) Reduction of the microbial population of apple cider by ultraviolet irradiation. Food Technol 22: 117–120.

Hoyer O (1998) Testing performance and monitoring of UV systems for drinking water disinfection. Water Supply 16: 419–424.

Huang YW, Toledo R (1982) Effect of high doses of high and low intensity UV irradiation on surface microbiological counts and storage life of fish. J Food Sci 47: 1667–1669, 1731.

Iwanami Y, Tateba H, Kodama N, Kishino K (1997) Changes of lemon flavor components in an aqueous solution during UV irradiation. J Agric Food Chem 45: 463–466.

Koutchma T, Parisi B (2004) Biodosimetry of Escherichia coli UV inactivation in model juices with regard to dose distribution in annular UV reactors. J Food Sci 69: 14–22.

Lu IY, Stevens C, Yakubu P, Loretan PA (1987) Gamma, electron beam and ultraviolet radiation on control and storage rots and quality of Walla Walla onions. J Food Process Preserv 12: 53–62.

Maharaj R, Arul J, Nadeau P (1999) Effect of photochemical treatment in the preservation of fresh tomato (*Lycopersicon esculentum* cv. Capello) by delaying senescence. Postharv Biol Technol 15: 13–23.

Moy IH, McElhandy T, Matsuzaki C (1977) Combined treatment of UV and gamma radiation of papaya for decay control. Food Preservation by Irradiation Proceedings at an IAEA, FAO, WHO Symposium, Wageningen, Netherlands.

Peleg M (1995) A model of microbial survival after exposure to pulsed electric fields. J Sci Food Agric 67: 93–99.

Ranganna B, Kushalappa AC, Raghavan GSV (1997) Ultraviolet irradiance to control dry rot and soft rot of potato in storage. Can J Plant Pathol 19: 30–35.

Rodov V, Ben-Yehoshua S, Kim JJ, Shapiro B, Ittah Y (1992) Ultraviolet illumination induces scoparone production in kumquat and orange fruit and improves decay resistance. J Am Soc Hortic Sci 117: 788–792.

Sastry SK, Datta AK, Worobo RW (2000) Ultraviolet light. J Food Sci Suppl, 65: 90–92.

Schneider IS, Frank HA, Willits CO (1960) Maple sirup XIV. Ultraviolet irradiation effects on the growth of some bacteria and yeasts. J Food Sci 25: 654–662.

US Food and Drug Administration (2000) 21 CFR Part 179. Irradiation in the Production, Processing and Handling of Food. Fed Regist 65: 71056–71058.

Chapter 10
Ionizing Radiation: Irradiation

10.1 Theoretical Aspects

Radiation is a form of energy traveling through space as radiant energy in a wave pattern. Such an energy form can occur naturally (e.g., from the sun or rocks) or can be produced by man-made objects, such as various electrical household appliances. The frequency or wavelength of the energy waves produced by different sources distinguishes the different types and functionality of the radiation, and they are classified in the electromagnetic spectrum, as shown in Table 9.1. High-frequency radiation, i.e., gamma rays, x-rays, or UV light, poses a risk to human health.

Radiation is called ionizing radiation when it has a sufficiently high frequency, such as that of gamma rays and x-rays, so that it results in the production of charged particles or ions in the material that it comes in contact with. Non-ionizing radiation, such as microwaves or infrared light, does not produce ions but can create heat under moist conditions and is routinely used for purposes such as cooking and reheating of foods.

Different types of radiation have the characteristic ability to ionize individual atoms or molecules and produce a positively charged ion and an electron, i.e.,

$$M \xrightarrow{h\nu} M^+ + e^-, \tag{10.1}$$

where M is an atom or molecule, $h\nu$ is a quantum of energy delivered by radiation, M^+ is the positively charged ion, and e^- is the electron released. Some types of ionizing radiation and their properties are listed in Table 10.1.

Various types of light, x-rays, and gamma rays are part of the electromagnetic spectrum and posses a dual nature according to quantum theory, in which they may be considered waves or packets or energy (photons). Energy and wave properties are related by

$$\nu = \frac{c}{\lambda} \tag{10.2}$$

E. Ortega-Rivas, *Non-thermal Food Engineering Operations*,
Food Engineering Series, DOI 10.1007/978-1-4614-2038-5_10,
© Springer Science+Business Media, LLC 2012

Table 10.1 Source and characteristics of different types of ionizing radiation

Radiation	Source	Typical energy (MeV)	Degree of penetration[a]
X-rays	Man-made photons	0.01–10	20–150
Alpha rays	He nuclei from isotopes	1–10	~0.005
Beta rays	Electrons from isotopes	0.01–1	0.1–0.5
Gamma rays	Photons from isotopes	0.1 MeV to several million electronvolts	30–100
Cathode rays	Electrons from accelerators	0.1–10	0.1–5
Protons	Protons from accelerators	10^6-eV range	0.003
Neutrons	Neutrons from fission or isotopes	10^6-eV range	~10

[a]Depth of absorber of unit density required to reduce radiation intensity by a factor of 100

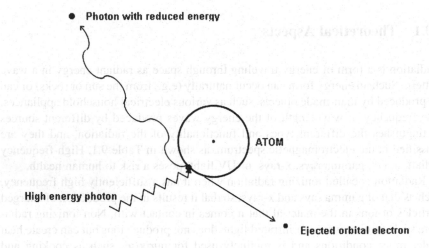

Fig. 10.1 The Compton effect

and

$$E = hv, \tag{10.3}$$

where v is the frequency, c is the velocity of light, λ is the wavelength, E is the energy of the photon, and h is the Planck constant. The energy of photons is usually measured in electronvolts (eV). An electronvolt is defined as the energy acquired by an electron in a potential fall of 1 V and has a value equal to 1.602×10^{19} J.

The total energy impinging a material being irradiated by photons equals the product of the number of photons and the energy of each photon. The types of interaction between matter and photons depend on the photon energy, and three known modes of interactions of significance are the Compton effect, pair production, and the photoelectric effect. In the Compton effect, the photon is considered as a particle that undergoes collision with an orbital electron, ejecting it with kinetic energy derived from the photon and reducing the energy of the photon (Fig. 10.1).

Table 10.2 Absorption coefficients for several materials (cm)

Material	Energy of photons (MeV)		
	0.5	1.0	4.0
Air	0.00008	0.00005	0.00004
Water	0.09	0.067	0.033
Aluminum	0.23	0.16	0.082
Iron	0.63	0.44	0.37
Lead	1.7	0.77	0.48

Pair production is possible when photons of high energy, such as high-energy x-rays, are converted into an electron and a positron. As for the case of the photoelectric effect, this happens when all the energy of the photon is absorbed by an atom, with the photon disappearing and an orbital electron being ejected. The energy of the photon is partially used up in overcoming the binding forces holding the electron in its orbit and is partially converted into the kinetic energy of the electron.

The interaction of photons with matter results in absorption of the energy described by Eq. 9.1 introduced in connection with the absorption of UV light by matter. Coefficients of absorption for different types of materials are given in Table 10.2.

Electrons emitted by a hot cathode can be accelerated to very high velocities by electron accelerators. High-energy electrons are also produced by disintegration of some radioactive atoms. Electrons produced by radioactive isotopes are called beta rays, in contrast with cathode rays, which are produced by machines. There is an important difference between these two types of electrons. Cathode rays can be produced with a uniform energy, whereas beta rays have an energy spectrum. Radioactive isotopes are also emitters of high-energy gamma rays with energies specific to a given isotope. Electrons lose energy when they interact with matter and slow down. As more and more electrons in a beam slow down, the number of fast electrons decreases with the depth of the absorber. If a uniform energy beam of electrons enters an absorber, the relative number of electrons with the initial energy decreases linearly with depth, but not all the electrons at a given depth have the same energy. As a result, the relative ionization at any point does not vary linearly with distance. In the case of beta rays, which are not monoenergetic, when they enter the absorber, the attenuation of the intensity is similar to that of gamma rays. The maximum range of an electron beam R_{max} is related to its energy by the Feather equation, which is valid for energies above 0.5 MeV, and is represented by

$$R_{max} = \frac{0.542E - 0.133}{\rho}, \tag{10.4}$$

where E is the energy and ρ is the absorber density.

10.2 Processing Considerations

Irradiation is performed in specially contained areas where a material is exposed to a determined amount of radiation in a continuous or batch process. The level of exposure is designed to take into account interdependent variables, such as the operation mode (batch or continuous), the optimum energy requirement to successfully safeguard the product, and the source of irradiation (gamma rays, x-rays, etc.).

10.2.1 Radiation Sources

Ionizing radiation can be produced by a number of sources, such as x-rays and electron beams generated by electron accelerators. On the other hand, radioactive materials decay and produce alpha rays, beta rays, or gamma rays, as well as x-rays and electrons. Commercial sources of gamma rays or, to a lesser extent, x-rays and electrons, are cobalt-60 (^{60}Co) and cesium-137 (^{137}Cs). Electron beams are the most cost-efficient form of irradiation, but they can only penetrate food to a limited depth, whereas x-rays are expensive, but are a penetrative form of radiation suitable for bulk operations. Gamma rays are relatively inexpensive and highly penetrative, making them a cost-effective alternative for food irradiation.

Radioactive cesium can be separated from spent nuclear fuel, whereas radioactive cobalt is obtained by neutron irradiation of the non-radioactive cobalt isotope ^{59}Co. Cobalt-60 units used for food processing are produced by machining cobalt metal into shapes suitable for radiation source design, encapsulating them into a 0material that does not become radioactive in a neutron flux, and exposing them to an intense neutron flux in a reactor. A second encapsulation then follows the irradiation. The operations involving handling of ^{60}Co are performed by remote control, because exposure to the now intensely radioactive material would be lethal to humans. Cobalt-60 emits two gamma rays per disintegration, with energies of 1.17 and 1.33 MeV, and a half-life of 5.3 years. The main characteristics of different irradiation sources are summarized in Table 10.3.

Table 10.3 Main characteristics of different irradiation sources

Radiation source	Advantages	Disadvantages
Cobalt-60	High penetrability	Source replenishment needed
	Permanent radiating source	Low throughput
	High efficiency	
Electron beams	High efficiency	Low penetrability
	High throughput	Power and cooling needed
	Switch on–switch off possibility	Technically complex
X-rays	High penetrability	Low efficiency
	Switch on–switch off possibility	Power and cooling needed
	High throughput	Technically complex

Table 10.4 Radiation units

Unit	Definition
Becquerel (Bq)	One unit of disintegration per second
Curie (Ci)	1 Ci = 3.7 × 10^{10} Bq
Half-life	The time taken for the radioactivity of a sample to fall to half its initial value
Electronvolt (eV)	Energy of radiation, usually as megaelectronvolts (MeV)
Radiological unit (rad)	Absorbed dose (where 1 rad is the absorption of 1×10^{-2} J of energy per kilogram of material)
Gray (Gy)	Absorbed dose (where 1 Gy is the absorption of 1 J of energy per kilogram of material)

10.2.2 Dosimetry of Irradiation

The amount of energy absorbed per unit mass of an irradiated food product is known as the "absorbed dose" or simply as the "dose." The absorbed dose is proportional to the ionizing radiation energy absorbed per unit mass of irradiated material, and the effects of the treatment are related to this quantity, which is the most important specification for any irradiation process (Cleland 2006). Currently, the international unit of absorbed dose is the gray (Gy), which is equivalent to 1 J of energy per kilogram of material. It is often more convenient to refer to dose in terms of thousands of grays (kGy). Units of different sorts have been used to characterize irradiation doses over the years. Table 10.4 lists various irradiation units.

Although the process of irradiation can be carefully controlled, it is necessary to confirm the absorbed dose received by food, and this is normally achieved using dosimetry systems (Ehlermann 2001). Dosimeters (or dose meters) are placed at different points within a selected number of packages. A range of dosimetry systems are available, but not all are suitable for measuring the low or high doses of irradiation used to treat some foods. It is, therefore, necessary to select a system appropriate for the purpose for which it is being used (Stevenson 1990).

Ionizing radiation is absorbed as it passes through a food, but not all parts will receive the same absorbed dose. A given part of the food will receive a maximum dose, whereas another part will receive a minimum dose. The uniformity of the dose distribution, also called the overdose ratio, can be improved by irradiating the product from both sides. Prior to treating a food with ionizing radiation, one must establish the dose distribution throughout the food package (Stevenson 1990) in order to take into account differences in the products being irradiated, including factors such as seasonal crop variations, anomalies in bulk density, uneven density variation, and random packaging of agricultural products (Diehl 1995). The dose distribution may be established by placing a number of dosimeters throughout the package or packages of food being treated. By doing this, one can measure the absorbed dose and define positions within the package which have received the maximum and minimum irradiation doses (Stevenson 1990). For the irradiation of any food, the minimum absorbed dose should be sufficient to achieve the intended technological purpose and the maximum absorbed dose should be less than that which would compromise consumer safety or wholesomeness or would adversely affect structural integrity, functional properties, or sensory attributes (CAC 2003). The maximum dose delivered to a food should not exceed 10 kGy, except when necessary to achieve a legitimate technological purpose (WHO 1999).

The dose and the distribution of the dose through a product are determined by the characteristics of the product and the source (Diehl 1995). Product characteristics are, mainly, the density of the food itself and the density of the packing of the individual food containers within the carrier in which the irradiation takes place. Source characteristics will differ depending on the type of irradiator used. For gamma radiators, relevant factors include the isotope, the source strength, and the geometry, whereas for machine sources these include factors such as the beam energy and the beam power.

10.2.3   Irradiation Plant

In designing a particular food irradiation process, one must ensure that a suitable radiation source will be provided at the lowest cost, the appropriate dose and dose distribution is given to each food item, and the operating personnel will not be subjected to harmful radiation either during normal operation or in the event of accidents or breakdowns. The irradiation process involves exposing the food, either packaged or in bulk, to a predetermined dose of ionizing radiation. In food preservation only gamma rays and high-energy electrons can be used in practice. The gamma rays are usually produced by irradiators consisting of packaged ^{60}Co, and occasionally by ^{137}Cs radionuclides. Irradiation installations using gamma-ray-type sources consist of a high-energy isotope source to produce the specific type of ray needed. An isotope source cannot be switched off, so it is shielded within a pool of water below the process area. In operation, the source is raised and packaged food is loaded onto conveyors or trucks, and is transported through the radiation field in a circular path. A typical installation may consist of nickel-plated cobalt wafers encased in stainless steel into groups, to resemble irradiation walls. The source is immersed in a pool of water when not in use, but during irradiation it is raised by remote control and a conveyor transports the material to be irradiated between the walls. A gamma irradiator with the flux entering the irradiated material from two opposite sides has the advantage of being capable of delivering a uniform dose to the material being irradiated. Since the half-thickness in water for gamma rays of the energy of the ^{60}Co irradiators is approximately 11 cm, it is possible to achieve efficient energy absorption patterns even for large, thick food packages. A basic diagram of a ^{60}Co radiation facility is illustrated in Fig. 10.2.

Machine generation of irradiation has the advantage of being controllable, i.e., the radiation stops when the machine is turned off. Additionally, there is no need to dispose of the radioactive material when it is no longer useful. Electron generators

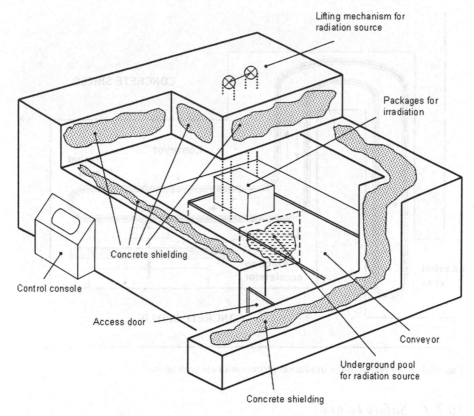

Lifting mechanism for radiation source

Packages for irradiation

Concrete shielding

Control console

Access door

Conveyor

Underground pool for radiation source

Concrete shielding

Fig. 10.2 Basic diagram of a ⁶⁰Co irradiation facility

are also capable of high-energy outputs, consequently being suitable for high-throughput applications. Their major disadvantage is the limited penetration by electrons. At the maximum energy level allowed by the FDA of 10 MeV, the penetration of electron beams is limited. Conversion of electron beams to x-rays increases the penetration capability, but such conversion is still inefficient with current technology, so the costs of the process increase dramatically. An alternative that has been suggested is the use of pulsed x-rays, whose delivery results in higher dose rates and the possibility of intermittent application. It has been also proposed that this mode of x-ray exposure may be useful in minimizing quality defects and maximizing bactericidal effects. Another suggested possibility is, of course, the use of combined treatments or hurdles. For example, specific doses of irradiation may be applied in combination with mild heat treatment. A typical layout of a process using machine-generated irradiation is given in Fig. 10.3.

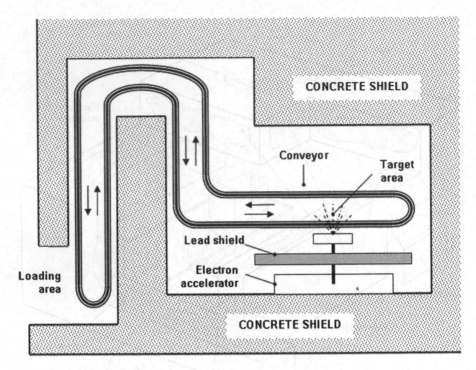

Fig. 10.3 Plant layout of the irradiation process with electron beams

10.2.4   Safety Issues

Extensive studies undertaken on the safety of irradiated food have indicated that the compounds formed in irradiated foodstuffs are generally the same as those produced using other food processing technologies such as cooking, canning, and pasteurization and that any differences are not a risk to the consumer (GAO 2000). A project in the field of food irradiation (IFIP), which ran from 1970 to 1982, involved a large number of animal feeding studies, with over 70 reports being generated. Overall, 100 compounds from irradiated beef, pork, ham, and chicken were examined. The data generated were reviewed at a series of international meetings organized by the World Health Organization (WHO), and these were often jointly held with the Food and Agriculture Organization of the United Nations (FAO) and the International Atomic Energy Agency (IAEA). In 1980, a joint FAO/IAEA/WHO Expert Committee on Food Irradiation (IJECFI) concluded that the maximum recommended dose for foods is 15 kGy, with the average dose not exceeding 10 kGy. At this dose, ^{60}Co and ^{137}Cs have insufficient emission energies to induce radioactivity in the food. According to the IJECFI, this dose presents no toxicological hazards and no special nutritional or microbial problems in foods (WHO 1981).

The Raltech studies were among the most extensive toxicology studies performed on irradiated food, the findings of which are summarized in a report by Thayer et al. (1987). These studies took 7 years to complete at a cost of $8 million and, undoubtedly, led to the most comprehensive safety evaluation ever undertaken on irradiated food (Diehl 1995). During these studies, a total of 230,000 broiler chickens were processed, producing the 134 t of chicken meat required for this work and used to produce four diets: frozen control, thermally processed control, gamma-ray-irradiated enzyme-inactivated chicken meat, and electron-irradiated enzyme-inactivated chicken meat. These diets were subject to nutritional studies, teratology studies, chronic toxicity studies, and genetic toxicology studies. With regard to the nutritional studies, the results showed that all the diets containing the chicken meat had higher protein efficiency ratios than the casein standard and were not significantly affected by any of the ways the chicken had been processed. In terms of the genetic toxicology studies, the Ames test (*Salmonella* microsomal mutagenicity test system) found that the way in which the chicken was processed, either irradiated or non-irradiated, did not affect the response of the test system to known mutagens and no positive results were observed for any of the chicken diets in the absence of the known mutagens. Canton-S *Drosophila melanogaster* was used to test for sex-linked recessive lethal mutations and it was demonstrated that none of the diets produced evidence of sex-linked recessive lethal mutations. However, it was observed that there was a significant reduction in the egg hatchability of cultures of *Drosophila melanogaster* reared on the gamma-ray-irradiated chicken meat diet. Additional testing was performed to confirm these results, and it was concluded that although the irradiated chicken meat was not mutagenic in the test system used, the number of offspring from *Drosophila melanogaster* being fed diets containing chicken was consistently reduced, particularly for those containing irradiated chicken.

Irradiation of food and agricultural products is currently allowed in about 40 countries, and around 60 irradiation facilities operate commercially in the USA (Sommers 2004). Commonly irradiated food products include spices and dry vegetable seasonings (Loaharanu and Murrell 1994), but many other products have been commercialized or are at an evaluation stage. The reasons for low commercialization of food-irradiated products, in spite of their having being well researched and developed for more than a century, include antinuclear activism, industry hesitation, time-consuming approval processes, and insufficient consumer education. Smith and Pillai (2004) identified the major issues expressed by anti-irradiation groups as misuse to avoid plant sanitation, and environmental concern. The volumes of irradiated food products are increasing, however, and the future of irradiated foods looks promising (Olson 2004).

10.3 Effects of Radiation

10.3.1 Chemical Effects of Ionizing Radiation

A primary effect of irradiation is the radiolysis of water. When pure water is irradiated, the following highly reactive entities are formed (Stewart 2001):

$$H_2O \rightarrow \cdot OH, \ e_{aq}^-, \ \cdot H, \ H_2, \ H_2O_2, \ H_3O^+,$$

where ·OH is a hydroxyl radical, e_{aq}^- is an aqueous (or solvated or hydrated) electron, ·H is a hydrogen atom, H_2 is a hydrogen molecule, H_2O_2 is hydrogen peroxide, and H_3O^+ is a solvated (or hydrated) proton.

The hydroxyl radical is a powerful oxidizing agent, whereas the hydrated electron is a strong reducing agent and the hydrogen atoms are slightly weaker reducing agents. Hydrogen (H_2) and hydrogen peroxide (H_2O_2) are the only stable end products of water radiolysis, being produced in a low yield even when the irradiation doses are high (Diehl 1995).

The presence of oxygen during irradiation can influence the course of radiolysis. Hydrogen atoms can reduce oxygen to the hydroperoxyl radical (·OOH), which is a mild oxidizing agent. Another oxidizing agent, the superoxide radical ($·O_2^-$), is formed from the reaction of the solvated electron with oxygen. Both the hydroperoxyl radical and the superoxide radical can give rise to hydrogen peroxide. Oxygen can also add to other radicals produced when water is irradiated, giving rise to peroxy radicals ($·RO_2$). Ozone (O_3), a powerful oxidant, can also be formed from oxygen during irradiation (Stewart 2001).

10.3.2   Effects on Foods

Since most foods contain a considerable amount of water, radiolysis caused by irradiation will have an effect on food-irradiated products. In radiolysis, the hydroxyl radical is a powerful oxidizing agent and the hydrated electron is a strong reducing agent, and both oxidation and reduction reactions take place when water-containing food is irradiated. Some other radiolysis products mentioned, such as the peroxy radical and ozone, can also be formed when foods are irradiated. Radiation can, therefore, affect all the major components of food, that is, carbohydrates, proteins, and lipids (Stewart 2001). It should, however, be noted that even at the high doses used for sterilization, the changes are small and similar to those produced by other food processing technologies such as pasteurization. Temperature can also influence radiolytic changes within a foodstuff. In deep-frozen food the reactive intermediates of water radiolysis are trapped and are thus kept from reacting with each other or with the substrate; thus, freezing can have a protective effect. During thawing of frozen food, there is an increase in the yield of radiolytic products. However, during transition from the frozen to the thawed state, the reactive intermediates react preferentially with each other rather than with the food components. Consequently, any damage to food components is much less than if the food had been irradiated in the unfrozen state (Diehl 1995).

As for any other food processing method, irradiation has advantages and disadvantages. Some foods are sensitive to ionizing radiation, and too high a dose may cause sensory changes rending the food unacceptable to the consumer. Therefore, the actual dose employed is a balance between what is needed and what can be tolerated by the product without unwanted changes (Farkas 2006). A considerable amount of research has been performed into effects of ionizing radiation on sensory, nutritive, and functional properties of different food products.

10.3.3   Effects on Microorganisms

The potentially lethal effects of ionizing radiation were known soon after the discovery of x-rays by Roentgen and of natural radioactivity by Becquerel, both in the late nineteenth century. The sensitivity of organisms to radiation increases with their increasing complexity. Biological effects of irradiation are usually divided into direct effects and indirect effects. In direct action, a sensitive target such as the DNA of a living organism is damaged directly by an ionizing particle or ray, whereas indirect action is caused mostly by the products of water radiolysis.

Similarly to the UV treatment kinetics discussed previously, ionizing radiation treatment is a dose–response method in which models are derived from kinetic data to predict the efficiency of variables of alternative processes. These models depict the relationship between the treatment intensity and a microbial population resistance parameter. The treatment intensity corresponds to the ionizing radiation dose, for this case. A generally accepted model of microbial inactivation kinetics for ionizing radiation is

$$\frac{N_t}{N} = e^{-D/D_0}, \tag{10.5}$$

where N_t is the microbial load after contact time t, N is the initial microbial load, D is the radiation dose, and D_0 is a lethality constant dependent on the organism and irradiation conditions.

D_0 may be conveniently considered the dose resulting in the destruction of approximately 63.3% of the microorganisms. A useful related constant is D_{10}, or decimal reduction dose, which is the dose that will reduce the number of microorganisms by a factor of 10. D_{10} is, of course, equal to $2.303/D_0$. Typical values of lethality constants for selected bacteria are listed in Table 10.5. A graphical representation of Eq. 10.5 for typical vegetative bacteria, spore-forming bacteria, and yeasts is given in Fig. 10.4.

The evident broad range of lethality for given microorganisms (Table 10.5) is due to the fact that many factors, apart from the type of microorganism, affect the survival of irradiated species. The most important factors affecting the survivability of microorganisms are the water content, the pH, the temperature, the dose rate, including type of radiation, and the use (or not) of some sort of pretreatment. In terms of water content, because of the release of radiolytic products, the more dilute the medium exposed to radiation, the lower the microbial survival. A decrease in pH has an effect by increasing the oxygen concentration in the medium,

Table 10.5 Lethality constant values for selected bacteria

Microorganism	D_{10} (Gy)
Vegetative bacteria (range)	100–2,000
Campylobacter species	10–300
Escherichia coli (including O157:H7)	200–500
Micrococcus radiurans	2,500–6,000
Pseudomonas fluorescens	50–200
Salmonella species	250–2,000
Staphylococcus aureus	200–600
Bacterial spores (range)	300–6,000
Bacillus subtilis	300–3,000
Clostridium botulinum (type A)	3,000–6,000
Clostridium botulinum (type E)	1,000–1,500
Clostridium perfringens	1,100–2,200
Clostridium sporogenes PA3679	~2,200

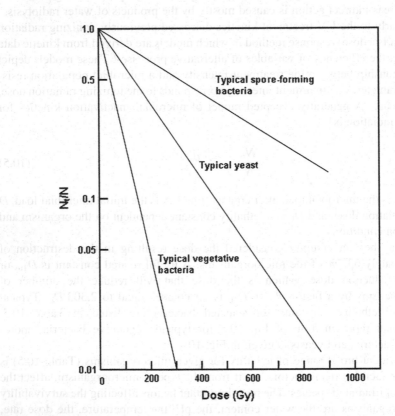

Fig. 10.4 Typical irradiation dose–response curves for different types of microbes

which usually sensitizes the microorganism to radiation-induced effects. Temperature has a very significant effect, in which the effect of radiation is generally increased by increasing temperature. The dose rate bears a direct relation with the model described by Eq. 10.5, i.e., the higher the dose, the most effective the inactivation. Also, as discussed previously, gamma rays are more penetrating that electron beams. Pretreatment processes may have an important effect, particularly when creating conditions to germinate spores in order to increase the sensitivity to radiation of spore-forming microorganisms.

The lethal effect of ionizing radiation on pathogenic bacteria does not necessarily eliminate the danger of toxins formed by these bacteria, for example, *Clostridium botulinum* and *Staphylococcus aureus*. The doses required to eliminate toxins greatly exceed the lethality requirements.

Yeast and mold measurements of individual cell survivors after irradiation treatment are not as convenient as for the case of bacteria and, therefore, Eq. 10.5 is not applicable. The approximate doses of radiation needed to efficiently eliminate the danger of sufficient cell survival are 5,000–20,000 Gy for yeasts and 1,000–10,000 Gy for molds.

Viruses are difficult to eliminate by irradiation and their D_{10} values may well be in excess of 5,000 Gy. Since viruses are normally quite sensitive to heat, radiation in combination with heat treatment may often be an effective option for controlling bacteria and viruses as well.

More complex biological organisms, such as parasites and even insects, can also be inactivated or eradicated by use of ionizing radiation. Among protozoans, *Toxoplasma* and *Entamoeba* are often food-borne, and can cause illness. They are readily inactivated by low radiation doses on the order of less than 500 Gy. Helminthic parasites, including *Trichinella* and several tapeworms, would require doses of up to 5,000 Gy. For the case of insects, whose infestation is the cause of major losses of plant-derived food products, it has been observed that immediate lethality requires substantial doses of radiation, but reproduction and development of the next insect stage (i.e., larvae from eggs, pupae from larva, and adults from pupae) could be prevented by doses of only 10–50 Gy. Studies reported by Urbain (1986) suggested that development of insects and mites could be completely arrested by doses of 100–1,000 Gy.

10.3.4 Effects on Enzymes

The resistance of enzymes to irradiation is much higher than the resistance shown by microorganisms. To be inactivated, enzymes generally require doses about 5–10 times the doses required to inactivate microbes. Enzymes in foods require, therefore, inactivation by other means, for example, by blanching, prior to any irradiation treatment. The D_{10} values of enzymes can be up to 50 kGy and almost four D_{10} values would be required for complete inactivation. Irradiated foods could, thus, be more prone to microbial activity during the shelf life than non-irradiated foods. High resistance of enzymes to irradiation has been demonstrated with milk phosphatase, which was not destroyed by irradiation doses sufficient to sterilize milk (Procter and Goldblith 1951).

Enzymes are affected by the indirect effects of free radicals formed in the solvent phase and, consequently, dilute enzyme solutions are relatively more sensitive to irradiation than concentrated solutions. Enzymes in their natural environments, as in foods, are relatively very resistant, so their activity is unaffected by normal doses, thus limiting the achievable shelf life extension of processed food products, including fruits and vegetables.

10.4   Applications

Ionizing radiation has a wide scope of applications, including disinfestation, decontamination, shelf life extension, quality improvement, and treatment of waste streams. Applications relevant to food processing can be broadly classified into (1) commercial sterilization in hermetically closed containers, known as "radappertization," (2) reduction of the concentration of spoilage organisms, equivalent in its effects to thermal pasteurization, and often referred to as "radurization," (3) inactivation of non-spore-forming pathogens, which has been termed "radicidation," (4) inhibition of sprouting and delaying the senescence of plant food materials, and (5) disinfestation of different crops.

10.4.1   Processed Foods

It is commonly known that ionizing radiation has been used for the purpose of preserving foods in different ways. The use of ionizing radiation for this purpose has been extensively studied for over 100 years, making it one of the most thoroughly researched means by which food can be treated to make it safer to eat and last longer (Diehl 1995). The two main drivers for the use of food irradiation are the enhancement of food safety and of trade in agricultural products (Borsa 2004). The process should, however, not be used as a substitute for good manufacturing practices but rather as a means of reducing risk. Although food irradiation is considered an authentic non-thermal process, and so it is supposed to maintain the quality attributes of processed foods, it has often been associated with development of off-flavors, as well as with softening and increasing permeability of tissue of some fresh fruits and vegetables (Stewart 2009). Some typical applications of irradiation of foods are listed in Table 10.6.

Owing to its bactericidal and antiparasitic properties, ionizing radiation is a highly effective means of enhancing the safety of muscle-based foods. The levels of enteric pathogens associated with meat and poultry products such as *Clostridium jejuni*, *Escherichia coli* O157:H7, *Staphylococcus aureus*, *Salmonella* spp., *Listeria*

Table 10.6 Scope of applications of irradiation in some food products

Application	Food product	Typical dose (kGy)
Pasteurization	Meat, poultry, shellfish, spices	2–7
Increase shelf life	Fruits, vegetables, meat, poultry, fish	0.5–5.0
Disinfestation	Grain, flour, dried fruits	0.2–0.8
Inhibition of sprouting	Onions, garlic, potatoes	0.03–0.14

monocytogenes, and *Aeromonas hydrophila* can be significantly decreased or eliminated at irradiation doses of less than 3 kGy. Only enteric viruses and endospores of the genera *Clostridium* and *Bacillus* are highly resistant to ionizing radiation, but even these are significantly affected (Thayer 1995; Ahn and Lee 2006). The sensitivity of microorganisms to ionizing radiation is affected by factors such as temperature during irradiation, the stage of growth, the presence of oxygen, availability of water, and the composition of the medium in which they are present (Molins 2001). Despite the recognizable advantages of treating meat and poultry products with ionizing radiation, there are some quality aspects which limit adoption of the technology by the meat industry. Irradiation can produce a characteristic aroma, accelerate lipid oxidation, and change the color of the meat (Ahn and Lee 2006). The characteristic odors have been described as "barbecued corn-like" or "bloody sweet." The odors have also been described as being metallic, sulfide, wet dog, goaty, or burned. The intensity of the characteristic odors developed in irradiated meat products depends on the dose of irradiation applied.

Fresh fish and shellfish are highly perishable products that can be irradiated for safe consumption. Pathogenic bacteria of concern in seafood are those found naturally in the freshwater or marine environment and include the genera *Vibrio* and *Aeromonas* and *Clostridium botulinum* type E. *Vibrio* spp. are relatively sensitive to low-dose gamma irradiation compared with other pathogens (Foley 2006). Use of ionizing radiation to eliminate the risk of *Vibrio vulnificus* is the major justification for using irradiation technology. *Vibrio parahaemolyticus* O3: K6 is the most process resistant of all the *Vibrio* pathogens tested to date, but is reduced to non-detectable levels following a dose of 1.5 kGy. It has been shown (Kilgen 2001) that a sublethal dose of 1.5 kGy is optimum for elimination of *Vibrio vulnificus* in oysters and that this dose can also extend the shelf life of commercially shucked and packaged oyster meat from 14 to 24 days when stored at 4°C in a commercial cooler.

A noteworthy review on the irradiation of fruits and vegetables undertaken by Thomas (2001a) stated that the main reasons for treating fruits and vegetables with ionizing radiation include (1) extension of shelf life by delaying the physiological and biochemical processes leading to maturation and ripening, (2) control of fungal pathogens causing postharvest rot, (3) inactivation of human pathogens to maintain microbiological safety, (4) as a quarantine treatment for commodities subject to infestation by insect pests of quarantine importance, and (5) to increase recovery of juice from berry fruits. Niemara (2007) investigated the relative efficacy of a sodium hypochlorite wash versus irradiation to inactivate *Escherichia coli* O157: H7 internalized in leaves of romaine lettuce and baby spinach. A cocktail mixture of three isolates of *Escherichia coli* O157:H7 were drawn into the leaves, after which the leaves were washed with a sodium hypochlorite sanitizing solution or treated with ionizing radiation (0.25–1.5 kGy). The results showed that treatment of the leaves with irradiation (but not the chemical sanitizers) effectively reduced the numbers of viable *Escherichia coli* O157:H7 cells internalized in the leafy green vegetables in a dose-dependent manner. A more complex response to irradiation was observed in the spinach leaves than in romaine lettuce leaves, with a marked tailing effect in spinach at higher doses as compared with a linear response in the lettuce. The specific doses to be used should be determined for each product on the basis of the patterns of antimicrobial efficacy and specific product sensory responses. Recent studies on some fruits have indicated that irradiation treatment can improve quality. Hussain et al. (2007) conducted work on peach fruits which after harvest were irradiated in the dose range of 1–2 kGy followed by storage under ambient or refrigerated conditions. Evaluation of the anthocyanin content of the fruits showed that irradiation enhanced the color development under both storage conditions and that doses of 1.2–1.4 kGy effectively maintained a higher total soluble solids concentration, reduced weight loss, and delayed decay of the fruit by 6 days under ambient conditions and by 20 days when stored under refrigeration. Fan et al. (2004) reviewed the effect of low-dose ionizing radiation on fruit juices. They found that irradiation at low doses effectively inactivated food-borne pathogens and reduced the levels of the mycotoxin patulin and brownness. However, irradiation did induce undesirable chemical changes such as the accumulation of malondialdehyde, formaldehyde, and tetrahydrofuran. They found the literature on the negative flavor changes of irradiated juice to be contradictory, although they did state that evidence did exist concerning the involvement of volatile sulfur compounds in the development of off-flavor. It was suggested that, as for other products, many of the undesirable effects of irradiation could be reduced by using low temperatures during treatment, by the addition of antioxidants to the product, and by combining irradiation with other techniques and treatments such as mild heating and use of antimicrobials.

10.4.2   Shelf Life Extension

One form of shelf life extension is to inhibit sprouting in bulbs and tubers. Bulb and tuber crops are important fruit vegetables cultivated and consumed in most areas of the world. Onions, shallots, and garlic constitute the major bulb crops, whereas potatoes and yams are the most economically important tuber crops (IAEA 1997). Commercial irradiation of potatoes has been performed in Japan since 1973, when the first industrial-scale irradiation facility for treatment of potatoes was set up by the Shihoro Agricultural Cooperative Association. Back in 2001, it was reported that the plant was processing 15,000–20,000 t annually (Thomas 2001b). The tubers are sold for industrial processing and for household use. Factors that determine the efficacy of radiation treatment include the time interval between harvest and irradiation, the radiation dose, the cultivar, and the initial product quality (IAEA 1997; Thomas 2001b). Good tuber and bulb crop handling and storage management practices are essential prerequisites for their successful irradiation on a commercial scale. Upon harvest, the crops must be dried well, cleaned of adhering soil, particularly in the case of tubers, and sorted to remove damaged and infected material. Generally, irradiation has no significant effect on the carbohydrate and sugar content of onions, whereas in garlic a lower content of water-soluble carbohydrates has been reported. For potatoes, the effect of irradiation on their sugar content is variable, and this could be attributed to variations in the cultivar, time of treatment, and storage temperature. It has been observed that there is a temporary rise in the level of both reducing and non-reducing sugars immediately after irradiation, which often returns to a normal level during storage, followed by an increase on prolonged storage or senescent sweetening. The postirradiation storage temperature may influence the rise in the levels of sucrose and reducing sugars as demonstrated in early work by Metlitsky et al. (1957). They showed that tubers stored at 1.5°C for 7 months had higher levels than their non-irradiated counterparts, whereas at a higher temperature the irradiated tubers had lower sugar levels than the controls. In a similar manner, increased levels of reducing sugars were observed for tubers treated at 100 and 250 Gy and stored at 0–4°C for 1 month compared with those stored at the higher temperature of 25°C.

Spices and herbs have been important items in international trade for centuries, most of which are produced in equatorial and tropical countries by small landholders and local farmers. Generally, microbial contamination is inevitable under the prevailing production, harvesting, and postharvest handling conditions; therefore, most dried food ingredients of vegetable origin may contain large numbers of microorganisms that may cause spoilage or defects in composite food products into which they are incorporated, or more rarely, may cause food poisoning (Farkas 2001). Until the early 1980s, the method most commonly used to destroy microorganisms in dried food ingredients was fumigation using ethylene oxide, or to a much lesser extent, using propylene oxide. However, for toxicological reasons, the use of ethylene oxide has been discouraged or even banned for food use, as is the case in the European Union. Consequently, the use of ionizing radiation was considered as alternative method for decontamination of dried food ingredients and it is now one of the main commercial applications of the technology (Farkas 2001). Irradiation has a strong antimicrobial effect, and depending on the initial number and type of microorganisms present, and on the chemical composition of the product, doses of 3–10 kGy can be used to reduce the total aerobic viable cell counts even in highly contaminated spices and other dried food ingredients, without having an adverse effect on their flavor, texture, or other properties (Farkas 1988).

Nuts represent a food commodity and product of economic interest in different parts of the world. One of the major problems affecting the shelf life, quality, and safety of nuts is contamination with molds, which results in aflatoxin formation. To date, the nut industry has relied on chemical fumigants such as methyl bromide, ethylene oxide, and propylene oxide for control of microbial growth (Kwakwa and Prakash 2006). The use of gamma radiation is being considered as an alternative to fumigation as it offers various advantages in the treatment of almonds. Ionizing radiation can penetrate through the shell and provide homogeneous treatment of the surface of the nut kernel, without altering the general characteristics of raw almonds or raising the kernel temperature. It can be used at doses that will control microbial pathogens, molds, and other spoilage organisms.

10.4.3   Disinfestation

One of the important postharvest treatments in food processing is disinfestation, and chemicals are often used for this purpose. Ionizing radiation has been considered a suitable alternative for disinfestation of various food commodities. As indicated previously, the doses required to eradicate insects and mites or to prevent their reproduction are relatively low. Use of ionizing radiation is a very suitable method of disinfestation and can be used in bulk shipments of grains, legumes, and other commodities, in dry foods after they have been packaged, and in treatment of fruits from areas harboring insects and in quarantine at the locations to which they are exported. Disinfestation doses range from 0.1 to 1 kGy.

References

Ahn DU, Lee EJ (2006) Mechanisms and prevention of quality changes in meat by irradiation. In: Sommers CH, Fan X (eds) Food Irradiation Research and Technology, pp 127–142. Blackwell Publishing, Ames, IA.

Borsa J. (2004) Outlook for food irradiation in the 21st century. In: Komolprasert V, Morehouse KM (eds) ACS Symposium Series 875: Irradiation of Food and Packaging: Recent Developments, pp 326–342. American Chemical Society, Washington DC.

CAC (2003) General Standard for Irradiated Foods (Codex Stan 106–1983, Rev 1–2003). Codex Alimentarius Commission (CAC), Rome, Italy.

Cleland MR (2006) Advances in gamma ray, electron beam and x-ray technologies for food irradiation. In: Sommers CH, Fan X (eds) Food Irradiation Research and Technology, pp11–35. Blackwell Publishing, Ames IA.

Diehl JF (1995) Safety of Irradiated Foods. Marcel Dekker, New York.

Ehlermann D (2001) Process control and dosimetry in food irradiation. In: Molins R (ed) Food Irradiation: Principles and Applications, pp 387–414. John Wiley & Sons, New York.

Fan X, Niemira BA, Thayer DW (2004) Low-dose ionizing radiation of fruit juices: benefits and concerns. In: Komolprasert V, Morehouse KM (eds) ACS Symposium Series 875: Irradiation of Food and Packaging–Recent Developments, pp 138–150. American Chemical Society, Washington DC.

Farkas J (1988) Irradiation of Dry Food Ingredients. CRC Press, Boca Raton FL.

Farkas J (2001) Radiation decontamination of spices, herbs, condiments, and other dried food ingredients. In: Molins R (ed) Food Irradiation: Principles and Applications, pp 291–312. John Wiley & Sons, New York.

Farkas J (2006) Irradiation for better foods. Trends Food Sci Technol 17: 148–152.

Foley DM (2006) Irradiation of seafood with a particular emphasis on *Listeria monocytogenes* in ready-to-eat products. In: Sommers CH, Fan X (eds) Food Irradiation Research and Technology, pp 185–197. Blackwell Publishing, Ames IA.

GAO (2000) Food Irradiation. Available Research Indicates that Benefits Outweigh Risks. United States General Accounting Office, Report to Congressional Requesters, GAO/RCED-00-217, August 2000.

Hussain PR, Meena RS, Dar MA, Wani AM (2007) Studies on enhancing the keeping quality of peach (Prunus persica Bausch) Cv. Elberta by gamma-irradiation. Radiat Phys Chem 77: 473–481.

IAEA (1997) Irradiation of Bulb and Tuber Crops: A Compilation of Technical Data for its Authorization and Control. IAEA-Tecdoc-937, April 2007, International Atomic Energy Agency, Vienna, Austria.

Kilgen MB (2001) Irradiation processing of fish and shellfish products. In: Molins R (ed) Food Irradiation: Principles and Applications, pp 193–212. John Wiley & Sons, New York.

Kwakwa A, Prakash A (2006) Irradiation of nuts. In: Sommers CH, Fan X (eds) Food Irradiation Research and Technology, pp 221–235. Blackwell Publishing, Ames IA.

Loaharanu P, Murrell D (1994) A role for irradiation in the control of food borne parasites. Trends Food Sci Nutr 5: 190–195.

Metlitsky LV, Rubin BA, Krushchev VG (1957) Use of γ-radiation in lengthening storage time of potatoes. In Proceedings of the All Union Conference on the Application of Radioactive and Stable Isotopes and Radiation in the National Economy and Science, USAEC Report–AEC-tr-2925.

Molins RA (2001) Irradiation of meats and poultry. In: Molins RA (ed) Food Irradiation: Principles and Applications, pp 131–191. John Wiley & Sons, New York.

Niemara BA (2007) Relative efficacy of sodium hypochlorite wash versus irradiation to inactivate *Escherichia coli* O157:H7 internalized in leaves of romaine lettuce and baby spinach. J Food Prot 70: 2526–2532.

Olson DG (2004) Food irradiation future still bright. Food Technol 58 (7): 112.

Procter BE, Goldblith SA (1951) Food processing with ionizing radiations. Food Technol 5 (9): 376–380.

Smith JS, Pillai S (2004) Irradiation and food safety. Food Technol 58 (11): 48–55.

Sommers CH (2004) Food irradiation is already here. Food Technol 58 (11): 22.

Stevenson MH (1990) The practicalities of food irradiation. In: Turner A (ed) Food Technology International Europe, pp 73–77. Sterling Publications International, London.

Stewart EM (2001) Food irradiation chemistry. In: Molins R (ed) Food Irradiation: Principles and applications, pp 37–76. John Wiley & Sons, New York.

Stewart EM (2009) Food irradiation. In: Process-Induced Food Toxicants, pp. 387–412. Stadler RH, Lineback DR (eds). John Wiley & Sons, New York.

Thayer DW (1995) Use of irradiation to kill enteric pathogens on meat and poultry. J Food Saf 15: 181–192.

Thayer DW, Christopher JP, Campbell LA, Ronning DC, Dahlgren RR, Thomson GM, Wierbicki E (1987) Toxicology studies of irradiation-sterilized chicken. J Food Prot 50: 278–288.

Thomas P (2001a) Irradiation of fruits and vegetables. In: Molins R (ed) Food Irradiation: Principles and applications, pp 213–240. John Wiley & Sons, New York.

Thomas P (2001b) Irradiation of tuber and bulb crops. In: Molins R (ed) Food Irradiation: Principles and applications, pp 241–272. John Wiley & Sons, New York.

Urbain W (1986) Food Irradiation. Academic Press, Orlando, FL.

WHO (1981) Wholesomeness of Irradiated Food: A Report of a Joint FAO/IAEA/WHO Expert Committee on Food Irradiation. WHO Technical Report Series, 659, World Health Organization, Geneva.

WHO (1999) High-Dose Irradiation: Wholesomeness of Food Irradiated with Doses above 10 kGy. Report of a Joint FAO/IAEA/WHO Study Group, WHO Technical Report Series 890. World Health Organization, Geneva.

Chapter 11
Ultrasound in Food Preservation

11.1 Introduction

Ultrasound waves are similar to sound waves but, having a frequency above 16 kHz, cannot be detected by the human ear. Bat and dolphins are said to use low-intensity ultrasound to locate prey, and some marine species make use of high-intensity ultrasound pulses to stun their prey. Low-intensity and high-intensity ultrasound have had some applications in food processing and preservation. Low-intensity ultrasound has been used as a non-destructive analytical method to assess the composition and structure of foods. High-intensity ultrasound has been employed to cause physical disruption of tissues, create emulsions, clean equipment, and promote chemical reactions. Ultrasound has the properties of sound waves, such as reflection, interference, adsorption, and scattering. Ultrasound can propagate through solids, liquids, and gases (McClements 1997; Povey and McClements 1988).

Sound waves can propagate parallel or perpendicular to the direction of travel through a material. Since the propagation of sound waves is normally associated with a liquid medium. Parallel waves are known as longitudinal waves and perpendicular waves are also known as shear waves (Fig. 11.1). The direction of particle motion in longitudinal waves is the same as the wave motion. Longitudinal waves are capable of traveling in solids, liquids, or gases, and have short wavelengths with respect to the transducer dimensions, producing sharply defined beams and high velocities. In shear waves, particle motion is perpendicular to the direction of wave propagation and, since liquids and gases do not support stress shear under normal conditions, shear waves can only propagate through solids. The velocity of shear waves depends on the material in which they propagate and such velocity is relatively low compared with that of longitudinal waves. Some velocities of transverse and longitudinal waves are listed in Table 11.1. The longitudinal velocity is dependent on the state of the material and can be used to follow processes such as

E. Ortega-Rivas, *Non-thermal Food Engineering Operations*,
Food Engineering Series, DOI 10.1007/978-1-4614-2038-5_11,
© Springer Science+Business Media, LLC 2012

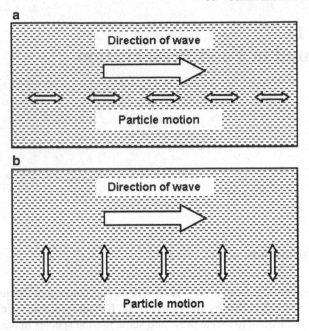

Fig. 11.1 Sound waves: (**a**) longitudinal wave, (**b**) transverse wave

Table 11.1 Velocities of longitudinal and transverse sound waves in different materials

Material	Type of wave	State	Velocity (m/s)
Air	Longitudinal	Gas, 20°C	344
Aluminum	Longitudinal	–	6,374
	Transverse	–	3,111
Water	Longitudinal	Water vapor	500
		Water, 25°C	1,498
		Ice	3,760
	Transverse	Ice	2,000
Orange juice	Transverse	Liquid, 20°C	1,540
		Frozen, –40°C	3,310
Beef[a]	Longitudinal	Warm, 37°C	
		Perpendicular	1,595
		Parallel	1,605
		Chilled, 0°C	
		Perpendicular	1,525
		Parallel	1,531
		Frozen, –9°C	
		Perpendicular	2,870
		Parallel	2,930
Fat content of meat mixtures	Longitudinal	100% lean meat	1,543
		50% lean meat/50% fat	1,584
		100% fat	1,617
Olive oil	Longitudinal	Liquid, 60°C	1,490
		Solid, –30°C	1,990

[a]Direction of ultrasound either parallel or perpendicular to muscle fiber orientation

freezing of orange juice (Lee et al. 2004) and meat (Miles and Cutting 1974). It is also sensitive to differences in structure, such as fiber orientation in meat, and comminuted meat composition.

11.2 Principles: Acoustic Cavitation and Sonochemistry

Ultrasound waves are characterized by velocity c, wavelength λ, and frequency f, by the following relationships:

$$c = \lambda f, \tag{11.1}$$

$$c = \sqrt{\frac{K}{\rho}}, \tag{11.2}$$

$$c = \sqrt{\frac{C_p P}{C_v \rho}}, \tag{11.3}$$

where ρ is the density of the medium, K is the Young's modulus for solids or the bulk modulus of elasticity for liquids, P *is the pressure (for gases)*, C_p is the specific heat at constant pressure, and C_v is the specific heat at constant volume.

The wavelength is the distance between adjacent wave crests, as illustrated in Fig. 11.2, and the frequency is the number of wave crests passing a point in a unit time. Frequency resembles the vibration of a wave generator. The wavelength of ultrasound at 20 kHz in water at 25°C is approximately 7.5 cm, but in a denser system, such as muscle or bone, it will be higher. The amplitude is the height of the wave (Fig. 11.2) and it determines the wave strength. Amplitude refers to the motion

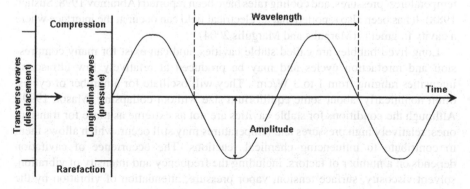

Fig. 11.2 Description of ultrasound wave characteristics

of the ultrasound source, the motion of the receiver of the sound, or the motion of the medium through which the sound wave is traveling. The amplitude A is related to the energy contained in the wave E by

$$E \propto A^2. \tag{11.4}$$

The intensity is a measure of the flow of acoustic energy through a unit area of the medium in unit time. If sound waves propagate through biological tissues, the intensity of the signal decreases with the distance of travel, owing to scattering of the sound waves, as well as owing to absorption of part of the sound energy by the material. Absorption of sound may be due to a number of mechanisms, such as viscous losses, heat condition losses, and losses due to molecular exchange of energy (Povey and McClements 1988). Scattering is caused by interfaces between materials with different acoustic impedances. Similarly to previous examples of energy absorption by radiating sources, the ultrasound intensity I follows an exponential pattern depending on the initial intensity of the transducer I_0, the distance from the transducer X, and the amplitude attenuation coefficient of the material α. An analogous expression to Eq. 9.1 for this case is, therefore,

$$I = I_0 e^{-2\alpha X}. \tag{11.5}$$

The amplitude attenuation coefficient differs between materials and with the frequency of the sound. In general, the absorption coefficient increases with increasing frequency, and in biological systems such a coefficient can differ markedly between types of materials (Povey 1989).

When an intense sound propagates through a liquid, it creates regions of compression (positive pressure) and rarefaction (negative pressure). If the negative pressure during rarefaction is sufficient, a cavity or bubble can form in the liquid, and two types of cavitation can result: transient or "inertial" and stable or "non-inertial." Transient cavitation occurs if a cavity experiencing vibration increases in size progressively over a number of compression and rarefaction cycles, until it reaches a size at which it collapses violently (Perkins 1988). On collapse of the cavity, very high, but localized, temperatures, pressures, and cooling rates have been reported (Abramov 1998; Suslick 1988). It has been also reported that an electrical field can occur at the interface where a cavity fragments (Margulis and Margulis 2004).

Long-lived bubbles are called stable cavities, and can exist for many compression and rarefaction cycles and may be produced at relatively low ultrasound intensities ranging from 1 to 3 W/cm^2. They will oscillate for a number of cycles, often nonlinearly, about some equilibrium size without collapsing (Mason 1999). Although the conditions for stable cavities are not as extreme as those for transient ones, relatively high pressures and temperatures may still occur, which allows them to contribute to influencing chemical reactions. The occurrence of cavitation depends on a number of factors, including the frequency and intensity of vibration, solvent viscosity, surface tension, vapor pressure, attenuation of vibration by the medium, presence of gas bubbles as cavitation nuclei, and ambient pressure as well as the ambient temperature (Mason 1999).

Ultrasound causes a range of effects on chemical reactions in treated materials, such as formation of free radicals, increase of reaction rates, reduction of the induction period, improvement of catalyst efficiency, alteration of the reaction pathway (Margulis 2004), and production of sonoluminescence, which is the light emitted during the collapse of either stable or transient cavities due to the pressures and temperatures generated (Yasui et al. 2004). Within the collapsing cavity there are extreme pressures, temperatures, and cooling rates, as well as gas or vapor from the liquid medium. All these conditions are capable of creating hydroxyl radicals from water, which can react with other chemicals in the cavity, or diffuse into the liquid medium, where they can react with some other compounds. Reactions involving volatile compounds are also possible when volatile compounds are present in the liquid medium, which may be a plausible situation for many liquid food systems. Volatile compounds can also diffuse into the cavity during expansion, and can undergo chemical reactions during collapse of the cavity. In addition to chemical reactions caused by the diffusion of the reactive species or the reaction products into the liquid medium, cavity collapse can also affect some chemical reactions in the liquid medium. The high shear created by cavity collapse may also cause breaks in polymer chains, increase reaction rates owing to an increase in the kinetic energy of molecules, or alter interaction with the solvent.

Ultrasound waves may be able to alter chemical processes at surfaces, causing different effects depending on whether they are small, large, or liquid surfaces. There can be mechanical damage to the solid material, with shock waves and microjets causing surface damage, such as pitting of solid surfaces, fragmentation of brittle materials, deaggregation of groups of particles, and high-velocity collisions between small particles accelerated by ultrasound that may promote abrasion or fusion. For the case of immiscible liquids, cavitation at their interfaces can create emulsions, greatly increasing the area of contact between the two materials.

Although the main focus of sonochemistry is on cavitation-related effects, the non-cavitational effects of ultrasound can still play a role in chemical reactions. Ultrasound agitation of liquids, for instance, improves mass transfer and heating of the material because the energy absorbed from the ultrasound waves can increase the rate of chemical reactions.

11.3 Ultrasound Treatment as a Preservation Technology

11.3.1 Inactivation of Microorganisms

Ultrasound treatment in itself has not been found to be efficient enough for microbial inactivation of foodstuffs. Although ultrasound treatment alone can destroy microorganisms, it is a relatively inefficient process, requiring quite long

Table 11.2 Effect of sonication combined with heating on inactivation of
two microorganisms, expressed as D_{10} values (min)

Product	Heat only	Presonication		Simultaneous treatment	
		20 kHz	800 kHz	20 kHz	800 kHz
Listeria monocytogenes at 60°C					
UHT milk	2.1	0.4	>10	0.3	1.4
Rice pudding	2.4		0.3		4.5
Zygosaccharomyces bailii at 55°C					
Orange juice	10.5	2.4	1.4	3.9	>10
Rice pudding	11.0	2.3		1.0	>10

processing times to produce a significant reduction in microbial counts. As a result,
research into ultrasound treatment as an antimicrobial method has mainly focused
on combining ultrasound treatment with other treatments, especially heat and
pressure treatments. Combination of ultrasound treatment with heat processing
has been referred to as "thermosonication," Ordonez et al. (1987) whereas combination with pressure processing has been termed "manosonication." The
advantages of combining ultrasound treatment, both as a pretreatment or simultaneously, with heat treatment can be observed from the examples shown in
Table 11.2. The shear forces and rapidly changing pressures created by ultrasound
waves are effective in microbial inactivation, but mostly if combined with other
methods including heating, pH modification, and chlorination (Lillard 1994). The
mechanisms of cell disruption and the effects on different microorganisms have
been reviewed by Rahman (1999). A combined heat and ultrasound treatment under
pressure, known as "manothermosonication" (MTS), was described by Sala et al.
(1995). Exploratory investigations indicate that the lethality of MTS treatments is
6–30 times greater than that of a corresponding heat treatment at the same temperature, and was greater for yeasts than for bacterial spores. The efficiency of MTS
depends on the intensity, amplitude, and time of electrosonication, and on the
pressure applied. Microbial inactivation kinetics for MTS follows the same pattern
as those for thermal treatments. It can, therefore, be hypothesized that ultrasound
reduces the heat resistance of microorganisms by physical damage to cell structures,
caused by extreme pressure changes and disruption of cellular protein molecules,
making them more sensitive to denaturation by heat.

11.3.2   Effects on Spores

The lengthy processing required to destroy some microorganisms and particularly
spores at temperatures below the temperatures used for conventional heat treatment, makes ultrasound treatment unattractive. The ability of certain species of

microorganisms, such as *Bacillus* spp., to produce spores provides protection against adverse factors. It has been observed that the resistance that those spore-forming microorganisms show against thermal processes is also shown against sonication. Ultrasound treatment of bacterial spores can have little effect at low temperature, with no reduction in *Bacillus steareothermophilus* spore numbers by an extended ultrasound treatment at 12°C for 30 min (Palacios et al. 1991). *Clostridium sporoegenes* and *Bacillus cereus* spores suspended in Ringer solution survived sonication for 30 min (Ahmed and Russell 1975). Ultrasound was also reported to have no significant effect on the survival of *Clostridium botulinum* spores in honey (Nakano et al. 1989).

11.4 Ultrasonic Equipment

For actual application of ultrasound waves, there is a need to use instruments known as transducers and suitable equipment, including whistle reactors, ultrasonic baths, and probe systems (Mason 1999). A transducer is a device needed to convert one form of energy into another. In ultrasonic applications, the transducers are design to convert mechanical or electrical energy into high-frequency sound. There are two main types of transducers: mechanical and electroacoustic. Mechanical transducers rely on the flow of a liquid or a gas through a siren, rotor, or turbine to generate ultrasound. Electromechanical transducers are widely used in modern ultrasonic applications, and are based on the inherent electrostrictive phenomenon in certain materials to produce piezoelectric or magnetostrictive transducers (Abramov 1998).

In a whistle reactor, a stream of liquid flows, and passes a metal blade, in order to produce vibration (Mason 1999). The frequency of the vibration depends on the liquid flow rate. The principle of this type of generator is shown in Fig. 11.3. A blade with wedge-shaped edges is positioned in front of a nozzle. Liquid is pumped through this nozzle and the jet emerging from it impinges on the leading edge of the blade and sets it vibrating. The blade is normally clamped at one or more nodal points and resonates at its natural frequency, propagating waves of ultrasonic frequency through the liquid. Whistle reactors are often used for high-power liquid applications, such as homogenization, emulsification, and dispersion.

Ultrasonic baths are cheap, simple, and versatile, and comprise a metal bath with one or more transducers attached to the walls of the tank (Mason 1999). Items to be treated can be directly immersed in the bath and subjected to ultrasound wave propagation. The maximum power input of ultrasonic baths is generally low, ranging from 1 to 5 W/cm^2. A diagram of an ultrasonic bath is illustrated in Fig. 11.4.

Probe systems (Fig. 11.5) consist of a metal horn coupled to an ultrasonic transducer. The horn is used to amplify the vibration produced by eletrostrictive material, which is generally a piezoelectric material, in the transducer (Perkins 1988). Amplifying the vibration produced by the transducer is necessary as the amplitude of the waves produced by the piezoelectric material is too small to

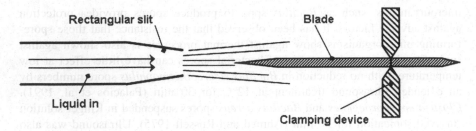

Fig. 11.3 Principle of the whistle reactor

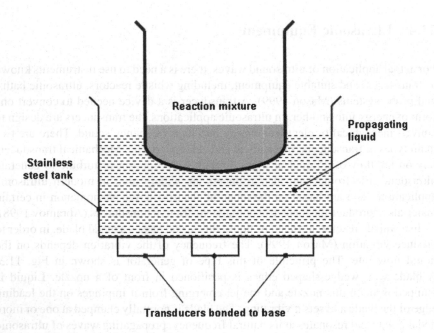

Fig. 11.4 An ultrasonic cleaning bath

produce a useful effect. An appropriate horn design will increase the amplitude of the vibration at the face of the horn. Probe horns have advantages such as the possibility of being placed directly in or against the material being processed, and the production of controlled ultrasound intensities of up to several hundred watts per square centimeter. Among their disadvantages are erosion of the tip of the horn by cavitation, free-radical formation, and heating of the material exposed to ultrasound. Tip erosion by cavitation can cause contamination of the material being processed by metal residues, as well as a gradual change in the horn length, which in turn will affect its efficiency.

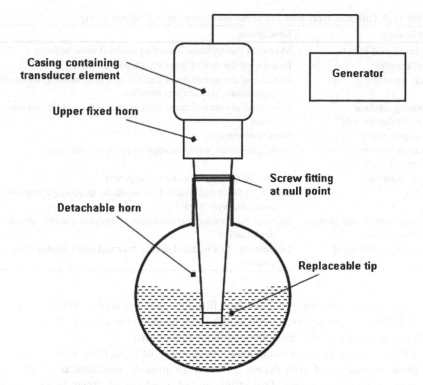

Casing containing transducer element

Upper fixed horn

Detachable horn

Generator

Screw fitting at null point

Replaceable tip

Fig. 11.5 Ultrasonic probe equipment

11.5 Applications

Ultrasound has a number of food safety and processing applications unrelated to destruction of microorganisms in food which can be cost-effective and more relevant. As can be seen in Table 11.3, most of the applications of ultrasound in the food industry are other than food-preserving operations.

Ultrasound can be used as a processing aid in speeding up heat transfer and mass transfer unit operations, which results in improved efficiencies in several applications. When ultrasound is applied to freezing liquids, very fine crystals can be generated by maximizing ice crystal formation with a limited degree of supercooling (Mason 1998). Faster freezing rates are possible, as are finer and more uniform crystal sizes, reducing food product damage and improving the texture of ice creams (Mortazabi and Tabatabaie 2008) and the quality of thawed frozen products (Li and Sun 2002).

Mass transfer plays an important role in many food process engineering operations, such as drying, dewatering, filtration, membrane separations, salting, and osmotic dehydration. Ultrasound has been shown to improve the efficiency of many mass transfer processes either by a direct involvement in the process or by supporting the process. There has been substantial research into industrial

Table 11.3 Different applications of ultrasound treatment in food processing

Application	Description
Antimicrobial effects	Microbial inactivation, microbial removal from surfaces
Heat transfer	Increase of the rate of freezing, thawing, cooking
Mass transfer	Increase of the rate of mass transfer in drying, brining, membrane separations, dewatering, filtration
Cleaning, surface decontamination	Loosening of contaminants, such as grit on vegetables and wax on fruits
Meat processing	Meat tenderization
Homogenization, emulsification	Homogenization, emulsification of milk, mayonnaise
Crystallization	Control of nucleation and crystal growth
Cutting	Cutting of fresh and frozen fruit products, including composite and multilayer foods
Enzyme activity and protein denaturation	Enzyme inactivation, enhancement of enzyme activity, protein denaturation
Defoaming, defrothing	Defoaming of carbonated drinks, beer and other liquids during canning

applications of ultrasound-assisted bed filtration and dewatering (Gallego-Juarez et al. 2003; Tarleton and Wakeman 1998) and it has been shown that ultrasound can substantially improve the efficiency of these operations.

Ultrasound can be used to enhance the cleaning of processing equipment with strategic mounting of transducers in order to provide mechanical disturbance in areas difficult to clean. The efficiency of washing of fresh foods can be improved by application of ultrasound, reducing or eliminating the need for chemical cleaners. Loosening of contaminants, such as grit on vegetables or egg surfaces and grease and wax on fruits, can be greatly improved by using ultrasound waves. The contaminants after loosening can be removed by conventional cleaning methods.

Tenderness is the most important consumer quality attribute in cooked meat (Miller et al. 2001). The tenderness of meat is influenced by multiple factors, including animal characteristics, preslaughter handling of animals, and postslaughter handling of carcasses (Tarrant 1998). There are myriad techniques available to tenderize meat, including conventional methods such as aging, protease treatment, salt solution injection, and mechanical treatment, and alternative approaches such as explosive shock, ultrahigh pressure, and injection of calcium salts (Tarrant 1998). Ultrasound treatment can be counted among these less conventional approaches, offering potential as a tenderizing treatment that can be applied without causing the changes in appearance that some other methods cause.

The use of ultrasound waves provides a satisfactory means of dispersing one immiscible liquid in another to produce an emulsion. In a system consisting of two immiscible liquids, if cavitation occurs at the interface between them, one phase will become dispersed in the other. Thus, if the internal phase is added to the external phase while the latter is subjected to ultrasound, an emulsion can be formed. From the three commonly employed methods of generating ultrasound waves, mechanical generators are finding increasing use in the food industry for emulsification applications. The most common form of mechanical ultrasonic generator used for food emulsification is the wedge resonator or whistle reactor previously described (Fig. 11.3). The intensity achieved by the whistle reactor is not very high, but suffices at the proximity of the blade to cause cavitation in the liquids to bring about emulsification.

When ultrasound is applied to solid media, it can induce molecular excitation, leading to changes in crystal structures. High local temperatures can be induced and this is useful for welding of some materials. At higher levels, ultrasound can cut material without the need for direct surface contact. The recent development of ultrasonic knives is of great potential value to the food industry for difficult-to-cut materials (Schneider et al. 2002), such as crumbly cheeses and confectionary.

The shearing and compression effects of ultrasound cause denaturation of proteins, which may result in reduced enzyme activity in some foods. However, a short burst of ultrasound may increase enzyme activity, possibly by breaking down large molecular structures and making the enzymes more accessible for reaction with substrates. The resistance of most enzymes to ultrasound is quite high, so the intensity and prolonged time to try to inactivate them will produce adverse changes to the texture and other physical properties of the food under treatment, and may substantially reduce its sensory attributes.

Ultrasound is a clean means for breaking foams. The mechanisms of acoustic defoaming, which are insufficiently known, may be a combination of high acoustic pressures, radiation pressure, resonance of the bubbles, cavitation, streaming, and atomization from the surface of the bubble. The potential use of high-intensity ultrasound for defoaming has been known for several decades, although it has not been successfully introduced in industry. This is probably because the acoustic defoamers used were generally based on aerodynamic acoustic sources, which presented many difficulties for practical applications. A new ultrasonic defoamer has been developed by using the stepped-plate transducer and it has been successfully applied to control foam in fermenting vessels and on high-speed canning lines for carbonic beverages (Gallego-Juarez 2002). The transducer rotates to increase the defoaming area. This technique was positively applied to control the foam in a beer fermenter of 6-m diameter. In high-speed canning of beverages, foam is produced on the top of the can, and to avoid liquid losses, the ultrasonic radiation is focused on it and the excess foam is quickly destroyed. The new system was applied to control foam in the filling of cans with commercial beverages at a speed of more than 20 cans/s.

References

Abramov OV (1998) High intensity-ultrasonics: theory and industrial applications. Gordon and Breach, Amsterdam.
Ahmed FIK, Russell C (1975) Synergism between ultrasonic waves and hydrogen peroxide in killing of microorganisms. J Appl Bacteriol 39: 31–40.

Gallego-Juarez JA (2002) Macrosonics: phenomena, transducers and applications. Keynote lecture KL-05 at Forum Acusticum Sevilla 2002, Seville, Spain, 16–20 September 2002.

Gallego-Juarez JA, Elvira-Segura L, Rodriguez Corral G (2003) A power ultrasonic technology for deliquoring. Ultrasonics 41: 255–259.

Lee S, Pyrak-Nolte MJ, Cornillon P, Campanella O (2004) Characterization of frozen orange juice by ultrasound and wavelet analysis. J Sci Food Agric 84: 405–410.

Li B, Sun DW (2002) Novel methods for rapid freezing and thawing of foods-a review. J Food Eng 54: 175–182.

Lillard HS (1994) Decontamination of poultry skin by sonication. Food Technol 48 (12): 72–73.

Margulis MA (2004) Sonochemistry as a new promising area of high energy chemistry. High Energy Chem 38: 135–142.

Margulis MA, Margulis IM (2004) Mechanism of sonochemical reactions and sonoluminescense. High Energy Chem 38: 285–294.

Mason TJ (1998) Power ultrasound in food processing: the way forward. In: Povey MJW, Mason TJ (eds) Ultrasound in Food Processing, pp 105–126. Blackie Academic and Professional, London.

Mason TJ (1999) Sonochemistry. Oxford University Press, Oxford.

McClements DJ (1997) Ultrasonic characterization of foods and drinks: principles, methods, and applications. Crit Rev Food Sci Nutr 37: 1–46.

Miles CA, Cutting CL (1974) Changes in the velocity of ultrasound in meat during freezing. J Food Technol 9: 119–122.

Miller MF, Carr MA, Ramsey CB, Crockett KL, Hoover LC (2001) Consumer thresholds for establishing the value of beef tenderness. J Anim Sci 79: 3062–3068.

Mortazabi A, Tabatabaie F (2008) Study of ice cream processing after treatment with ultrasound. World App Sci J 4: 188–190.

Nakano H, Okabe T, Hashimoto H, Yoshikuni Y, Sakagushi G (1989) Changes in *Clostridium botulinum* spores in honey during long-term storage and mild heating. Jpn J Food Microb 6: 97–101.

Ordonez JA, Aguilera MA, Garcia ML, Sanz B (1987) Effect of combined ultrasonic and heat treatment (thermoultrasonication) on the survival of a strain of *Staphylococcus aureus*. J Dairy Res 54: 61–67.

Palacios P, Burgos J, Hoz L, Sanz B, Ordonez JA (1991) Study of substances released by ultrasonic treatment from *Bacillus steareothermophilus* spores. J Appl Bacteriol 71: 445–451.

Perkins JP (1988) Power ultrasound. In: Mason TJ (ed) The Uses of Ultrasound in Chemistry, pp 47–59. The Royal Society of Chemistry, Cambridge.

Povey MJW (1989) Ultrasonics in food engineering II. Applications. J Food Eng 9: 1–20.

Povey MJW, McClements DJ (1988) Ultrasonics in food engineering I. Introduction and experimental methods. J Food Eng 8: 217–245.

Rahman MS (1999) Light and sound in food preservation. In: Rahman MS (ed) Handbook of Food Preservation, pp 669–686. Marcel Dekker Inc, New York.

Sala FJ, Burgos J, Condon S, Lopez P, Raso J (1995) Effect of heat and ultrasound on microorganisms and enzymes. In: Gould GW (ed) New Methods of Food Preservation, pp 176–204. Blackie Academic and Professional, London.

Schneider Y, Zahn S, Linke L (2002) Qualitative process evaluation for ultrasonic cutting of food. Eng Life Sci 2: 153–157.

Suslick KS (1988) Homogeneous sonochemistry. In: Suslick KS (ed) Ultrasound–Its Chemical, Physical and Biological Effects, pp 123–163. VCH Publishers, New York.

Tarleton ES, Wakeman RJ (1998) Ultrasonically assisted separation processes. In: Povey MJW, Mason TJ (eds) Ultrasound in Food Processing, pp 193–218. Blackie Academic and Professional, London.

Tarrant PV (1998) Some recent advances and future priorities in research for the meat industry. Meat Sci 49: S1–S16

Yasui K, Tuziuti T, Sivakumar M, Iida Y (2004) Sonoluminescence. Appl Spectrosc Rev 39: 399–436.

Chapter 12
Pulsed Light Technology

12.1 Introduction

Pulsed light contains a broad spectrum of "white" light from UV wavelengths of 200 nm to infrared wavelengths of 1,000 nm. A peak of pulsed light is observed between 400 and 500 nm, as shown in Fig. 12.1. This light is the non-ionizing part of the electromagnetic spectrum, and, in contrast to gamma rays and x-rays, does not cause ionization of small molecules. Pulsed light has a spectrum similar to that of sunlight (Fig. 12.1), except that it also contains some UV wavelengths that are filtered out of sunlight by Earth's atmosphere. The light is produced in short high-intensity pulses that are approximately 20,000 times the intensity of sunlight at sea level. They last for a few hundred microseconds and the energy they impart is measured as "fluence," often quoted as joules per square centimeter. The intense flashes of light produced by the pulses of light can be used for the destruction of microorganisms. Pulsed light treatment involves the use of a flash of high-intensity light for the purpose of killing microorganisms on the surface of either food or packaging materials.

Pulsed light inactivates microorganisms by a combination of photothermal and photochemical effects. The antimicrobial effects of light at UV wavelengths are due to absorption of the energy by highly conjugated carbon double bonds in proteins and nucleic acids, which disrupt cellular metabolism, and are well documented (Jagger 1967; Koller 1965; Smith 1977). As already mentioned, pulsed light contains a broad spectrum of white light from UV to infrared wavelengths. Of that spectrum, 45% lies in the visible range, 30% in the infrared range, and 25% in the UV range. The UV component of pulsed light has a photochemical effect but, since most of the energy is in the visual spectrum, the effect is mostly photothermal. A large amount of energy is transferred rapidly to the surface of the food, raising the temperature of a thin surface layer sufficiently enough to destroy vegetative cells. The energy penetrating the surface of the treated material is absorbed in accordance with the Beer–Lambert law, whose different expressions have been already

E. Ortega-Rivas, *Non-thermal Food Engineering Operations*,
Food Engineering Series, DOI 10.1007/978-1-4614-2038-5_12,
© Springer Science+Business Media, LLC 2012

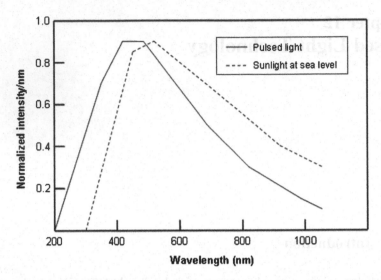

Fig. 12.1 Wavelength distributions of pulsed light and sunlight

described (Chaps. 9 and 11). For the specific case of pulsed UV light, an appropriate relationship is as follows:

$$E(X) = (1 - r)E_0 e^{-\alpha X}, \tag{12.1}$$

where $E(X)$ is the energy of light transmitted to a distance X below the surface of the material, r is the reflection coefficient of the material, E_0 is the radiating light hitting the surface of the material, and α is the extinction coefficient, which is a measure the transparency (or opacity) of the material at a given wavelength λ. Most solids are opaque ($\alpha \rightarrow \infty$), not transmitting radiation, whereas many liquids and gases are transparent ($\alpha \rightarrow 0$), not absorbing any part of the energy. In most materials, including foods, light intensity rapidly decreases when light penetrates into the bulk.

The absorbed energy is converted primarily into heat, but some is also used for excitation of different molecules. The energy absorbed E_d is

$$E_d = E(X)\left(1 - e^{-(\alpha)d}\right), \tag{12.2}$$

where d is the depth of a layer below the distance X.

The absorbed light energy is usually dissipated into heat and results in an increase in temperature ΔT of

$$\Delta T = \frac{E_d}{\rho c_p A d}, \tag{12.3}$$

where ρ is the density of the material, c_p is its specific heat, and A is its surface area.

Table 12.1 Main features of continuous UV light and pulsed light

Feature	Continuous UV light	Pulsed light
Wavelength	254 nm	100–1,100 nm (typical)
Instantaneous energy	Less	Magnified several thousand fold
Inactivation mechanism	Photochemical DNA damage by thymine dimer formation	Damage to cells by photochemical changes and by localized heating
Natural cooling of lamp	No	Enables the lamp to cool between pulses
Mercury	Commonly used as source	Provides a mercury-free alternative
Inactivation efficiency	Normal	Up to four times increased inactivation efficiency compared with continuous UV light
Temperature increase	Not significant	Significant temperature increase due to infrared component

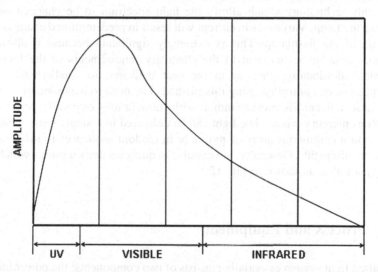

Fig. 12.2 A single pulse of light showing its spectral components

For satisfactory disinfection, the UV component is especially important, with lethal effects as discussed in Chap. 9. In water and some transparent fluids, sufficiently intense penetration of the UV components to depths of 0.25 cm is possible. Most food materials are opaque, however, so penetration of the UV component is extremely low, limiting the applications of this technology to treatment of surfaces and transparent fluids.

Pulsed UV light, as compared with continuous UV light, presents some advantages, which are summarized in Table 12.1. A single pulse of light contains a dynamic spectrum and presents many opportunities for matching the light to the specific requirement of the application. For example, as illustrated in Fig. 12.2, the spectrum of a single pulse is time-dependent, starting with the shortest UV

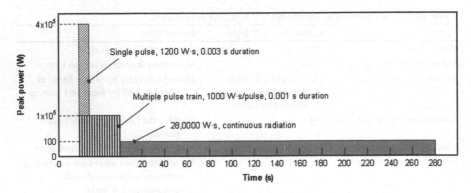

Fig. 12.3 Comparison of pulsed and continuous light

wavelength and ending with the longest infrared wavelength. Pulsed UV treatment is the only technology which allows the light spectrum to be changed without changing the lamp. Variations in current will result in predetermined changes in the spectrum of the flashlamp. This is extremely significant because it allows the flashlamp light to be matched to the chemistry requirements of the biological substrate. Additionally, changes in the heat delivered to smaller and thinner substrates can be controlled using this method. The need to make minor increases in the heat delivered is most common with formulations originally designed for continuous mercury lamps. The light can be delivered in a single pulse, a burst of pulses, and a continuous array of pulses or in random sequences to meet specific process requirements. Generally, lower pulse frequencies deliver pulses with higher energy per pulse, as shown in Fig. 12.3.

12.2   Process and Equipment

The pulsed light system essentially consists of two components: the power unit and the lamp. The power unit is used to generate a high voltage, whose resulting high-current pulses are employed in the lamp. As shown in Fig. 12.4, AC power is first converted to high-voltage DC power, which is then used to charge a capacitor. After the capacitor has been charged to certain voltage, a high-voltage switch discharges the charge in the capacitor into a lamp. The system is properly contained to protect personnel from high voltages. Cooling water is used to minimize heating of the treated product. The treatment unit contains one or more inert gas lamps. When a high-current pulse is applied to the lamp, gas contained within the lamp emits an intense pulse of light. The frequency of flashing, the number of lamps, and the flashing configuration depend on the treatment application. Monitoring of the lighting system is extremely important to ensure that the treatment area is properly treated. The system uses two types of diagnostic monitors, the lamp output (fluence) and the lamp current. Lamp fluence is measured to ensure that sufficient UV

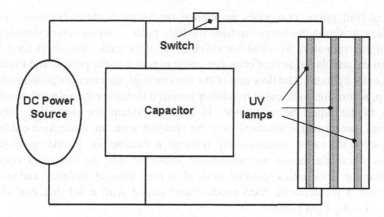

Fig. 12.4 Equipment for generating pulsed light

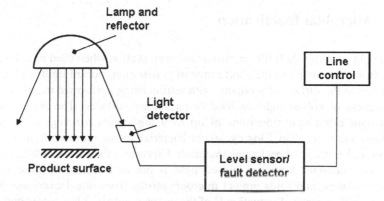

Fig. 12.5 A monitoring system for pulsed light energy

radiation is generated by the lamp as to inactivate microorganisms. This is accomplished with the use of a silicon photodiode that is capable of detecting whether the lamp has the required output of UV light. A decreasing output would signal the need for replacement of the lamp. This control is also required to shut down the operation in cases where objects do not receive treatment above some predetermined threshold (Fig. 12.5). A second monitoring control used in the system measures the lamp current for every flash. The current is an indication of the intensity and the spectrum of the radiation. If the current falls below a preselected threshold, the operation is shut down. The system involves illuminating the desired treatment area with $0.1-3$ J/cm^2 per flash, with total accumulated fluences of $0.1-12$ J/cm^2. Flashes are, in general, applied at a rate of $0.5-10$ Hz for several hundred microseconds.

Although several companies already produce pulsed light systems for applications other than food processing and preservation, pulsed light technology is not yet applied at an industrial scale in the food sector. Some processing plant schemes for use in the food industry have, however, been patented (Dunn et al. 1989). In one of these patents, a system for sterilizing pumpable foods, such as water of fruit juices, comprises an annular treatment chamber between an inner cylinder containing the lamps emitting the light pulses, and an outer cylinder made of a highly reflective material for driving the light back. The fluid food flows through the annular treatment chamber and is exposed to the pulsing light reflected by the outer cylinder. The flow rate of the feed through the chamber is controlled by a pump, according to the pulse frequency required to deliver the selected number of pulses to the liquid being treated. In another system for sterilizing foods in packages, preformed containers may be sprayed with an absorption-enhancing agent solution, move progressively through a number of flashlamp treatment stations where the lamps are introduced above or into the container openings providing the light pulses required to sterilize their internal surfaces, and are then filled with a presterilized food product and sealed with a lid that can also be sterilized using light pulses.

12.3 Microbial Inactivation

Pulsed light provides shelf life extension and preservation when used in a variety of foods. However, owing to the short range of penetration, pulsed light applications are limited to the surface of products, with testing being performed mainly into the effectiveness of pulsed light on food or packaging surfaces. The application of pulsed light can cause reductions of up to 9 log cycles for vegetative micro-organisms and more than 7 log cycles for bacterial spores on smooth, non-porous surfaces such as those of packaging materials. On complex surfaces such as those of most food materials, however, pulsed light is not as effective because surface recesses, fissures, and folds protect microorganisms from direct exposure. When this is the case, microbial reduction is of the order of only 2–3 log cycles but, since the pulsed light systems affect only the surface of the product being treated, fewer photoproducts result from this treatment than are produced by thermal treatments, thus minimizing product degradation.

The kinetics of microbial inactivation for pulsed light acting on surfaces follows a first-order logarithmic decrease, in which inactivation is a direct function of the treatment intensity (fluence) and the treatment time (number of pulses or flashes), and may be generally represented by (Severin et al. 1984)

$$\frac{N_t}{N} = e^{-kIt}, \tag{12.4}$$

where N and N_t are the concentrations of viable microorganisms before and after treatment respectively, k is the first-order inactivation coefficient, I is the intensity of pulsed light, and t is the time treatment.

A typical graph for microbial inactivation for a given intensity and different pulses is shown in Fig. 12.6. The fluence required to reduce the microbial population by 1 log cycle using a single light pulse can be defined as "decimal reduction fluence" D_{F1}, and can be used to compare the resistances of different microorganisms to pulsed light.

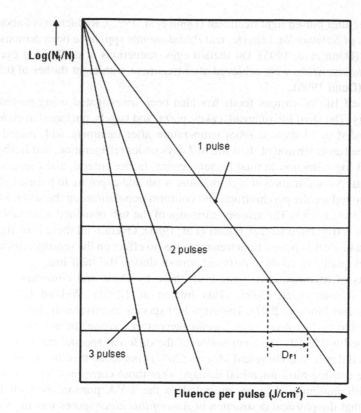

Fig. 12.6 Typical graph of microbial inactivation on a surface by pulsed light treatment

The higher its value, the more resistant the microorganism. Dunn et al. (1989) reported D_{F1} values of 1.2, 2.6, 2.8, and 2.9 kJ/m^2 for spores of *Aspergillus niger*, *Bacillus pumilis*, *Bacillus subtilis*, and *Bacillus stearothermophilus*, respectively.

Most testing on pulsed light microbial inactivation was conducted by Dunn et al. (1995), who own a patent for a broad-spectrum pure light process (Dunn et al. 1989) based on an acquired patent by Hiramoto (1984). The concentration of mold spores of *Aspergillus niger* on the surface was reduced by 7 log cycles with several flashes of 1 J/cm^2 intensity (Dunn et al. 1991). Some other microorganisms, including *Escherichia coli*, *Staphylococcus aureus*, *Bacillus subtilis*, and *Saccharomyces cerevisiae*, were inactivated using 1–35 pulses with an intensity ranging from 1 to 2 J/cm^2. Reduction in the concentration of an inoculum of *Pseudomonas aeruginosa* by 1.5 log cycles was achieved by applying two flashes with an energy density of 16 J/cm^2 in curds of commercial dry cottage cheese. The surface temperature of the cheese was only increased by 5°C and the sensory quality was reported to be impaired (Dunn et al. 1991). Successful disinfection of surfaces was reported for fish and shrimp. *Salmonella* serovars were reduced by 2 log cycles on chicken wings in samples inoculated with either 2 or 5 log/cm^2. *Listeria innocua* was reported to be reduced by 2 log cycles on hot dogs inoculated with 3 or 5 log per frankfurter after pulsed light treatment (Dunn et al. 1995). Reductions of about 2–3 log cycles of *Salmonella, Listeria,* and *Pseudomonas* spp. have been demonstrated on meats (Dunn et al. 1995). On shelled eggs, reductions of up to 8 log cycles of *Salmonella enteritidis* were achieved after treatment with eight flashes of 0.5 J/m^2 per flash (Dunn 1996).

The shelf life of various foods has also been investigated using pulsed light technology. The shelf life of bread, cakes, pizza, and bagels, packaged in clear film, was extended to 11 days at room temperature after treatment with pulsed light. Shrimps had an extension of shelf life to 7 days under refrigeration, and fresh meats had a 1–3 log reduction in total bacteria count, lactic, enteric, and *Pseudomonas* spp. counts. A combination of high-pressure wash and exposure to pulsed light was reported to reduce the psychrotroph and coliform populations on the surface of fish tissue by 3 log cycles. The sensory attributes of the fish remained acceptable after 15 days of refrigerated storage (Dunn et al. 1988). Overall, all these investigations seem to indicate that pulsed light treatment has no effect on the sensory attributes or nutritional quality of foods at different periods during the shelf life.

Reports of investigations, apart from those by Dunn and coworkers, include those by Rowan et al. (2000), Takeshita et al. (2003), Wekhof (2003), and Woodling and Moraru (2007). The microbial species inactivated included *Listeria innocua, Aspergillus niger,* and *Saccharomyces cerevisiae.* In general, all these studies tried to identify the contribution of the different spectral ranges of pulsed light to cell death. Woodling and Moraru (2007) concluded that the portions of the UV range causing most microbial damage were those corresponding to UVB and UVC, with minimum damage attributed to the UVA portion. Wekhof (2003) reported that the physical destruction of *Aspergillus niger* spores was the result of structural collapse, whereas Takeshita et al. (2003) found enlarged vacuoles in *Saccharomyces cerevisiae* after exposure to pulsed light, attributing the effect to intracellular heating. Rowan et al. (2000) as well as Woodling and Moraru (2007) reported only minimal heating effects after treating a variety of food-related microorganisms.

In applications other than treatment of surfaces of foods and packaging materials, kinetics data on inactivation of microorganisms with pulsed light is somewhat scarce and too limited to allow the development of quantitative models of inactivation. Uesugi et al. (2007) reported, however, that a Weibull model was successfully applied to describe the survival of *Listeria inoccua* and *Escherichia coli* after exposure to pulsed light, both on clear and on translucent liquid substrates. The Weibull model is a non-mechanistic model that uses a power function to describe the variation of survivor ratio as a function of treatment intensity (van Boekel and Martinus 2002). An attempt to use this model for surface inactivation was not successful, possibly because of the influence of surface properties and initial inoculum levels on inactivation (Uesugi et al. 2007).

12.4 Other Effects

Apparently, neither the effect of pulsed light treatments on nutritional components of vegetables nor the potential formation of toxic by-products has been studied yet. Since the wavelengths used for pulsed light treatment are too long to cause ionization of small molecules (Dunn et al. 1995), the formation of radioactive by-products is not really expected. Pulsed light treatment of foods has been approved by the FDA (1996) under code 21CFR179.41. According to Dunn et al. (1997), in assessing the safety of foods treated with all forms of radiation, the agency considers changes in the chemical composition of the food that may be induced by the proposed treatment, including any potential changes in nutrient levels. The legal status of pulsed light technologies in the European Union (EU) has been considered by a different approach. The legislation in the EU is not technology-oriented but is food-oriented or food-ingredient-oriented, so pulsed light technology falls within the scope of regulation 258/97 on novel foods and novel food ingredients, article 1, item f (European Union 1997). Among other categories, this legislation applies to foods and food ingredients to which a production process not currently used has been applied, and evaluates possible changes in nutritional value, in metabolism, and in level of undesirable substances. The EU would not approve, therefore, pulsed light technology as such. It may approve (or not approve) specific foods and food ingredients treated with pulsed light technology. It has been proved that continuous-wave UV light treatment can increase the concentration of phytochemicals in fruits. Cantos et al. (2001) found that UV treatment can increase by more than tenfold the levels of resveratrol in grapes. Given the similarities between both techniques, pulsed light might also have the same effect.

12.5 Applications

Short-duration high-power light pulses have been tested sufficiently to demonstrate that they are a viable alternative to inactivate microorganisms on food products, by a combination of photochemical and photothermal effects. Compared with thermal sterilization, pulsed light technology achieves effective microbial inactivation with a much lower processing time, thus preserving more food nutritional and sensory properties. Pulsed light technology implies, however, high investment costs, which complicates possible applications to high-value added products and particular market situations. The major limitation of pulsed light technology is, however, its low penetrating power and the requirement for transparency and smoothness of the food items to be treated. For this reason, pulsed light technology may be successfully used only for bulk treatments of extremely transparent materials, or for surface treatments of less transparent ones, as suggested in Table 12.2. A quite interesting application is in sterilizing films or packaging materials, particularly those used in

Table 12.2   Potential applications of pulsed light technology based on its limitations

Unpacked solid foods requiring only a primary surface sterilization or decontamination

Packaged solid food requiring terminal sterilization or decontamination using pulsed-light-compatible packages

Unpacked liquid foods flowing through treatment chambers

Liquid foods packaged into pulsed-light-compatible packages

aseptic technology, as a practical alternative to the use of hydrogen peroxide, whose residuals are not fully acceptable in foods.

Actual applications in which pulsed light technology has been tested successfully are treatment of baked products, including breadsticks, chocolate cupcakes, pizza, tortillas, and bagels, resulting in providing a longer shelf life by retarding mold growth. Also, as already mentioned, fish filets and several meats, including chicken wings and frankfurters, had a significantly longer shelf life when treated with high-intensity pulsed light.

References

Cantos E, Espín JC, Tomás-Barberán FA (2001) Postharvest induction modeling method using UV irradiation pulses for obtaining resveratrol-enriched table grapes: a new "functional" fruit? J Agric Food Chem 49: 5052–5058.

Dunn J, Bushnell A, Ott T, Clarke W (1997) Pulsed white light food processing. Cereals Food World 42: 502–515.

Dunn J, Clark W, Ott P (1995) Pulsed-light treatment of food and packaging. Food Technol 49 (9): 95–98.

Dunn JE (1996) Pulsed light and pulsed electric fields for foods and eggs. Poult Sci 75: 1133–1136.

Dunn JE, Clark RW, Asmus JF, Pearlman JS, Boyer K, Painchaud F, Hoffamn GA (1988) Methods and apparatus for preservation of foodstuffs. US International Patent Application No WO 88/03369.

Dunn JE, Clark RW, Asmus JF, Pearlman JS, Boyer K, Painchaud F, Hoffamn GA (1989) Methods for preservation of foodstuffs. US Patent 4,871,559.

Dunn JE, Clark RW, Asmus JF, Pearlman JS, Boyer K, Painchaud F, Hoffamn GA (1991) Methods for preservation of foodstuffs. Maxwell Laboratories Inc. US Patent 5,034,235.

European Union (1997) Regulation EC No 258/97 of the European Parliaments and of the Council of 27 January 1997 concerning novel foods and novel food ingredients.

FDA (1996) Code of Federal Regulations. 21CFR179.41.

Hiramoto T (1984) Method of sterilization. US Patent 4,464,366.

Jagger J (1967) Ultraviolet Photobiology. Prentice Hall, Upper Saddle River.

Koller L (1965) Ultraviolet Radiation. John Wiley & Sons, New York.

Rowan NJ, Anderson JG, MacGregor SJ, Fouracre RA (2000) Inactivation of food-borne enteropathogenic bacteria and spoilage fungi using pulsed light. IEEE Trans Plasma Sci 28: 83–88.

Severin BF, Suidan MT, Rittmann BE, Engelbrecht RS (1984) Inactivation kinetics in a flow-through UV reactor. J Water Pollut Control Fed 56: 164–169.

Smith K (1977) The Science of Photobiology. Plenum Press, New York.

Takeshita K, Shibato J, Sameshima T, Fukunaga S, Isobe S, Arihara K, Itoh M (2003) Damage of yeast cells induced by pulsed light irradiation. Int J Food Microbiol 85: 151–158.

Uesugi AR, Woodling SE, Moraru CI (2007) Inactivation kinetics and factors of variability in the pulsed light treatment of *Listeria innocua* cells. J Food Prot 70: 2518–2525.

van Boekel MAJS, Martinus AJS (2002) On the use of the Weibull model to describe thermal inactivation of microbial vegetative cells. Int J Food Microbiol 74: 139–159.

Wekhof A (2003) Sterilization of packaged pharmaceutical solutions, packaging and surgical tools with pulsed UV light. In: Proceedings of the Second International Congress on UV Technologies, Vienna, Austria, 9-11 July 2003.

Woodling SE, Moraru CI (2007) Effect of spectral range in surface inactivation of *Listeria innocua* using broad spectrum pulsed light. J Food Prot 70: 906–916.

Chapter 13
High-Voltage Pulsed Electric Fields

13.1 Introduction

With increasing demand to obtain processed foods with better attributes than have
been available to date, food researchers have pursued the discovery and development of improved preservation processes with minimal impact on the fresh taste,
texture, and nutritional value of food products. Different processed food products
may present a series of undesirable effects when treated by conventional, thermal
methods of pasteurization. Alternatives to traditional treatment which do not involve
direct heat have been investigated in order to obtain safe processed food products,
but with sensory attributes resembling those of the fresh raw material. Several
processing techniques have been investigated recently and include ultraviolet irradiation, gamma irradiation, ultrasound treatment, treatment with non-conventional
chemical reagents, application of high-intensity magnetic fields, ultra-high-pressure
processing (HPP), use of membrane technology, and application of high-voltage
pulsed electric fields (PEFs). Commercial applications of some alternative
technologies are varied and include purification of water, pasteurization of fruit
juices, and processing of milk. Application of these non-thermal technologies offers
interesting opportunities for mildly processed safe products with preserved sensory
and nutritional qualities. Although non-thermal technologies such as PEFs have
been demonstrated to have some advantages over the conventional thermal
technologies, they have only recently started gaining interest in the food industry.

13.2 Theoretical Background

13.2.1 Principle

PEF processing involves the application of an externally generated electric field across
a food product, with the intent of inactivating pathogenic microorganisms, modifying
enzymes, intensifying some processes, or achieving some specific transformation

E. Ortega-Rivas, *Non-thermal Food Engineering Operations*,
Food Engineering Series, DOI 10.1007/978-1-4614-2038-5_13,
© Springer Science+Business Media, LLC 2012

in the product. The technology has long been used for cell hybridization and electrofusion in genetic engineering and biotechnology. Its application is based on the transformation or rupture of cells under a sufficiently high external electric field, resulting in increased permeability and electrical conductivity of the cellular material. When a cell is exposed to high-voltage electrical pulses, their lipid bilayer and proteins are temporarily destabilized and perforation ensues, but if the electric field is removed, the pores reseal themselves. If the intensity of the electric fields is too high, however, the microorganism will not be able to repair itself and will start leaking small compounds or even undergo lysis (Amiali et al. 2006). The magnitude of the critical breakdown voltage depends on the type of biological cell, its size, its diameter, its shape, the growth phase of the microorganisms, and the medium in which the microorganisms are present. Successful application of PEFs for a product depends on rational and innovative optimization of process variables. PEF applications allow better control of electric power input and effective permeabilization of cellular membranes without significant temperature elevation (Weaver and Chizmadzhev 1996). In general, the transmembrane voltage u_m induced on the cell membrane owing to an external electric field is given as (Zimmermann 1986)

$$u_m \approx \alpha d_c E \cos \theta, \tag{13.1}$$

where α is a parameter depending on the cell shape ($\alpha = 0.75$ for a spherical cell and $\alpha = 1$ for a rectangular cell), d_c is the cell diameter, E is the electric field strength, and θ is the angle between a point on the membrane surface and the direction of electric field. The smaller the exposed cells are in the electric field, the higher the field strength required for creating a critical transmembrane potential needed for plasmolysis of the cell membrane. The mean diameters of microorganisms and biological tissue cells are in the ranges of 10 nm to 1 μm and 10 μm to 1 mm, respectively (Aguilera et al. 2000). High electric field pulses with strengths in the range of 20–50 kV/cm are used to kill microorganisms in PEF pasteurization.

13.2.2 Mechanisms of Microbial Inactivation

PEF processing involves a short burst of high voltage to a food placed between two electrodes (Qin et al. 1995). When a high electric voltage is applied, a large flux of electric current flows through food materials, which may act as electrical conductors because of the presence of electric charge carriers such as a large concentration of ions (Barbosa-Cánovas et al. 1999). The mechanism of inactivation of microorganisms exposed to PEFs has not been fully clarified yet. Two typical theories have been proposed over the years: the dielectric breakdown theory (Zimmermann et al. 1976) and the electroporation theory (Castro et al. 1993). In a suspension of cells, an electric field causes a potential difference across the

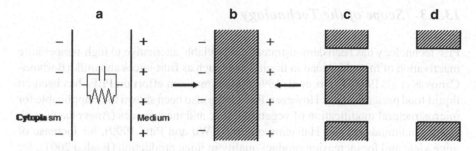

Fig. 13.1 Dielectric breakdown mechanism on a cell membrane: (**a**) cell membrane with induced potential, (**b**) membrane compression, (**c**) reversible pores on the membrane, (**d**) irreversible pores on the membrane

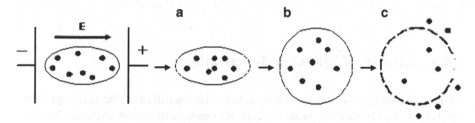

Fig. 13.2 Electroporation mechanism on a cell membrane: (**a**) osmotic unbalance, (**b**) swelling, (**c**) membrane rupture

membrane and induces a sharp increase in membrane conductivity and permeability. Membrane destruction occurs when the induced membrane potential exceeds a critical value of 1 V for a short duration (2 μs) in many cellular systems, corresponding to an external field of roughly 10 kV/cm for *Escherichia coli*, the commonly accepted standard for food pasteurization (Castro et al. 1993; Coster and Zimmermann 1975; Kinosita and Tsong 1977; Zimmermann 1986; Zimmermann et al. 1974). The dielectric breakdown or electroplasmolysis (McLellan et al. 1991) can also be explained by the mechanism known as electroporation, which is the formation of pores in the lipid or the protein domains of the cell membrane. Lipid bilayers are susceptible to applied electric fields because of the electric charges of the lipid molecule and the permeability for the bilayer to ions. Electric charges will cause lipid molecules to reorient themselves under an intense electric field, thus creating hydrophilic pores and impairing the bilayer barrier against ions. Besides these, the physiological impact and electroosmotic effect may also influence the efficiency of electroplasmolysis (Weaver and Chizmadzhev 1996). Figures 13.1 and 13.2 present diagrammatic schemes of the dielectric breakdown and the electroporation mechanisms, respectively.

13.2.3   Scope of the Technology

PEF technology has been demonstrated to be a viable alternative to high-temperature inactivation of microbial load in liquid foods such as fruit juices and milk (Barbosa-Canovas et al. 1999; Knorr et al. 1994). Most research effort on PEFs has been on liquid food pasteurization. However, PEFs have also been shown to be applicable for microstructural modification of vegetable, fish, and meat tissues (Angersbach et al. 2000; Gudmundsson and Hafsteinsson 2001; Wu and Pitts 1999), for increase of juice yield and for increasing product quality in juice production (Bazhal 2001), for processing of vegetable raw materials (Papchenko et al. 1988), and for winemaking and sugar production (Gulyi et al. 1994). PEF treatment significantly enhances certain food processing unit operations such as pressing (Bazhal and Vorobiev 2000), diffusion (Jemai 1997), osmotic dehydration (Rastogi et al. 1999), and drying (Ade-Omowaye et al. 2000).

13.3   Kinetics of Microbial Inactivation

When microorganisms are treated with heat, the logarithm of the cell population decreases linearly with the treatment time for constant treatment intensity. Alternative technologies are also believed to inactivate microorganisms logarithmically. Dose–response models are derived from kinetic data to predict the efficiency of variables of alternative processes (Van Gerwen and Zwietering 1998). These models depict the relationship between the treatment intensity and a microbial population resistance parameter. The treatment intensity corresponds to the electric field intensity for PEF technology. Graphically, the models described can be represented by sigmoid plots. Survivor plots for alternative preservation technologies commonly exhibit a shoulder or a tail (Ohshima et al. 1997). As shown in Fig. 13.3, the initial delay in inactivation of *Saccharomyces cerevisiae* in apple juice during PEF treatment with up to 600 pulses per second (pps) is a typical shoulder, whereas the constant number of *Listeria brevis* during extended PEF treatment in apple juice corresponds to the tailing effect. The intensity of the deleterious agent (i.e., electrical discharge) must exceed a threshold value before microbial inactivation happens, which would explain the shoulder. A mechanistic concept considers the shouldering and the tailing effects as artifacts due to factors such as lethal injury, cell clumping, and heterogeneous treatment zones.

The microbial survival fraction in terms of the applied field intensity E and the treatment time t can be represented by (Grahl and Märkl 1996; Hülsheger et al. 1981)

$$\frac{N_t}{N} = \left(\frac{t}{t_c}\right) e^{-\frac{(E-E_c)}{k}},\tag{13.2}$$

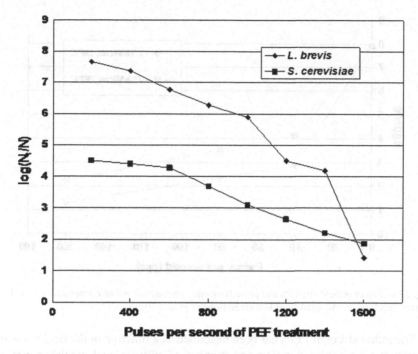

Fig. 13.3 Inactivation of *Lactobacillus brevis* and *Saccharomyces cerevisiae* by pulsed electric field (PEF) pasteurization of apple juice applying 1,200 pulses per second (pps) at 35 kV/cm (Adapted from Aguilar-Rosas et al. 2007)

where N_t is the microbial load after contact time t, N is the initial microbial load, and t_c, E_c, and k are the treatment time threshold value, the field intensity threshold value, and an independent constant factor determined for a specific microorganism, respectively.

If logarithms to base 10 on both sides of Eq. 13.2 are taken, the left-hand side of such an equation will represent the inactivation ratio or log reduction, which refers to 90% reduction in the initial microorganism's population. The transformed equation will also indicate that the inactivation ratio depends linearly on the applied field strength and logarithmically on the treatment time. Although the electric field strength will represent a more pronounced effect than the treatment time, both are important elements (see, e.g., Fig. 13.4).

Considering the kinetic equations, smaller values of E_c or t_c indicate greater susceptibility of the particular microorganism to PEF processing. The above first-order kinetics may be simplistic for most PEF applications. Tailing phenomena have been observed in microbial inactivation kinetics (Sensoy et al. 1997). A two-phase kinetic model may be required to adequately describe PEF inactivation kinetics. Numerous models have been developed to fit sigmoid patterns of inactivation (Van Gerwen and Zwietering 1998). Some other, more complicated inactivation kinetics models have been reported (Raso et al. 2000). The search for a most robust kinetic model for PEF inactivation of microorganisms is currently ongoing.

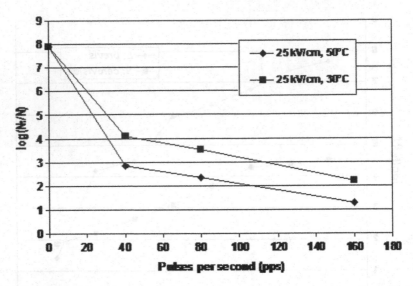

Fig. 13.4 Effect of field strength and pulse frequency on inactivation of *Escherichia coli* in PEF-treated peach juice (Adapted from Gutierrez-Becerra et al. 2002)

The lethal effect of PEFs has been described as a function of the field intensity, treatment time (pulse duration and number of pulses), and a model constant determined by the microorganism and its physiological status (Hülsheger et al. 1981). Sale and Hamilton (1967) demonstrated the non-thermal lethal effect of high-voltage electric fields on bacteria such as *Escherichia coli, Staphylococcus aureus, Micrococcus lysodeikticus, Sternbergia lutea, Bacillus subtilis, Bacillus cereus, Bacillus megaterium*, and *Clostridium welchii* as well as on yeasts such as *Saccharomyces cerevisiae* and *Candida utilis*. Similar results were reported by Mizuno and Hori (1988), Pothakamury et al. (1993), Sato et al. (1988), and Zhang et al. (1994). Different scientific papers and reviews (Knorr et al. 1994; Pothakamury et al. 1996; Raso et al. 2000) have dealt with the effect of PEFs on microorganisms. The inactivation of *Saccharomyces cerevisiae, Escherichia coli, Listeria monocytogenes, Listeria innocua, Staphylococcus aureus, Salmonella* spp., *Bacillus subtilis*, and *Bacillus cereus* has been extensively studied (Cserhalmi et al. 2006).

Microorganisms differ in their sensitivity to PEF exposure. It is generally accepted that yeasts are more sensitive to PEF treatment than vegetative bacteria. Also, Gram-negative bacteria are more susceptible to PEF treatment than Gram-positive bacteria. These trends were reported by Qin et al. (1995), who investigated the inactivation of *Escherichia coli* and *Saccharomyces cerevisiae* by PEF treatment. Yeasts cells are larger than bacterial cells, so they may exhibit a lower breakdown transmembrane potential and, therefore, they will be more sensitive to PEF processing. In terms of the resistance of Gram-positive bacteria, as compared with the sensitivity of Gram-negative bacteria, it can be argued that the more rigid and thicker cell wall of Gram-positive bacteria constitutes a protection against the lethality of PEFs (Aronsson 2002). Taking into account that PEF treatment inactivates microorganisms on the basis of their electromechanical instability, the lethality should differ not only for different species but also for different growth phases of each species. Hülsheger et al. (1981) reported that cells harvested from the logarithmic growth phase were more sensitive to PEFs than those from the stationary growth phase. It has also been reported that *Escherichia coli* cells in the logarithmic growth phase were more sensitive to PEFs than cells in the stationary phase (Pothakamury et al. 1996).

13.4 Effects on Enzymes

Information on the inactivation of enzymes by PEF treatment is very limited compared with that on the effects of PEF treatment on microorganisms. The results presented in the literature over the years are, also, quite contradictory. The sensitivity of enzymes to PEF treatment is very different from that of microorganisms. Process variables, medium characteristics, and the enzyme structure significantly influence the enzymic inactivation in different applications (Van Loey et al. 2002; Yeom et al. 2002). For example, the activity of α-amylase from *Bacillus licheniformins* decreased significantly after 30 exponential decay pulses using an intensity of 26 kV/cm in different applications. After PEF treatment, the remaining activity was only 15% (Ho et al. 1997). However, PEF treatment of alkaline phosphatase resulted in inconsistent results that ranged from 5% to 96% inactivation (Barbosa-Cánovas et al. 1998; Castro 1994; Ho et al. 1997). Contrastingly, Grahl and Märkl (1996) applying exponential decay pulses at 21.5 kV/cm did not detect any decrease in the activity of alkaline phosphatase in raw milk. Studies on enzyme inactivation of PEF-treated apple juice have also been reported. Apple juice was pasteurized by an ultrahigh temperature (UHT) treatment and the results were compared with those obtained using a PEF process (Sanchez-Vega et al. 2008). Enzyme inactivation and the physicochemical properties of the treated juices were compared, using a non-treated sample as the control. The UHT treatment was more efficient in enzyme inactivation, reducing by 95% the residual activity of polyphenoloxidase (PPO) at the maximum temperature and time. However a PEF treatment at 38.5 kV/cm and 300 pps combined with a temperature of 50°C achieved a 70% reduction of residual PPO activity (Fig. 13.5).

13.5 Effects on Nutritive and Sensory Attributes

13.5.1 Milk and Dairy Products

PEF treatment may influence the physical and chemical properties of products. The nature and extent of influence of PEF treatment on quality changes are still being actively discussed. In recent years, with the demand for high-quality milk and

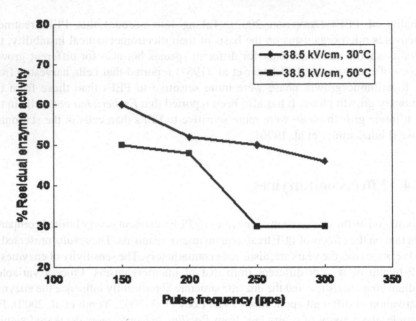

Fig. 13.5 Effects of PEF-treated apple juice on polyphenoloxidase inactivation caused by applying 300 pps at 38.5 kV/cm (Adapted from Sanchez-Vega et al. 2008)

milk products, more and more researchers have focused on studies of loss of sensory and physicochemical characteristics in milk and dairy products following treatment with PEFs. Barsotti et al. (2002) indicated that PEF treatment of model emulsions and liquid dairy cream may result in dispersal of oil droplets and dissociation of fat globule aggregates. Dunn (1996) reported that milk treated with PEFs suffered less flavor degradation than thermally processed milk. Dunn proposed the possibility of manufacturing dairy products such as cheese, butter, and ice cream using PEF-treated milk, although no detailed information was given in his report. Qin et al. (1995) conducted a study of shelf life, physicochemical properties, and sensory attributes of milk with 2% milk fat, treated with an electric field of 40 kV/cm and six to seven pulses. No physicochemical or sensory changes were observed after treatment in comparison with a sample treated by thermal pasteurization. Bendicho et al. (1999) studied the destruction of riboflavin, thiamine (water soluble), and tocopherol (liposoluble) in milk by treatment with PEFs. The vitamin concentrations before and after treatment were determined by high-performance liquid chromatography. They observed no destruction of vitamins by treatment with pulses. Michalac et al. (2003) studied the variation in color, pH, proteins, moisture, and particle size of UHT skim milk subjected to treatment with PEFs. They found no differences in the parameters studied before and after treatment. Shin et al. (2007) applied PEFs with square-wave pulses to whole milk inoculated with *Escherichia coli*, *Pseudomonas fluorescens*, and *Bacillus stearothermophilus*. The samples were exposed to an electric field of 30–60 kV/cm with 1-μs pulse width and 26–210 μs treatment time in a continuous PEF treatment system. Eight log reductions were obtained for *Escherichia coli* and *Pseudomonas fluorescens*, and 3 log reductions were obtained for *Bacillus stearothermophilus* with 210-μs treatment time and a pulse intensity of 60 kV/cm. There was no significant change in pH and titration acidity of milk samples after PEF treatment.

Sepúlveda-Ahumada et al. (2000) compared the textural properties and sensory attributes of Cheddar cheese made with heat-treated milk, PEF-treated milk, and untreated milk. In the hardness and springiness study, the cheeses made from milk pasteurized by any method were harder than those made from untreated milk. In the sensory evaluation, the panelists also found differences between the cheeses made from untreated milk and milk treated with PEFs or heat. Regardless of the differences, Sepúlveda-Ahumada et al. still considered using PEF-treated milk to obtain cheese as a feasible option to improve the product quality.

13.5.2 *Fruit and Vegetable Juices and Beverages*

PEF treatment of various liquid foods, including apple juice and orange juice, has not shown any significant physicochemical changes (Charles-Rodriguez et al. 2007; Jia et al. 1999). There was a slight decrease in vitamin C content in PEF-treated orange juice compared with heat-treated orange juice (Zhang et al. 1997). Gallardo-Reyes et al. (2008) conducted a comparative study of orange juice pasteurized by UHT treatment (processing at 110°C, 120°C, and 130°C for 2 and 4 s) and PEF treatment (20 and 25 kV/cm for 2 ms). They concluded that although there was no difference in the pH and soluble solids obtained with both treatments and freshly squeezed control samples, the color of the PEF-treated sample was closer to that of the control.

The flavor of orange is due to more than 200 chemical compounds (Maarse 1991), and is caused by hydrocarbons, aldehydes, esters, ketones, and alcohols. Limonene is the most important flavor compound in terms of quantity, although not in terms of quality (Sizer et al. 1988). It has been reported (Ahmed et al. 1978) that acetaldehyde, citral, ethyl butyrate, limonene, linalool, octanal, and α-pinene are the major contributors to orange juice flavor. The development of off-flavors in orange juice has been attributed to the degradation of limonene to α-terpineol and other compounds (Tatum et al. 1975). The effects of PEF processing on specific flavor compounds of orange juice have been studied (Jia et al. 1999; Yeom et al. 2000). It was found that 40% of decanal was lost by heat treatment at 90°C for 3 min, whereas no loss was observed by PEF treatment at 30 kV/cm, for either 240 or 480 μs. Octanal showed a loss of 9.9% for the heat treatment and 0% for either of the two PEF treatments. Some compounds suffered losses for the PEF treatments, but always in lower proportion than in the heat-pasteurized juice. For example, 5.1% and 9.7% of ethyl butyrate was lost with the 240-μs and 480-μs treatments, respectively, but 22.4% was lost in the thermal process. The loss of these volatile compounds in orange juice treated by PEFs was attributed to the vacuum degassing system of the PEF unit (Jia et al. 1999). The advantage of PEF compared with thermal processing was also observed in nutritive aspects. PEF-treated orange juice retained a significantly higher content of ascorbic acid than heat-pasteurized juice during storage at 4°C (Yeom et al. 2000). Although more research needs to be completed before PEF can be considered as the sole treatment to retain completely all flavor and color components of orange juice, it can be stated that PEF-treated juice retains more flavor and shows less browning than conventionally pasteurized juice. Under certain conditions, PEF-treated orange juice retains ascorbic acid better than heat-treated juice. All these findings are important and may prove invaluable for the adoption of PEF treatment as a real alternative for orange juice pasteurization.

The flavor components in apple juice are numerous, so flavor identification may be considered more complex than for orange juice owing to the aromatic nature of apples. Eight odor-active volatiles have, however, been identified as the most important contributors to the aroma-flavor authenticity of apple juice (Cunningham et al. 1986). Zarate-Rodriguez et al. (2000) compared PEF treatment and ultrafiltration for pasteurization of apple juice. No significant changes were observed in variables such as pH, sugar content (°Brix), and acidity, expressed as malic acid, for the PEF-treated juice and the ultrafiltered one. Color was the quality attribute that showed a change for membrane treatments. The trend observed was for the juice to become darker as a function of transmembrane pressure applied. Similarly to ultrafiltration treatments, relative color changes were observed but the effect registered was the opposite, i.e., the treated juices became paler as a function of the applied field strength. Color changes were independent of the number of pulses but were dependent on the field strength. The different color ratio perception in ultrafiltration-treated and PEF-treated juices could be due to haze formation, which may be caused by tannins, proteins, and carbohydrate polysaccharides. The direct effects of PEFs on volatiles of apple juice, and comparison with a conventional thermal treatment, have also been investigated (Aguilar-Rosas et al. 2007). PEF treatment and high-temperature, short-time (HTST) treatment were tested to determine the decrease in concentration of eight volatiles responsible for odor. In general terms, PEF treatment resulted in better retention of most of the volatile compounds responsible for the color and flavor of the treated apple juice. Also, important biochemical substances in apple juice, such as phenol compounds, were better retained by PEF treatment than by HTST treatment.

The effect of a PEF treatment applied in a continuous system on the physical and chemical properties of freshly squeezed citrus juices (grapefruit, lemon, orange, tangerine), was studied by Cserhalmi et al. (2006). The aim of the work was to investigate the effect of PEF technology on pH, sugar content (°Brix), electric conductivity, viscosity, non-enzymic browning index, hydroxymethylfurfural (HMF) content, color, organic acid content, and volatile flavor compounds of the citrus juices mentioned. The juices were treated with 50 pulses at 28 kV/cm. The treatment temperature was below 34°C. The pH, sugar content (°Brix), electric conductivity, viscosity, non-enzymic browning index, and HMF content of citrus juices practically did not change. The color also remained unchanged for all samples. In all cases, the absorption spectra of treated and untreated samples were similar to each other. There was no significant change in organic acid content of the juices. The volatile flavor compounds of the treated juices were essentially the same as those present in unprocessed juice.

The influence of different PEF intensities and conventional HTST treatment on quality characteristics [pH, sugar content (°Brix), total acidity, turbidity, HMF content, color, microbial flora, pectin methylesterase (PME) activity, and sensory analysis] of blended orange and carrot juice was investigated (Rivas et al. 2006). HMF content and color parameters did not vary with any of the treatments. Total acidity and turbidity were slightly higher after HTST treatment. The sensory characteristics of the PEF-treated juice were more similar to the those of the untreated juice than to those of the HTST-pasteurized juice. Heat pasteurization was more efficient in inactivating microbial flora and PME and preventing the growth of microbial flora and reactivation of PME at 2°C and 12°C for 10 weeks.

Torregrosa et al. (2006) reported higher vitamin A contents of orange–carrot juices, compared with untreated control samples, when the product was treated at 25 kV/cm. The effects of PEFs on oxidative enzymes and the color of fresh carrot juice were studied by Quitão-Teixeira et al. (2008). A response surface methodology was used to evaluate the effect of pulse polarity (monopolar or bipolar mode), pulse width (from 1 to 7 μs), and pulse frequency (from 50 to 250 Hz) on the color and peroxidase inactivation of PEF-treated carrot juice. The total treatment time and the electric field strength were set at 1,000 μs and 35 kV/cm, respectively, at a temperature below 35°C. The physicochemical characteristics of carrot juice were measured. There was a linear relationship between the electrical conductivity and the temperature of the carrot juice. The results showed that PEF-treated carrot juice (35 kV/cm for 1,000 μs applying a 6-μs pulse width at 200 Hz in bipolar mode) had 73.0% inactivation of peroxidase. The color coordinates did not change significantly.

A study of the effects of PEF treatments on the bioactive components (polyphenols, catechins, and free amino acids), the color, and the flavor of green tea infusions has been reported (Zhao et al. 2009). PEF treatment efficiently retained polyphenols, catechins, and the original color of green tea infusions with an electric field strength from 20 to 40 kV/cm for 200 μs. PEF treatments also caused a significant increase in the total free amino acids of green tea infusions. The total free amino acids increased by 7.5% after PEF treatment at 40 kV/cm. The increase in total amino acids induced by PEF treatment, especially for threonine, is beneficial for the quality of commercial ready-to-drink green tea infusion products. There was no significant effect of PEF treatment at 20 or 30 kV/cm on flavor compounds of green tea infusions. However, PEF treatment caused losses of volatiles in green tea infusions to different extents when the PEF dose was higher than a critical level. The total concentration of volatiles lost was approximately 10% after PEF treatment at 40 kV/cm for 200 μs.

Sampedro et al. (2009) presented the results of a project aimed at comparing thermal treatment, PEF treatment, and HPP on PME activity and the concentration of volatile compounds in an orange juice–milk beverage. Thermal treatment (85°C, 1 min), PEF treatment (25 kV/cm, 65°C), or HPP (650 MPa, 50°C) was needed to inactivate 90% of PME. Twelve volatile compounds were extracted and selected for quantification by instruments following the application of the different treatments. The average loss in concentration of volatile compounds was between 16.0% and 43.0% after thermal treatment. After PEF treatment, the average loss was between −13.7% and 8.3% at 25°C, between 5.8% and 21.0% at 45°C, and between 11.6% and 30.5% at 65°C. After HPP, the average loss was between −14.2% and 7.5% at 30°C and between 22.9% and 42.3% at 50°C. The results showed the potential of the non-thermal technologies in providing food with a higher standard of quality compared with thermal processing.

The degradation of ascorbic acid was determined in a ready-to-drink orange juice–milk beverage treated by PEFs (Zulueta et al. 2010). The effects of PEF treatment were compared with those of heat pasteurization (90°C, 20 s). Four electric field strengths (15, 25, 35, and 40 kV/cm) and six treatment times (from 40 to 700 μs) for each field were studied. Ascorbic acid degradation was adjusted to an exponential model. For the shelf-life study a 25 kV/cm treatment for 280 μs was applied and the beverages were stored at 4°C and 10°C. The ascorbic acid degradation rate during storage was adjusted to zero-order kinetics, showing that beverages stored at 4°C had better ascorbic acid retention than beverages stored at 10°C. No significant differences were found between heat pasteurization and PEF treatments during storage.

13.5.3 Solid Food Materials

There have been attempts to improve juice extraction from plants and fruit products using PEFs (Bazhal and Vorobiev 2000; Schilling et al. 2007). PEFs may enhance the extraction of higher amounts of valuable compounds into the extraction juice, resulting in high quality. Apple juice contains several phenolic compounds, including chlorogenic acid, catechins, procyanidins, quercetin glycosides, and phloridzin. Schilling et al. (2007) monitored the phenolic compositions of juices obtained from PEF-treated mash and untreated samples. The PEF treatments were at field intensities of 1, 3, and 5 kV/cm. There was no significant difference between treated and untreated samples. However, enzymic maceration of the apple mash resulted in a marked increase in the content of quercetin glycosides. Schilling et al. also reported that there was no difference in the antioxidative capacity of the apples juices. It was then postulated that the fact that the antioxidative potential was not affected indicates that radical formation had not taken place during PEF treatment.

Table 13.1 shows the influence of PEF treatment on the quality of juice expressed from apple and sugar beet. The data in Table 13.1 were obtained from Lazarenko et al. (1977), Bazhal and Vorobiev (2000), and Bazhal (2001). Bazhal and Vorobiev (2000) reported changes in the color of juices treated at different

Table 13.1 Qualitative characteristics of apple juice from control and pulsed electric field (*PEF*)-treated samples

Variable	Apple juice		Sugar beet juice	
	PEF	Control	PEF	Control
Density (kg/m^3)	1059.4	1057.7	1.0392	1.038
Sugar content (°Brix)	13.8	13.1	9.8	9.5
pH	3.91	3.84	5.6	5.65
Pectin (mg/L)	290	517		
Kinematic viscosity (10^{-6} m^2/s)	5.917	6.747	2.284	2.595
Absorbance (wavelength of 520 nm)				
Filtered	0.02	0.39		
Non-filtered	0.03	1.18		
Transmittance	0.67	0.33		

intensities. The significant change in juice color may be attributed to the inhibition of PPO by electric fields (Giner et al. 2001). Lazarenko et al. (1977) suggested that an electric field can break the chains of pectin molecules, resulting in deceased pectin concentrations and thus reducing the kinematic viscosity of the extracted juices. Lower viscosity improves juice filtration. The reduction in the transmittance of juice after electroplasmolysis indicates a reduction in suspended particle contents because of improved tissue filtration properties resulting from the additional pores formed in the cell walls after PEF treatment. PEF treatment of sugar beet resulted in increased purity of the extracted juice compared with that obtained by traditional processing by thermal plasmolysis. Despite lower temperature leaching for PEF-treated samples, the sucrose loss in pulp decreased from 0.62% (thermally treated beet) to 0.57% (Knorr et al. 2001). It is uncertain if all necessary chemical analyses have been performed to fully ascertain the effect of PEF treatment on the quality of processed foods. However, the clear consensus is that liquid food products generally retain their fresh-like quality after PEF treatment.

Solid food products undergo significant changes when treated with PEFs. Changes in the electrical conductivity of the treated vegetable samples indicated increasing cell permeability (Lebovka et al. 2000, 2001). Table 13.2 shows that the diffusion coefficient of sugar from beetroot increases from 0.68×10^{-9} to 1.2×10^{-9} m^2/s after PEF treatment (Jemai 1997; Gulyi et al. 1994). The elastic modulus of sugar beet decreased after PEF treatment (Bazhal 2001). The microstructure of salmon and chicken changed considerably because of PEF treatment as the muscle cells decreased in size and gaping occurred (Gudmundsson and Hafsteinsson 2001). Electric field treatment generally affects biological cell membranes, whereas heating destroys the cell walls (Calderón-Miranda et al. 1999). There is the potential of inducing rheological changes in a product as a result of PEF treatment. This phenomenon depends on the type of product involved and requires detailed investigations.

Table 13.2 Some properties of vegetable tissues estimated for untreated (control) and PEF-treated samples

Property	Value of property		Material	Operation	Reference
	Control	PEF treatment			
Electrical conductivity (S/m)	0.003–0.007	0.035–0.070	Apple	Conductivity measurement	Lebovka et al. (2001)
	0.03	0.41	Carrot	Conductivity measurement	Rastogi et al. (1999)
	0.06	0.53	Potato	Conductivity measurement	Knorr and Angersbach (1998)
Porosity (%)	67	75	Apple	Using of penetrometer	Bazhal et al. (2003)
Water diffusion coefficient (m^2/s)	0.98×10^{-9}	1.55×10^{-9}	Carrot	Osmotic dehydration	Rastogi et al. (1999)
Sugar diffusion coefficient (m^2/s)	0.68×10^{-9}	1.2×10^{-9}	Sugar beet	Leaching	Gulyi et al. (1994)
Mass transfer coefficient (kg/m^2 s)	0.043	0.058	Paprika	Drying	Ade-Omowaye et al. (2002)
Constant drying rate (kg/m^2 s)	9.68×10^{-4}	13.02×10^{-4}	Paprika	Drying	Ade-Omowaye et al. (2002)
Heat transfer coefficient (W/m^2 s)	73.13	98.36	Paprika	Drying	Ade-Omowaye et al. (2002)
Elastic modulus (MPa)	12.5	6.5	Sugar beet	Compression test	Matvienko (1996)
	1.53	0.32	Apple	Compression test	Bazhal et al. (2003)
Failure stress (MPa)	1.26	0.53	Apple	Compression test	Bazhal et al. (2003)

13.6   Combination with Other Preservation Techniques

PEF technology is mainly applied for processing of liquid foods. The physicochemical characteristics of the medium, such as temperature, electrical conductivity, ionic strength and pH, are all influential for the efficiency of the technology. An important aspect, possibly reported first by Hülsheger et al. (1981), is that of the synergistic effect of the medium temperature with PEF treatment on the inactivation ratio. An increase in the medium temperature decreases the breakdown transmembrane potential of the cell membrane, boosting the combined thermal–electrical effect on the inactivation ratio (Pothakamury et al. 1996). The key issue is to maintain the medium temperature within low ranges in order to avoid thermal degradation of the product.

Application of the PEF method is restricted to food products that can withstand high electric fields. The dielectric property of a given food product is closely related to its physical structure and chemical composition. Homogeneous liquids with low electrical conductivity provide ideal conditions for continuous treatment with the PEF method. Jayaram et al. (1993) reported reduction in the resistance of the treatment chamber with an increase in the fluid medium. Lowering the conductivity of the medium possibly increased the ionic concentration difference between the cytoplasm and the medium, which could cause a drain on cell energy reserves, weakening the membrane structure.

The microbial inactivation ratio on PEF treatment is highly influenced by ionic strength and pH. The inactivation ratio is normally reduced in higher ionic strength solutions. In this case, as the pH is reduced from neutral, the inactivation ratio is increased. Vega-Mercado et al. (1996) studied the effects of the pH and ionic strength of a liquid medium in PEF treatments. They reported that that ionic strength was responsible for poration and compression of the cell, whereas the pH affected the cytoplasm when the poration was completed. According to Vega-Mercado et al. (1996), all these factors disturbed the homeostasis of the microorganisms, thereby leading to an increase in the inactivation ratio.

Evrendilek et al. (2001) studied a yogurt drink prepared using combined PEF treatment (30 kV/cm for 32 μs) and heat treatment (60°C for 32 s). They reported no significant differences between the control sample and the treated samples in terms of color, soluble solids, and pH. However, when milk was subjected to long-duration pulses (Perez and Pilosof 2004), or high-intensity electric fields (45–55 kV/cm) as described by Floury et al. (2005), the structure of milk protein was apparently modified.

Charles-Rodriguez et al. (2007) reported a direct increase in pH as a function of temperature in HTST-pasteurized apple juice when comparing it with PEF-treated juice, whose pH did not vary. They believed that this trend could be explained in terms of the evaporative effect of organic acids caused by a temperature increase. This variation of pH with temperature could have an effect on the shelf life, since a higher pH could increase the possibility of growth of yeasts or enzyme activity.

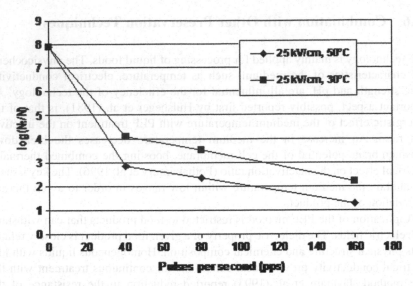

Fig. 13.6 Effects of PEF-treated orange juice on inactivation of *Escherichia coli* (Adapted from Gallardo-Reyes et al. 2008)

A synergistic effect was observed when pasteurizing orange juice by PEF treatment coupled with a mild increase in temperature (Gallardo-Reyes et al. 2008). These authors reported 6 log reductions of *Escherichia coli* in PEF-pasteurized orange juice when applying an electric field of 25 kV/cm and 160 pps, enhanced by exposure to 50°C. The inactivation was improved with this temperature when compared with use of a lower (30°C) temperature (Fig. 13.6).

The hurdle approach has also been used to combine factors, apart from heat and electric voltage, in microbial inactivation of PEF-treated orange juice. The effect of thermosonication and PEF treatment on inactivation of *Staphylococcus aureus* and selected quality aspects in orange juice was investigated (Walkling-Ribeiro et al. 2009). Thermosonication (10 min at 55°C) applied in combination with PEF treatment (40 kV/cm for 150 μs) resulted in inactivation of *Staphylococcus aureus* comparable to that achieved by conventional HTST treatment. Hodgins et al. (2002) investigated variations in temperature, acidity, and the number of pulses to maximize microbial inactivation in orange juice. The effect of PEF treatment combined with the addition of nisin, lysozyme, or a combination of both to orange juice was also investigated. Combinations of different hurdles, including moderately high temperatures (below 60°C) and antimicrobial compounds to reduce *Salmonella* spp. in pasteurized and freshly squeezed orange juices have been explored (Liang et al. 2002). The combination of different obstacles such as moderately high temperatures (below 50°C), antimicrobial compounds and PEF treatment to reduce the levels of naturally occurring microbes in red and white grape juices was explored (Wu et al. 2005). The microbial count decreased with the increase in electric field and temperature treatment when a constant number of pulses were applied. In red grape juice, up to 5.9 log reductions in counts were achieved by applying 20 pulses of 65 kV/cm (peak to peak) at 50°C with 2 h incubation of the juice containing lysozyme and nisin (0.4 g/100 mL).

13.7 Equipment: Design and Development

In general, a PEF system consists of a high-voltage power source, an energy storage capacitor bank, a charging current limiting resistor, a switch to discharge energy from the capacitor across the food, and a treatment chamber. The bank of capacitors is charged by a direct current power source obtained from an amplified and rectified regular alternating current main source. An electrical switch is used to discharge energy (instantaneously in a millionth of a second) stored in the capacitor storage bank across the food held in the treatment chamber. Apart from those major components, some adjunct parts are also necessary. In the case of continuous systems, a pump is used to convey the food through the treatment chamber. A chamber cooling system may be used to decrease the ohmic heating effect and control the food temperature during treatment. High-voltage and high-current probes are used to measure the voltage and current delivered to the chamber (Barbosa-Cánovas et al. 1999; Floury et al. 2005; Amiali et al. 2004, 2006). Figure 13.7 shows a basic PEF treatment unit, and Fig. 13.8 presents different chamber designs.

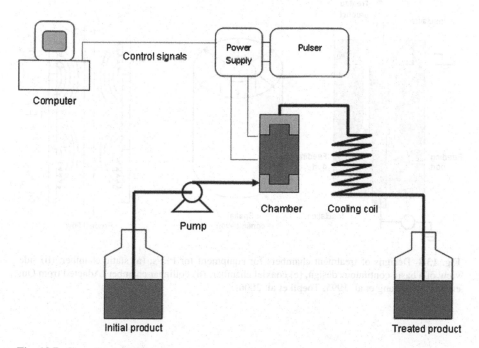

Fig. 13.7 Operation of PEFs

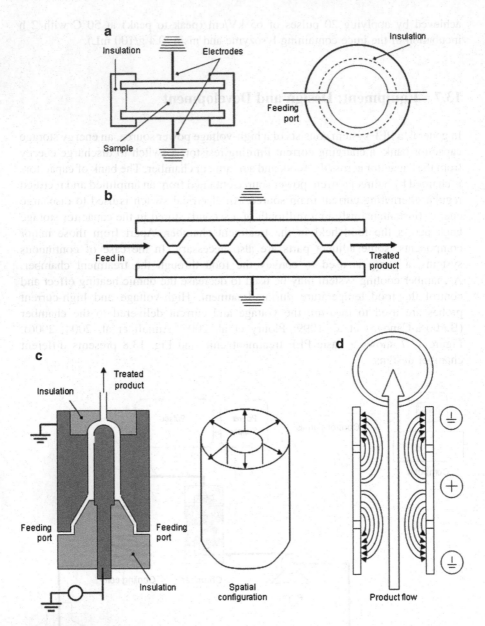

Fig. 13.8 Designs of treatment chambers for equipment for PEFs: (**a**) static chamber, (**b**) side view of a basic continuous design, (**c**) coaxial chamber, (**d**) collinear chamber (Adapted from Qin et al. 1995; Zhang et al. 1995; Toepfl et al. 2006)

The type of electric field waveform applied is one of the important descriptive characteristics of a PEF treatment system. Exponentially decaying and square waves are among the most common waveforms used. To generate an exponentially decaying voltage wave, a direct current power supply charges the bank of capacitors, which are connected in series with a charging resistor. When a trigger signal is applied, the charge stored in the capacitor flows through the food in the treatment chamber. Exponential waveforms are easier to generate from the generator point of view. Generation of square waveforms generally requires a pulse-forming network consisting of an array of capacitors and inductors. It is more challenging to design a square waveform system than an exponential waveform system. However, square waveforms may be more lethal and energy efficient than exponentially decaying pulses (Amiali et al. 2006; Evrendilek and Zhang 2005; Zhang et al. 1995). To produce an effective square waveform using a pulse-forming network, the resistance of the food must be matched with the impedance of the pulse-forming network. Therefore, it is important to determine the resistance of the food in order to treat it properly.

The discharging switch also plays a critical role in the efficiency of the PEF system. The type of switch used will determine how fast it can perform and how much current and voltage it can withstand. In increasing order of service life, suitable switches for PEF systems include ignitrons, spark gaps, trigatrons, thyratrons, and semiconductors. Solid-state semiconductor switches are considered as the future of high-power switching. They have better performance and are easier to handle, require fewer components, allow faster switching times, and are more economically sound (Góngora-Nieto et al. 2002).

13.8 Applications

As previously mentioned, PEF technology has been found suitable for pasteurization of liquid foods, such as milk and dairy products, as well as fruit juices and beverages. Some other applications include increase of juice yield, processing of vegetable raw materials, winemaking, sugar production, and microstructural modification of vegetable, fish, and meat tissues. Application of the PEF method is restricted to food products that can withstand high electric fields. The dielectric property of a given food product is closely related to its physical structure and chemical composition. Homogeneous liquids with low electrical conductivity provide ideal conditions for continuous treatment with the PEF method. Non-fluid foods can also be processed by PEF treatment in batch mode operation as long as dielectric breakdown in the foods is prevented. Air bubbles in the food must be removed when using this method because the bubbles cannot withstand high electric fields. In general, PEF treatment is not suitable for most solid food products containing air bubbles when placed in the treatment chamber. Another limitation of the PEF method is the particulate size in the liquid food in both static and flow treatment modes. The maximum particle size in the liquid must be smaller than the gap of the treatment region in the chamber in order to maintain proper processing operation.

Most of the earlier studies with PEF treatment were conducted using only small sample amounts in static chambers. To continue the development of non-thermal pasteurization of foods it was necessary to design treatment systems that closely resembled commercial applications, both in treatment capacity and in treatment effectiveness. The treatment chamber is considered to be the crucial component of a PEF system since it is directly related to capacity. The first step in trying to take PEF processing to an industrial, commercial scale was the testing and development of different chamber configurations. Several designs of treatment chambers have been tested over the years, and Qin et al. (1995) presented an extensive review of the progress in chamber design for PEF utilization.

In spite of all the research and development, most of the PEF applications lie at the experimental level and in the pilot plant range, with limited commercialization yet. A large portion of work on PEF treatment have been focused on reducing the microbial load in liquid or semisolid foods in order to extend their shelf life and ensure their safety. The products that have been mostly studied include milk (Dunn and Pearlman 1987; Dutreux et al. 2000; Evrendilek and Zhang 2005; Fleischman et al. 2004; Grahl and Märkl 1996; Reina et al. 1998; Sensoy et al. 1997), apple juice (Ortega-Rivas et al. 1998; Vega-Mercado et al. 1997; Zarate-Rodriguez et al. 2000), orange juice (Jia et al. 1999; Yeom et al. 2000; Zhang et al. 1997), and liquid egg (Amiali et al. 2004, 2006; Hermawan et al. 2004; Jeantet et al. 1999, 2004). These studies and others have reported successful PEF-induced inactivation of pathogenic and food spoilage microorganisms, including *Listeria innocua* (Sepúlveda-Ahumada et al. 2005), *Staphylococcus aureus* (Sobrino-Lopez et al. 2006), and *Escherichia coli* (Charles-Rodriguez et al. 2007) as well as selected enzymes such as PPO (Giner et al. 2002), PME (Espachs-Barroso et al. 2006), and peroxidase (Elez-Martínez et al. 2006), resulting in better retention of flavors and nutrients and fresher taste compared with heat-pasteurized products. Reviews on the effects of PEFs on enzyme activity (Van Loey et al. 2002) and combined enzymic and microbial effects (Espachs-Barroso et al. 2003) are available in the literature.

As compared with many other technologies in any field, the development of PEF treatment as a feasible and viable industrial process has faced troubles and hurdles, but it was commercialized briefly (Clark 2006). PEF treatment has gone through all stages of an adaptable industrial process. In the 1960s, research into PEF treatment started with fundamental studies to derive a theoretical basis to understand it. The pioneering experiments were performed in a batch manner and with small samples. In the years following those initial efforts, researchers studied the effects of PEF treatment on different target microorganisms to prove its microbial inactivation efficiency. Once this aspect had been thoroughly demonstrated, investigations were devoted to the quality of processed products, both sensory and nutritive. After that stage, and approaching the turn of the century, efforts were focused on adapting the many designs of chambers and devices into modular-type units of equipment to provide laboratory-scale, pilot-scale, and industrial-scale machinery. It can be seen, considering the many review and research papers referred to in this chapter, that research activity is taking place aimed at different aspects, such as further understanding of enzyme-related effects and the influence of PEF treatment on the structure and composition of biochemical components responsible for sensory and nutritive attributes of processed foods.

In summary, PEF treatment can be considered a real alternative for pasteurization of liquid foods. However, the only documented commercial application of PEF technology in fruit juice pasteurization is that mentioned earlier (Clark 2006). The Genesis Juice Corporation, a maker of premium refrigerated fruit juices, installed and used for a limited time in 2005–2006 a 200 L/h PEF processing unit in its Eugene (Oregon, USA) plant. Genesis juice was briefly commercialized and the company reported that the main motivation of using PEF treatment was the avoidance of loss of flavor associated with conventional heat pasteurization methods. The shelf life of Genesis juices processed using PEF treatments was 4 weeks. Genesis juices complied with FDA regulations and the case was presented at professional events as a successful experience (Serrano 2006; Toepfl et al. 2006). The prices of Genesis PEF-pasteurized fruit juices were higher than those of conventional processed fruit juices, but the company promoted them as being from organic fruit and having fresher flavor than traditionally processed juices.

References

Ade-Omowaye BIO, Angersbach A, Eshtiaghi NM, Knorr D (2000) Impact of high intensity electric field pulses on cell permeabilisation and as pre-processing step in coconut processing. Innov Food Sci Emerg Technol 1: 203–209.

Ade-Omowaye BIO, Rastogi NK, Angersbach A, Knorr D (2002) Osmotic dehydration of bell peppers: influence of high intensity electric field pulses and elevated temperature treatment. J Food Eng 54: 35–43.

Aguilar-Rosas SF, Ballinas-Casarrubias M, Nevarez-Moorillon GV, Martin-Belloso O, Ortega-Rivas E (2007) Thermal and pulsed electric fields pasteurization of apple juice: effects on physicochemical properties and flavour compounds. J Food Eng 83: 41–46.

Aguilera JM, Stanley DW, Baker KW (2000) New dimensions in microstructure of food products. Trends Food Sci Technol 11: 3–9.

Ahmed EM, Dennison RA, Shaw PE (1978) Effects of selected oil and essence volatile components on flavor quality of pumpout orange juice. J Agric Food Chem 26: 368–372.

Amiali M, Ngadi MO, Raghavan GSV, Smith JP (2006) Inactivation of *Escherichia coli* O157:H7 and *Salmonella enteritidis* in liquid egg white using pulsed electric field. J Food Sci 71: 88–94.

Amiali M, Ngadi MO, Smith JP, Raghavan GSV (2004) Inactivation of *Escherichia coli* O157:H7 in liquid dialyzed egg using pulsed electric fields. Food Bioprod Process 82C: 151–156.

Angersbach A, Heinz V, Knorr D (2000) Effects of pulsed electric fields on cell membranes in real food systems. Innov Food Sci Emerg Technol 1: 135–149.

Aronsson K (2002). Inactivation, cell injury and growth of microorganisms exposed to pulsed electric fields using continuous process. PhD thesis. Chalmers University of Technology, Göteborg.

Barbosa-Cánovas GV, Góngora-Nieto MM, Pothakamury UR, Swanson BG (1999) Preservation of Foods with Pulsed Electric Fields. Academic Press, San Diego CA.

Barbosa-Cánovas GV, Pothakamury UR, Palou E, Swanson BG (1998). Nonthermal Preservation of Foods. Marcel Dekker Inc, New York.

Barsotti L, Dumay E, Mu TH, Diaz MDF, Cheftel JC (2002) Effects of high voltage electric pulses on protein-based food constituents and structures. Trends Food Sci Technol 12: 136–144.

Bazhal MI (2001) Etude du mécanisme d'électroperméabilisation des tissus végétaux. Application à l'extraction du jus des pommes. Doctoral thesis, Université de Technologie de Compiègne, Compiègne.

Bazhal MI, Ngadi MO, Raghavan GSV, Nguyen DH (2003) Textural changes in apple tissue during pulsed electric field treatment. J Food Sci 68: 249–253.

Bazhal MI, Vorobiev E (2000) Electrical treatment of apple cossettes for intensifying juice pressing. J Sci Food Agric 80: 1668–1674.

Bendicho S, Espachs A, Arantegui J, Martin O (1999) Effect of high intensity pulsed electric fields and heat treatments on vitamins of milk. J Dairy Res 69: 113–123.

Calderón-Miranda ML, Barbosa-Cánovas GV, Swanson BG (1999) Transmission electron microscopy of *Listeria innocua* treated by pulsed electric fields and nisin skimmed milk. Int J Food Microbiol 51: 31–38.

Castro AJ (1994). Pulsed electric field modification of activity and denaturation of alkaline phosphatase. PhD thesis. Washington State University, Pullman. WA.

Castro AJ, Barbosa-Cánovas GV, Swanson BG (1993) Microbial inactivation of foods by pulsed electric fields. J Food Process Preserv 17: 47–73.

Charles-Rodríguez AV, Nevárez-Moorillón GV, Zhang QH, Ortega-Rivas E (2007) Comparison of thermal processing and pulsed electric fields treatment in pasteurization of apple juice. Food Bioprod Process 85C: 93–97.

Clark JP (2006) Pulsed electric field processing. Food Technol 60: 66–67.

Coster HGL, Zimmermann U (1975) The mechanism of electrical breakdown in the membranes of *Valonia utricularis*. J Membr Biol 22: 73–90.

Cserhalmi Zc, Sass-Kiss A, Tóth-Marks M, Lechner N (2006) Study of pulsed electric field treated citrus juices. Innov Food Sci Emerg Technol 7: 49–54.

Cunningham DG, Acree TE, Barnard J, Butts RM, Braell PA (1986) Charm analysis of apple volatiles. Food Chem 19: 137–147.

Dunn J (1996) Pulsed light and pulsed electric field for foods and eggs. Poult Sci 75: 1133–1136.

Dunn JE, Pearlman JS (1987) Methods and apparatus for extending the shelf life of fluid food products. US Patent 4,695,472.

Dutreux N, Notermans WT, Gongóra-Nieto MM, Barbosa-Canovas GV, Swanson BG (2000) Pulsed electric fields inactivation of attached and free-living *Escherichia coli* and *Listeria innocua* under several conditions. Int J Food Microbiol 54: 91–98.

Elez-Martínez P, Aguiló-Aguayo I, Martín-Belloso O (2006) Inactivation of orange juice peroxidase by high-intensity pulsed electric fields as influenced by process parameters. J Sci Food Agric 86: 71–81.

Espachs-Barroso A, Barbosa-Cánovas GV, Martín-Belloso O (2003) Microbial and enzymatic changes in fruit juice induced by high-intensity pulsed electric fields. Food Rev Int 19: 253–273.

Espachs-Barroso A, Van Loey A, Hendrickx M, Martín-Belloso O (2006) Inactivation of plant methylesterase by thermal or high intensity electric field treatments. Innov Food Sci Emerg Technol 7: 40–48.

Evrendilek GA, Dantzer WR, Streaker CB, Ratanatriwong P, Zhang QH (2001) Shelf-life evaluations of liquid foods treated by pilot plant pulsed electric field system. J Food Process Preserv 25: 283–297.

Evrendilek GA, Zhang QH (2005) Effects of pulse polarity and pulse delaying time on pulsed electric fields-induced pasteurization of *E. coli* O157:H7. J Food Eng 68: 271–276.

Fleischman GJ, Ravishankar S, Balasubramaniam VM (2004) The inactivation of *Listeria monocytogenes* by pulsed electric field (PEF) treatment in static chamber. Food Microbiol 21: 91–95.

Floury J, Grosset N, Leconte N, Pasco M, Madec M, Jeantet R (2005) Continuous raw skim milk processing by pulsed electric field at non-lethal temperature: effect on microbial inactivation and functional properties. Lait 86: 43–57.

Gallardo-Reyes ED, Valdez-Fragoso A, Nevarez-Moorillon GV, Ngadi MO, Ortega-Rivas E (2008) Comparative quality of orange juice as treated by pulsed electric fields and ultra high temperature. AgroFood Ind Hi-Tech 19: 35–36.

Giner J, Gimeno V, Barbosa-Cánovas GV, Martín O (2001) Effects of pulsed electric field processing on apple and pear polyphenoloxidases. Food Sci Technol Int 7: 339–345.

Giner J, Ortega M, Mesegué M, Jimeno V, Barbosa-Cánovas GV, Martín O (2002) Inactivation of peach polyphenoloxidase by exposure to pulsed electric fields. J Food Sci 67: 1467–1472.

Góngora-Nieto MM, Sepúlveda-Ahumada DR, Pedrow P, Barbosa-Cánovas GV, Swanson BG (2002) Food Processing by Pulsed Electric Fields: Treatment Delivery, Inactivation Level and Regulatory Aspects. LWT-Food Sci Technol 35: 375–388.

Grahl T, Märkl H (1996) Killing of microorganisms by pulsed electric fields. Appl Microbiol Biotechnol 45: 148–157.

Gudmundsson M, Hafsteinsson H (2001) Effect of electric field pulses on microstructure of muscle foods and roes. Trends Food Sci Technol 12: 122–128.

Gulyi IS, Lebovka NI, Mank VV, Kupchik MP, Bazhal MI, Matvienko AB, Papchenko AY (1994) Scientific and Practical Principles of Electrical Treatment of Food Products and Materials. UkrINTEI, Kiev.

Gutierrez-Becerra LE, Li S, Ortega-Rivas E, Zhang QH (2002) Cold pasteurization of peach nectar using pulsed electric fields. IFT Annual Meeting, Anaheim, CA, June 2002.

Hermawan GA, Evrendilek WR, Zhang QH, Richter ER (2004) Pulsed electric field treatment of liquid whole egg inoculated with Salmonella enteritidis. J Food Saf 24: 1–85.

Ho SY, Mittal GS, Cross JD (1997) Effects of high electric pulses on the activity of selected enzymes. J Food Eng 31: 69–84.

Hodgins AM, Mittal GS, Griffiths MW (2002) Pasteurization of fresh orange juice using low-energy pulsed electric field. J Food Sci 67: 2294–2299.

Hülsheger H, Potel J, Niemann EG (1981) Killing of bacteria with electric pulses of high field strength. Radiat Environ Biophys 20: 53–65.

Jayaram S, Castle GSP, Margaritis A (1993) The effects of high field DC pulse and liquid medium conductivity on survivability of Lactobacillus brevis. Appl Microbiol Biotechnol 40: 119–122.

Jeantet R, Baron F, Nau F, Roignant M, Brulé G (1999) High intensity pulsed electric fields applied to egg white: Effect on Salmonella enteritidis inactivation and protein denaturation. J Food Prot 62: 1381–1386.

Jeantet R, Mc Keag JR, Fernández JC, Gosset N, Baron F, Korolczuk J (2004) Pulsed electric field continuous treatment of egg products. Sci Aliments 24: 137–158.

Jemai AB (1997) Contribution a l'étude de l'effet d'un traitement électrique sur les cossettes de betterave a sucre. Incidence sur le procède d'extraction. Doctoral thesis, Université de Technologie de Compiègne, Compiègne.

Jia M, Zhang QH, Min DB (1999) Pulsed electric field processing effects on flavor compounds and microorganisms of orange juice. Food Chem 65: 445–451.

Kinosita K Jr, Tsong TY (1977) Voltage-induced pore formation and hemolysis of human erythrocytes. Biochim Biophys Acta 471: 227–242.

Knorr D, Angersbach A (1998) Impact of high intensity electric field pulses on plant membrane permeabilization. Trends Food Sci Technol 9: 185–191.

Knorr D, Angersbach A, Eshtiaghi MN, Heinz V, Lee DU (2001) Processing concepts based on high intensity electric field pulses. Trends Food Sci Technol 12: 129–135.

Knorr D, Geulen M, Grahl T, Sitzmann W (1994) Food application of high electric Field pulses. Trends Food Sci Technol 5: 71–75.

Lazarenko BR, Fursov SP, Scheglov YA, Bordiyan VV, Chebanu VG (1977) Electroplasmolysis. Karta Moldovenaske, Kishinev.

Lebovka NI, Bazhal MI, Vorobiev E (2000) Simulation and experimental investigation of food material breakage using pulsed electric field treatment. J Food Eng 44: 213–223.

Lebovka NI, Bazhal MI, Vorobiev E (2001) Pulsed electric field breakage of cellular tissues: visualization of percolative properties. Innov Food Sci Emerg Technol 2: 113–125.

Liang Z, Mittal GS, Griffiths MW (2002) Inactivation of *Salmonella typhimurium* in orange juice containing antimicrobial agents by pulsed electric field. J Food Prot 65: 1623–1627.

Maarse H (1991) Volatile Compounds in Foods and Beverages. Marcel Dekker, New York.

Matvienko AB (1996) Intensification of the extraction of soluble substances by electrical treatment of plant materials and water. PhD thesis, Ukrainian State University of Food Technologies, Kiev.

McLellan MR, Kime RL, Lind LR (1991) Electroplasmolysis and other treatments to improve apple juice yield. J Sci Food Agric 57: 303–306.

Michalac SL, Alvarez VB, Zhang QH (2003) Inactivation of selected microorganisms and properties of pulsed electric field processed milk. J Food Process Preserv 27: 137–151.

Mizuno A, Hori Y (1988) Destruction of living cells by pulsed high-voltage application. IEEE Trans Ind Appl 24: 387–394.

Ohshima T, Sato K, Terauchi H, Sato M (1997) Physical and chemical modifications of high-voltage pulse sterilization. J Electrostat 42: 159–166.

Ortega-Rivas E, Zárate-Rodríguez E, Barbosa-Cánovas GV (1998) Apple juice pasteurization using ultrafiltration and pulsed electric fields. Food Bioprod Process 76C: 193–198.

Papchenko AY, Bologa MK, Berzoi SE (1988) Apparatus for processing vegetable raw material. US Patent 4,787,303.

Perez OE, Pilosof AMR (2004) Pulsed electric fields effects on the molecular structure and gelation of β-lactoglobulin concentrate and egg white. Food Res Int 37: 102–110.

Pothakamury UR, Barbosa-Cánovas GV, Swanson BG (1993) Magnetic field inactivation of microorganisms and generation of biological changes. Food Technol 47: 85–93.

Pothakamury UR, Vega H, Zhang Q, Barbosa-Cánovas GV, Swanson BG (1996). Effect of growth stage and processing temperature on the inactivation of *Escherichia coli* by pulsed electric fields. J Food Prot 59: 1167–1171.

Qin B, Zhang Q, Barbosa-Cánovas GV, Swanson BG, Pedrow PD (1995) Pulsed electric field treatment chamber design for liquid food pasteurization using a finite element method. Trans Am Soc Agric Eng 38: 557–565.

Quitão-Teixeira LJ, Aguiló-Aguayo I, Ramos AM, Martín-Belloso O (2008) Inactivation of oxidative enzymes by high-intensity pulsed electric field for retention of color in carrot juice. Food Bioproc Technol 1: 364–373.

Raso J, Alvarez I, Condon S, Sala FJ (2000) Predicting inactivation of *Salmonella seftenberg* by pulsed electric fields. Innov Food Sci Emerg Technol 1: 21–30.

Rastogi NK, Eshtiaghi MN, Knorr D (1999) Accelerated mass transfer during osmotic dehydration of high intensity electrical field pulse pretreated carrots. J Food Sci 64: 1020–1023.

Reina LD, Jin ZT, Yousef AE, Zhang QH (1998) Inactivation of *Listeria monocytogenes* in milk by pulsed electric field. J Food Prot 61: 1203–1206.

Rivas A, Rodrigo D, Martínez A, Barbosa-Cánovas GV, Rodrigo M (2006) Effect of PEF and heat pasteurization on the physical-chemical characteristics of blended orange and carrot juice. LWT-Food Sci Technol 39: 1163–1170.

Sale AJH, Hamilton WA (1967) Effect of high electric fields on microorganisms I. Killing of bacteria and yeast. Biochim Biophys Acta 148: 781–788.

Sampedro F, Geveke DJ, Fan X, Zhang HQ (2009) Effect of PEF, HHP and thermal treatment on PME inactivation and volatile compounds concentration of an orange juice-milk based beverage. Innov Food Sci Emerg Technol 10: 463–469.

Sanchez-Vega R, Mujica-Paz H, Marquez-Melendez R, Ngadi MO, Ortega-Rivas E (2008) Comparative study on enzyme inactivation and physicochemical properties of apple juice treated by ultrapasteurisation and pulsed electric fields. Ital Food Beverage Technol 52: 30–35.

Sato M, Tokita K, Sadakata M, Sakai T (1988) Sterilization of microorganisms by high-voltage pulsed discharge under water. Kagaku Kogaku Ronbunshu 4: 556–559.

Schilling S, Alber T, Toepfl S, Neidhart S, Knorr D, Andreas Schieber A, Reinhold C (2007) Effects of pulsed electric field treatment of apple mash on juice yield and quality attributes of apple juices. Innov Food Sci Emerg Technol 8: 127–134.

Sensoy I, Zhang QH, Sastry S (1997) Inactivation kinetics of *Salmonella dublin* by pulsed electric field. J Food Proc Eng 20: 367–381.

Sepúlveda-Ahumada DR, Gongóra-Nieto MF, San-Martin JA, Barbosa-Canovas GV (2005) Influence of treatment temperature on the inactivation of *Listeria innocua* by pulsed electric fields. LWT-Food Sci Technol 38: 167–172.

Sepúlveda-Ahumada DR, Ortega-Rivas E, Barbosa-Cánovas GV (2000) Quality aspects of cheddar cheese obtained with milk pasteurized by pulsed electric fields. Food Bioprod Process 78C: 65–71.

Serrano R (2006) Commercial juice processing using pulsed electric fields. Paper presented at the Workshop on Applications of Novel Technologies in Food and Biotechnology, Cork, Ireland, 11–13 September 2006.

Shin JK, Jung KJ, Pyun YR, Chung MS (2007) Application of pulsed electric fields with square wave pulse to milk inoculated with *E. coli*, *P. fluorescens*, and *B. stearothermophilus*. Food Sci Biotechnol 16: 1082–1084.

Sizer CE, Waugh PL, Edstam S, Ackermann P (1988) Maintaining flavor and nutrient quality of aseptic orange juice. Food Technol 42 (6): 152–159.

Sobrino-Lopez A, Raybaudi-Massilia R, Martin-Belloso O (2006) High-Intensity pulsed electric field variables affecting *Staphylococcus aureus* inoculated in milk. J Dairy Sci 89: 3739–3748.

Tatum JH, Steven N, Roberts B (1975) Degradation products formed in canned single-strength orange juice during storage. J Food Sci 40: 707–709.

Toepfl S, Heinz V, Knorr D (2006) Application of pulsed electric fields in liquid processing. Paper presented at the EFFOST Conference: Processing Developments for Liquids. Cologne, Germany, 3 April, 2006.

Torregrosa F, Esteve MJ, Frïogola A, Cortes C (2006) Ascorbic acid stability during refrigerated storage of orange- carrot juice treated by high pulsed electric field and comparison with pasteurized juice. J Food Eng 73: 245–339.

Van Gerwen SJC, Zwietering MH (1998) Growth and inactivation models to be used in quantitative risk assessments. J Food Prot 61: 1541–1549.

Van Loey A, Verachtert B & Hendrickx M (2002) Effects of high electric field pulses on enzymes. Trends Food Sci Technol 60: 1143–1146.

Vega-Mercado H, Martín-Belloso O, Qin BL, Chang FJ, Góngora-Nieto MM, Barbosa-Cánovas GV, Swanson BG (1997) Non-thermal food preservation: Pulsed electric fields. Trends Food Sci Technol 8: 151–157.

Vega-Mercado H, Pothakamury UP, Chang FJ, Barbosa-Cánovas GV, Swanson BG (1996) Inactivation of *Escherichia coli* by combining pH, ionic strength, and pulsed electric fields hurdles. Food Res Int 29: 119–199.

Walkling-Ribeiro M, Noci F, Riener J, Cronin DA, Lyng JG, Morgan DJ (2009) The impact of thermosonication and pulsed electric fields on *Staphylococcus aureus* inactivation and selected quality parameters on orange juice. Food Bioprocess Technol 2: 422–430.

Weaver JC, Chizmadzhev YA (1996) Theory of electroporation: a review. Bioelectrochem Bioenerg 41: 135–160.

Wu H, Pitts MJ (1999) Development and validation of a finite element model of an apple fruit cell. Postharvest Biol Technol 16: 1–8.

Wu Y, Mittal GS, Griffiths MW (2005) Effect of pulsed electric field on the inactivation of microorganisms in grape juices with and without antimicrobials. Biosyst Eng 90: 1–7.

Yeom HW, Streaker CB, Zhang QH, Min DB (2000) Effects of pulsed electric fields on the quality of orange juice and comparison with heat pasteurization. J Agric Food Chem 48: 4597–4605.

Yeom HW, Zhang QH, Chism GW (2002) Inactivation of pectin methyl esterase in orange juice by pulsed electric fields. J Food Sci 67: 2154–2159.

Zárate-Rodríguez E, Ortega-Rivas E, Barbosa-Cánovas GV (2000) Quality changes in apple juice as related to nonthermal processing. J Food Qual 23: 337–349.

Zhang QH, Barbosa-Cánovas GV, Swanson BG (1995) Engineering aspects of pulsed electric field pasteurization. J Food Eng 25: 261–281.

Zhang QH, Monsalve-González A, Barbosa-Cánovas GV, Swanson BG (1994) Inactivation of E. coli and S. cerevisiae by pulsed electric fields under controlled temperature conditions. Trans Am Soc Agric Eng 37: 581–587.

Zhang QH, Qiu X, Sharma SK (1997) Recent development in pulsed processing. New Technologies Yearbook. National Food Processors Association, Washington, DC.

Zhao W, Yang R, Wang M, Lu R (2009) Effects of pulsed electric fields on bioactive components, colour and flavour of green tea infusions. Int J Food Sci Technol 44: 312–321.

Zimmermann U (1986) Electrical breakdown, electropermeabilisation and electrofusion. Rev Physiol Biochem Pharmacol 105: 176–256.

Zimmermann U, Pilwat G, Riemann F (1974) Dielectric breakdown on cell membranes. Biophys J 14: 881–889.

Zimmermann U, Pilwat G, Beckers F, Rieman F (1976) Effects of external electric fields on cell membranes. Bioelectrochem Bioenerg 3: 58–83.

Zulueta A, Esteve MJ, Frígola A (2010) Ascorbic acid in orange juice-milk beverage treated by high intensity pulsed electric fields and its stability during storage. Innov Food Sci Emerg Technol 11: 84–90.

Chapter 14
Ultrahigh Hydrostatic Pressure

14.1 Introduction

High-pressure processing (HPP) technology has been commercially applied in the production of ceramic materials, carbides, and synthetic quartz for some time but is relatively new to the food industry (Torres and Velazquez 2008). In this alternative technology of food processing, food items are subjected to elevated pressures (up to about 600 MPa) to achieve microbial inactivation or to alter the food attributes in order to obtain consumer-desired qualities. When foodstuffs are subjected to extremely high pressures, protein and starch are denatured, and enzymes and microorganisms are deactivated in the same way as when heat is applied. Since no thermochemical changes occur, sensory characteristics are not affected and microorganisms are inactivated without causing significant flavor and nutritional changes to foods (Berlin et al. 1999). HPP retains food quality, maintains natural freshness, and extends microbiological shelf life, causing minimal changes in the fresh characteristics of foods by eliminating thermal degradation. As compared with thermal processing, HPP produces foods with better appearance, texture, and nutritive features. It can be conducted at ambient or refrigerated temperatures, thereby eliminating thermally induced cooked off-flavors. This technology is especially beneficial for heat-sensitive products and provides an alternative means of killing bacteria that can cause spoilage or food-borne disease without a loss of sensory quality or nutrients.

Early research (about the turn of the twentieth century) into the use of HPP to pasteurize foods found its potential limited because enzymes were largely unaffected, particularly in milk. There were, also, difficulties in manufacturing high-pressure units and in the inadequacy of packaging materials to withstand pressure during processing. Owing to the significant advances by the 1970s in the design of presses and the development of functional food packaging materials, research on use of HPP for food processing gained momentum, particularly in Japan. It was in Japan where the first commercial food products treated by HPP went on sale by 1990. One company introduced a range of pressure-processed jams, including apple, kiwi, strawberry, and raspberry, in flexible-sealed plastic packs. Two other

E. Ortega-Rivas, *Non-thermal Food Engineering Operations*,
Food Engineering Series, DOI 10.1007/978-1-4614-2038-5_14,
© Springer Science+Business Media, LLC 2012

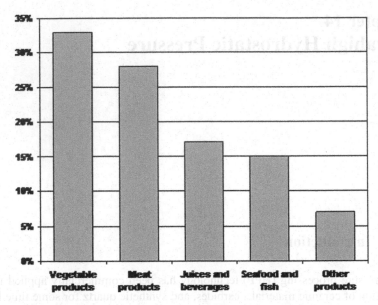

Fig. 14.1 Utilization of high-pressure processing preservation in different segments of the food industry

companies started production of bulk orange and grapefruit juices. Other products included fruit jellies, sauces, fruit yoghurts, and salad dressings. HPP-treated foodstuffs have been marketed in Europe and the USA since 1996. Over recent decades, process development in applications of HPP in food treatment has progressed rapidly. Among the several alternative food processing technologies (UV, pulsed light, ultrasound, pulsed electric fields, etc.) HPP has, undoubtedly, been a success in terms of the route from basic science to applied science to technology transfer and to industrialization. HPP-treated food products normally sell at three to four times the cost of conventional products, but the higher sensory quality has so far ensured sufficient demand for commercial viability. It is also a reality that, without doubt, the preservation of foods is by far the largest commercial application of HPP related to biological systems, and the application has steadily increased during recent years. In 2010 a total annual production of more than 200,000 t of HPP-treated products was reported in the food industry, with the approximate distribution shown in Fig. 14.1.

14.2 Theoretical Aspects

Hydrostatic pressure is generated by increasing the free energy, e.g., by heating in closed systems or by mechanical volume reduction. From the chemical industry, where the pressure is widely used to increase the reaction yield, the technology has been transferred to food and biotechnology, where pressures higher than 400 MPa

Table 14.1 Specific characteristics of different food preservation technologies

	Process parameters	Process intensity	Lethality	Structure impact
Drying	T, t	+	$--$	+ +
Heat	T, t	+ + +	+ + +	+ + +
Irradiation	$\int w$	+ +	+ + +	+ +
PEF	E, $\int w$	+ +	+ +	+
HPP	P, T, t	+	+ +	+

PEF pulsed electric field, *HPP* high-pressure processing, *T* temperature, *t* time, $\int w$ specific total energy input, *E* electric field strength, *P* pressure

can lead to a reversible and irreversible cleavage of intermolecular and intramolecular bonds (Knorr et al. 2006). In this way structural changes in membranes as well as the inactivation of enzymes involved in vital biochemical reactions are the key targets of microbial eradication by high pressure. The inactivation of a virus is supposed to depend on the denaturation of capsid proteins essential for host cell attachment. HPP is a relatively young technology for food preservation and its specific features are not as well understood as those characteristics of traditional technologies in the food industry, such as drying and heating. The high-pressure process itself is characterized by three parameters: temperature T, pressure P, and pressure exposure time t. Compared with other processes such as heat preservation, which is based on two parameters only (T, t), the three-parameter HPP offers a broad variability for process design. Table 14.1 shows typical processing parameters for traditional and novel preservation treatments.

In a qualitative approach, process efficiency is assessed in terms of the lethality of the treatment and its structural impact on the food matrix. Evidently, those treatments which are powerful in killing microbes usually have a strong destructive effect on the integrity of the food matrix, with severe consequences for quality and consumer acceptance. Traditional products are usually preserved by conventional technologies, which, to a large extent, meet the expectations of the consumers of those foods. On the other hand, traditional preservation strategies fail or are not applicable when new product developments are based on innovative or uncommon ingredient compositions. In those situations, non-thermal technologies such as irradiation, pulsed electric field treatment and HPP came into focus. The justification for applying novel preservation concepts is high safety margins, superior quality, and reasonable costs. Within the last 20 years, considerable knowledge on the impact of high pressure on microbes, viruses, food constituents, and food structures has been accumulated, and many practical applications of high-pressure technology in the food industry and biotechnology took advantage of the substantial advances in biochemistry and biophysics which led to an improved understanding of the mechanistic background.

In complex matrices, such as those in food materials, the desired effect of microbial inactivation may also produce biochemical changes which may affect the product properties in a negative manner. The suitable selection of the processing parameters temperature, time, and pressure can ensure that the processing goal is reached without extensive detrimental effects. Proteins, which play a major role in

Fig. 14.2 The principle
of isostatic pressure

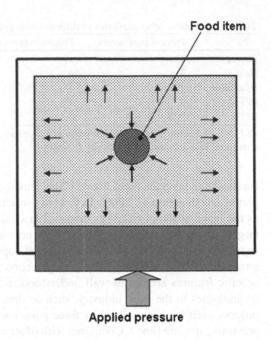

the metabolic activity of all living cells, are extremely susceptible to changes in the
environment. The stability of the molecular configuration of the protein in its
functional form is determined by a narrow band of parameter settings which impact
mainly on how it interacts with the solvent. In the case of water, which is the natural
protein environment, a hydration shell is formed which itself influences the intramolecular interaction. In the same manner, ionizing groups in lateral positions
produce conformational changes driven by the actual proton concentration and
ionic strength. The losses in functionality of proteins in response to those
perturbations are thus related to intramolecular reorientations or complete
unfolding leaving the polypeptide chain in a random-coiled state. In many cases,
the "functional" state is considered as the native state, whereas the "non-functional"
state is referred to as the denatured state no matter which particular molecular
structure is formed.

The governing principles of HPP are based on the assumption that foods
experiencing high pressure in a vessel follow the isostatic rule regardless of the
size or shape of the food. Such a rule states that pressure is instantaneously and
uniformly transmitted throughout a sample whether the sample is in direct contact
with the pressure medium or hermetically sealed in a flexible package (Fig. 14.2).
As compared with thermal processing, therefore, the time necessary for HPP should
be independent of the sample size (Rastogi et al. 2007). The effect of high pressure
on food chemistry and microbiology is governed by Le Chatelier's principle. This
principle states that when a system at equilibrium is disturbed, the system responds
in a way that tends to minimize the disturbance (Pauling 1964). In other words, high
pressure stimulates some phenomena, such as phase transition, chemical reactivity,

changes in molecular configuration, and chemical reactions that are accompanied by a decrease in volume, but opposes reactions that involve an increase in volume. The effects of pressure on protein stabilization are also governed by this principle, i.e., the negative changes in volume with an increase in pressure cause an equilibrium shift toward bond formation. Alongside this, the breaking of ions is also enhanced by high pressure, as this leads to a volume decrease caused by the electrostriction of water. Moreover, as hydrogen bonds are stabilized by high pressure, with their formation involving a volume decrease, pressure does not generally affect covalent bonds. Consequently, high pressure can disrupt large molecules or microbial cell structures, such as enzymes, proteins, lipids, and cell membranes, and leave small molecules such as vitamins and flavor components unaffected (Linton and Patterson 2000). Owing to the work of compression, HPP causes temperatures to rise inside the high-pressure vessel. This phenomenon is known as adiabatic heating and should be given due consideration during the preservation process. The values of the temperature increments in the food and the pressure-transmitting medium will be different, as they depend on the food composition as well as the processing temperature and pressure and the rate of pressurization (Otero et al. 2007). In food sterilization, adiabatic heating can be used advantageously to provide heating without the presence of sharp thermal gradients at the process boundaries (Toepfl et al. 2006). Pressure changes the physical properties of materials by affecting reaction rates. The dependence of reaction rate constants on pressure may be expressed in an equation analogous to the Arrhenius equation, which describes the dependence on temperature. The dependence on pressure is given by

$$-RT\left(\frac{d(\ln k)}{dP}\right)_T = \Delta V^+, \tag{14.1}$$

where R is the universal gas constant, T is the absolute temperature, k is a constant, and ΔV^+ is the activation volume, analogous to the inactivation energy in the Arrhenius equation. In the range of pressures used in HPP, their effects are less dramatic than the effects of temperature used in heat processing. A pressure increase of 100 MPa causes an increase on the order of that produced by a temperature increase of 10°C (Tauscher 1995).

14.3 Kinetics of Inactivation

The potentially lethal effect of high pressure on microorganisms has been documented for 100 years, at least. Applications to food processing are, however, more recent (Earnshaw 1995) and can be broadly classified into (1) effects on membranes, (2) effects on biochemical and physiology processes, and (3) effects on genetic mechanisms. The kinetics of inactivation by pressure, the sensitivity of different organisms of process significance, and the effect of various factors on sensitivity to pressure are the main factors to consider when designing equipment and operating HPP technology for use in food pasteurization and sterilization. To assess the efficiency of high pressure in microbial inactivation, it is necessary to understand the relation between the intensive and extensive parameters of the process, and the lethality effect. In the case of HPP, the intensive factor is the pressure and the extensive factor is the time (temperature and time are the corresponding factors for thermal processing). The models proposed for microbial inactivation by HPP follow a first-order equation as follows:

$$\ln \frac{N_t}{N} = -kt = -2.303 \left(\frac{t}{D_{10}} \right), \tag{14.2}$$

where N_t is the microbial load after contact time t, N is the initial microbial load, k is the reaction rate constant, and D_{10} the decimal reduction time.

Another model formulated from work with yeasts assumes that the change of D_{10} with pressure follow the equation

$$\log \frac{D_{10}}{D_{10R}} = \frac{P - P_R}{z(P)}, \tag{14.3}$$

where the subscripts R indicate a reference pressure condition and $z(P)$ is the pressure increase required to accomplish a one log reduction cycle in D_{10}.

An alternative model has been suggested as follows:

$$\ln \frac{k}{k_R} = \frac{V'(P - P_R)}{RT}, \tag{14.4}$$

where V' is an activation volume constant obtained from the slope of $\ln k$ plotted versus P.

14.4   Effects on Microorganisms

The effect of high pressure on microorganisms has been observed to be similar to that of high temperature. Examples of the effects of HPP on several micro-organisms are shown in Table 14.2. As mentioned before, HPP enables transmittance of pressure rapidly and uniformly throughout the food. The problems of spatial variations in preservation treatments associated with heat, microwave, or radiation penetration are, therefore, not evident in food products treated by HPP.

Table 14.2 Effects of ultrahigh hydrostatic pressure on various microorganisms

| Microorganism | Conditions | | | |
	Pressure (MPa)	Time (min)	Decimal reduction	Media
Saccharomyces cerevisiae	300	5	5	Satsuma mandarin juice
Aspergillus awamori	300	5	5	Satsuma mandarin juice
Listeria innocua	360	5	1	Minced beef muscle
Listeria monocytogenes	350	10	4	Phosphate buffer saline
Vibrio parahaemolyticus	172	10	2.5	Phosphate buffer saline
Salmonella typhimurium	345	10	1.8	Phosphate buffer
Total plate count	340	5	1.9	Fresh cut pineapple

14.4.1   Vegetative Cells

Vegetative cells, including yeast and molds, are pressure-sensitive, i.e., they can be inactivated by pressures of 300–600 MPa (Knorr 1994; Patterson et al. 1995). At high pressures, microbial death is considered to be due to permeabilization of the cell membrane. *Saccharomyces cerevisiae*, subjected to pressures of about 400 MPa, had its structure and cytoplasmic organelles grossly deformed and large quantities of intracellular material leaked out, whereas at 500 MPa, the nucleus could no longer be recognized and the loss of intracellular material was almost complete (Farr 1990). Changes that are induced in the cell morphology of the microorganisms are reversible at low pressures, but are irreversible at higher pressures, where microbial death occurs owing to permeabilization of the cell membrane. The application of high pressure increases the temperature of the liquid component of the food by approximately 3°C for every 100-MPa increase. When the food contains a significant amount of fat, the temperature rise is greater, on the order 8–9°C for every 100 MPa (Rasanayagam et al. 2003). Foods cool down to their original temperature on decompression if no heat is lost to, or gained from, the walls of the pressure vessel during the holding stage. The temperature distribution during the pressure-holding period can change depending on heat transfer across the walls of the pressure vessel, which must be held at the desired temperature to achieve truly isothermal conditions. High compression may also shift the pH, depending on the imposed pressure. For example, a decrease of pH in apple juice by 0.2 units for a pressure increment of 100 MPa has been reported (Heremans 1995).

14.4.2   Spores

Attempts to completely inactivate spores using high pressure have been unsuccessful so far. Even at the beginning of pioneering work applying high hydrostatic pressure to food processing there were indications that spores would be difficult to eradicate with HPP. Hite (1899) subjected milk to high hydrostatic pressure as opposed to high temperature, which causes undesirable burned flavor notes, in order to prevent it from turning sour. With use of pressures ranging from 400 to 700 MPa at room temperature, four log reduction cycles in microbial counts were achieved, while maintaining product freshness. Hite had some success in creating shelf-stable HPP-treated fruit products with a pH that prevented growth of spores, but he could never achieve shelf-stability of HPP-treated milk because of its neutral pH and the presence of spores (Hite et al. 1914). The ability to make low-acid pressure-treated products as safe as foods subjected to a botulinum cook (thermal sterilization equivalent to treatment at 121°C for 3 min) by pressure alone has not yet been possible. An early study by Timson and Short (1965) showed that spores of *Bacillus subtilis* and *Bacillus alvei* survived in milk treated at 1,034 MPa for 90 min at 35°C. The resistance to pressure of the bacterial spores is probably due to the structure and thickness of the coats of the bacterial spores that protect them against solvation and ionization (Sale et al. 1970). Microbial spores suspended in foods have been inactivated by HPP, but compared with the requirements for vegetative cells, the treatment conditions must be extreme, using as high pressure as possible for prolonged times and combining the treatment with elevated temperatures (Hoover 1993). Since HPP has the potential to inactivate spores in foods (Gould and Sale 1970; Heinz and Knorr 2001), it is important to understand the physiological properties of spores, especially factors that are pertinent to inactivation of spores by HPP. Therefore, in the attempt to achieve inactivation of spores using HPP as an alternative method of food preservation, a number of coactive preservative or synergistic factors to enhance the effects of high pressure on spores have been studied. These factors include pressure cycling and combinations of pressure, high and low temperatures, and antimicrobial agents.

The resistance to pressure and the resistance to heat of bacterial endospores appear to be unrelated. From a study of a number of *Bacillus* spp., Nakayama et al. (1996) showed that *Bacillus subtilis* spores were very pressure resistant, but their heat resistance was much lower than that of *Bacillus stearothermophilus* spores. A number of studies have compared the efficacy of inactivation of bacterial spores by pressure at ambient temperature with that at higher temperatures. Elevation of the pressure-processing temperature from room temperature to more than 50°C enhances inactivation of spores of both *Bacillus* and *Clostridium* spp. in buffer and meat (Reddy et al. 2003). Stewart et al. (2000) found that the spore numbers of *Bacillus subtilis* in McIlvaine citrate phosphate buffer were reduced more effectively at high temperatures than at ambient temperature, with five log cycle reductions in spores achieved at 404 MPa at 70°C, whereas only a 0.5 log cycle reduction could be achieved at 25°C at the same pressure. Mills et al. (1998) reported little or no inactivation of spores of *Clostridium sporogenes* following treatment at 600 MPa for 30 min at 20°C. However, combining heat and pressure simultaneously (400 MPa for 30 min at 60°C) or sequentially (heating at 80°C for 10 min at ambient pressure followed by treatment at 400 MPa for 30 min at 40°C or 60°C) was more effective, resulting in a three log cycle reduction.

Another aspect of HPP and its relation to spore inactivation that has been investigated is the ability of high pressure to induce spore germination. Some authors have suggested a high-pressure treatment to cause spore germination, followed by a subsequent treatment to inactivate the germinated microorganisms (Knorr 1994). The initiation of germination or inhibition of germinated bacterial spores and inactivation of piezoresistive microorganisms can be achieved in combination with moderate heating or other pretreatments such as ultrasound treatment. Gould and Sale (1970) studied the germination of *Bacillus* spores and demonstrated that treatments at 25 MPa and 50°C for 3 min caused the germination of 50–64% of the initially inoculated spores. Crawford et al. (1996) reported on the decrease of viable *Clostridium sporogenes* spores with increasing pressure up to 690 MPa. Higher pressures were inefficient in reducing further the spore counts. They postulated that spore germination possibly took place at pressures lower than the reported 690 MPa, and the germinated cells were inactivated at this maximum operating pressure.

14.4.3 Synergistic Effects

The use of antimicrobial factors in combination with HPP (hurdles) has been used not only for the case of spore inactivation, but also for different applications. Although the combination of high pressure with moderate to high temperatures would negate the "non-thermal" designation of HPP, pressure-assisted thermal sterilization treatments are still attractive alternatives to conventional thermal sterilization or retorting, as this type of process has been proven to be just as effective as and less damaging to flavor, texture, and nutritional attributes of food than thermal processes (Meyer et al. 2000). The combination of HPP treatment and temperature is frequently considered to be most appropriate for pasteurization and sterilization in food processing (Farr 1990; Patterson et al. 1995). An increase in process temperature above ambient temperature, and to a lesser extent, a decrease below ambient temperature, increases the inactivation rates of microorganisms during HPP. Temperatures in the range 45–50°C appear to increase the rate of inactivation of pathogens and spoilage microorganisms. Studies have shown that the antimicrobial effect of high pressure can be enhanced by using heat, low pH, carbon dioxide, organic acids, and bacteriocins, such as nisin. Mackey et al. (1995) reported on increased sensitiveness of *Listeria monocytogenes* cells to HPP caused by butylated hydroxyanisole, potassium sorbate, and acid conditions. Knorr (1994) as well as Papineau et al. (1991) reported on enhanced pressure inactivation of microorganisms when combining HPP treatments with additives such as different types of organic acids, sulfites, polyphenols, or chitosan. All these reported combinations allowed a lower processing pressure, temperature and time of exposure to be used.

14.5   Effects on Enzymes

Pressure can influence most biochemical reactions, since they often involve a change in volume. High pressure controls certain enzymic reactions. The effect of HPP on proteins/enzymes is reversible (unlike that of temperature) in the range 100–400 MPa and is probably due to conformational changes and subunit dissociation and association process (Morild 1981). High-pressure application leads to the effective reduction of the activity of food-quality-related enzymes (oxidases), which ensures the products are of high quality and shelf-stable. High pressure reduces the rate of the browning reaction known as the Maillard reaction, which actually consists of two reactions: condensation reaction of amino compounds with carbonyl compounds, and successive browning reactions including melanoidin formation and polymerization processes. The condensation reaction shows no acceleration by high pressure (5–50 MPa at 50°C), because it suppresses the generation of stable free radicals derived from melanoidin, which are responsible for the browning reaction (Tamaoka and Hayashi 1991).

Enzymes found in fruits, vegetables, milk, fish, and meat products include polyphenoloxidase (PPO), pectin methylesterase (PME), peroxidase (POD), lipases, and proteases. A recent review on the pressure-processing effects on enzymes present in fruits and vegetables can be found in Ludikhuyze et al. (2002). The thermal resistance of enzymes cannot be used to predict their pressure resistance (Hendrickx et al. 1998). Enzymes such as PPO and lipoxigenase (LOX) are inactivated at 300 MPa, whereas others such as PME and POD are very difficult to inactivate within the pressure range of the commercial units available today.

14.5.1   Polyphenoloxidase

PPO is considered a moderately heat stable enzyme (Yemenicioglu et al. 1997), and its inactivation is highly desirable as it causes enzymic browning of fruits and vegetables. In fruits such as apples, apricots, pears, plums, strawberry, bananas, peppers, and grapes and in vegetables such as mushrooms, onions, avocados, and potatoes, the pressures necessary to induce inactivation of PPO range from 200 to 1,000 MPa, depending on food composition, the pH, and the use of additives (Weemaes et al. 1999). PPO is pressure-sensitive in apples, strawberries, apricots, and grapes and pressure-resistant in other fruits, such as pears and plums. A low pressure may protect PPO from thermal inactivation and even enhance its activity in apple (Anese et al. 1995), pear (Asaka et al. 1994), and onion (Butz et al. 1994). However, apricot PPO and strawberry PPO are inactivated by pressures exceeding 100–400 MPa, respectively (Amati et al. 1996). Depending on the pH, pressures from 100 to 700 MPa are required for the inactivation of apple PPO (Anese et al. 1995). At a lower pH, pressure induced a faster inactivation of PPO (Jolibert et al. 1994). Apart from the pH, pressure inactivation is also influenced by the addition of salts, sugars, and compounds such as benzoic acid, glutathione, and $CaCl_2$. For example, pressure inactivation of apple PPO and mushroom PPO is enhanced by the addition of $CaCl_2$ (Jolibert et al. 1994) and the addition of benzoic acid or glutathione, respectively (Weemaes et al. 1997). PPO in white grape must is only partly inactivated at 300–600 MPa for 2–10 min (Castellari et al. 1997), with a clearly lower enzymic activity only at 900 MPa. Since this pressure exceeds the capacity of commercial units, complete inactivation can be achieved by pressure treatments combined with a mild thermal treatment (~40–50°C). Studies on the residual PPO activity in fruit purees suggest that inhibition is possible by pressure treatments combined with pH reduction, blanching, or a low refrigeration storage temperature (Ludikhuyze et al. 2000; Palou et al. 2000). For example, the PPO activity in banana puree adjusted to pH 3.4 and a water activity of 0.97 was reduced after steam blanching and then further reduced by a pressure treatment (Palou et al. 1999). However, extrapolating from findings in one matrix to another is not possible. Studies on the combination of thermal blanching and 100–200 MPa for 10–20 min showed that the PPO in red pepper fruits is more stable to pressure and temperature than the PPO in green peppers (Castro et al. 2008). Mushroom PPO, potato PPO, and avocado PPO are pressure-resistant as treatments at 800–900 MPa are necessary to reduce enzyme activity at room temperature (Weemaes et al. 1997). However, the effect of pressure on the PPO activity in mushrooms and potatoes is different (Gomes and Ledward 1996). In mushrooms, considerable browning is observed immediately after pressurization to 200 MPa, whereas in potatoes, limited browning is observed at 600 MPa and even at 800 MPa. In another study, treatments at 600 MPa and 35°C (including adiabatic compression heat) for 10 min decreased PPO activity in mushroom by 7%, but extending the treatment to 20 min resulted in no further PPO activity reduction (Sun et al. 2002). However, PPO activity decreased to 28% and 43% after 800-MPa treatments for 10–20 min, respectively. Furthermore, PPO isolated from apple, grape, avocado, and pear is inactivated slowly at 600 MPa but is completely inactivated at 900 MPa (Weemaes et al. 1998).

14.5.2 Pectin Methylesterase

PME is responsible for cloud destabilization in juices and the loss of consistency in many foods. This enzyme, one of the most abundant pectinases in plants, has high thermoresistance. Temperatures between 80°C and 95°C are required to induce significant inactivation and even then PME may remain active. This resistance reflects the presence of heat-labile and heat-stable PME isoforms (Wicker and Temelli 1988; van den Broeck et al. 2000). The pressure required for the inactivation at room temperature of PME from different sources has been reported to range significantly from ~150 to 1,200 MPa. Most studies report only partial inactivation of PME and this reflects the presence of pressure-resistant isoforms. Pressure-inactivation studies of PME in orange, grapefruit, guava, and tomato in the 0.1–800-MPa and 15–65°C range revealed a slight antagonistic effect of low pressure and high temperature (van den Broeck et al. 2000). However, pressure-treated orange juice resulted in a stable and higher-quality product as approximately 600 MPa reduces PME activity irreversibly by 90% (Irwe and Olson 1994; Ogawa et al. 1990). Lower-pressure treatments at room temperature (200–400 MPa) induced PME activation in freshly squeezed orange juice (Cano et al. 1997). Compared with orange PME, tomato PME is more pressure resistant. PME inactivation in tomato occurs at lower rates than in orange at high pressure (Weemaes et al. 1999; van den Broeck et al. 2000). Tomato PME is less pressure stable in the presence of Ca^{2+} ions or in citric acid buffer (pH 3.5–4.5) than in water. The antagonistic effect, i.e., lower inactivation at elevated pressure than at atmospheric pressure, was less pronounced in citric acid buffer (pH 3.8–4.5) and in the presence of $CaCl_2$ than in water. In the absence or presence of Ca^{+2} ions, the optimal pressure for enzyme activation was 100 and 400 MPa, respectively (van den Broeck et al. 2000). A study of PME in carrot extract (purified form), carrot juice, and carrot pieces showed that PME in carrot pieces is much more resistant to thermal and pressure treatments than PME in the carrot juice or the extract. This was attributed to the presence of a stabilizing factor in the carrot matrix and the fact that PME in carrot pieces is bound to the cell wall (Balogh et al. 2004). PME in carrots is inactivated by temperatures above 50°C but resists pressures up to 600 MPa (Balogh et al. 2004; Sila et al. 2007). For example, to reduce PME activity by 90% it is necessary to treat carrot juice at 800 MPa and 10°C for 36 min.

14.5.3   Peroxidase

This enzyme induces off-flavors during storage and is considered to be the most heat stable vegetable enzyme, and it is also extremely pressure resistant. For example, an applied pressure of 900 MPa in green peas at room temperature induced only a slight inactivation. Significant inactivation was achieved at 600 MPa only at 60°C (Quaglia et al. 1996). POD in strawberry puree and orange juice at 20°C can, however, be inactivated in 15 min at 300 and 400 MPa, respectively (Cano et al. 1997). The activity of POD in orange treated at room temperature for 15 min and up to 400 MPa decreased continuously, with the highest inactivation rate (50%) being observed at 32°C.

14.5.4   Lipases and Proteases

Lipases are responsible for the hydrolysis of animal and vegetable fats and oils. In pressure-treated foods, large differences in lipase activity have been reported (Macheboeuf and Basset 1934; Weemaes et al. 1998). In some cases, lipase has been reported to be stable at 1,100 MPa (Macheboeuf and Basset 1934), whereas in others inactivation has been achieved at 600 MPa (Seyderhelm et al. 1996).

A protease is any enzyme that conducts proteolysis, i.e., begins protein catabolism by hydrolysis of the peptide bonds that link amino acids together in the polypeptide chain forming the protein. The effect of HPP on the proteolytic enzymic activity in milk, meat, and fish products has been reported (Lakshmanan et al. 2005; Quiros et al. 2007). Inactivation of proteolytic enzymes at pressures over 800 MPa suggests that HPP may have a beneficial effect on the sensory properties of cheese (Reps et al. 1998). Also, the enzyme activity in crude enzyme extracts prepared from cold-smoked salmon muscles were reduced by high pressure. Proteolytic enzyme activity in cold-smoked salmon muscles was decreased by 20-min treatments at 9°C and pressures equal to or higher than 300 MPa. Also, proteolytic activity was substantially reduced by pressure treatments of protease extract solutions (Lakshmanan et al. 2005).

14.6 Effects on Nutritive and Quality Attributes

The primary structure of low molecular weight molecules such as vitamins, peptides, lipids, and saccharides is rarely affected by high pressure because of the very low compressibility of covalent bonds at pressures below 2,000 MPa (Cheftel and Culioli 1997; Oey et al. 2006). Certain macromolecules such as starches can, however, change their native structure during HPP in a manner analogous to the change during thermal treatments (Cheftel 1992; Heremans 1982). For example, starch granule solutions can form very smooth starch pastes, which can be used to replace fat in reduced-energy foods. Starch granule solutions can form a week gel owing to pressure-induced swelling of the granule (Stolt et al. 2000). Therefore, the occurrence of intermediate degradation levels of the lamellar crystalline regions of the starch granule can be anticipated, which is a possible reason for the significant difference in viscosity between starch gels formed at different pressures or temperatures, or both. The pressure range in which gelatinization occurs is specific for each starch and is partly dependent on its crystalline structure, e.g., B-type starches are more resistant to pressure than A- and C-type starches (Bauer and Knorr 2005), and the proportion of amylose and amylopectin (Blaszczak et al. 2007). Usually, the extent of gelatinization reached depends on the pressure, the treatment temperature, and the processing time (Stolt et al. 2000). Starch molecules are opened and partially degraded to produce increased sweetness and susceptibility to amylase activity. For example, the appearance, odor, texture, and taste of soybeans and rice did not change after HPP, whereas root vegetables, such as potato and sweet potato, became softer, more pliable, sweeter, and more transparent (Galazka and Leward 1995). Also, some HPP-treated fruits retained the flavor, texture, and color of the fresh product (Knorr 1995).

In proteins, high pressure induces unfolding of the molecular structure and then aggregation with different proteins in the food, or into a different form. Both possibilities will result in textural changes in the food material. Gel formation is observed in soy, meat, fish, and egg albumin proteins. Pressure-induced gels tend to maintain their natural color and flavor, and are described as smooth, glossy, and soft, when compared with gels formed by use of heat treatment. These results have been evaluated in relation to surimi products on an experimental scale. Interest is also focused on the possibility of unfolding the structure of lower-quality proteins, using HPP, to improve functional properties such as emulsifying and gelling capacity of high-protein food products. The effects of high pressures on protein structure have been discussed by Hendrickx et al. (1998), and the effects on protein functionality have been reviewed by Messens et al. (1997).

Another functional attribute of HPP that has been reported is its suitability for meat tenderization. High-pressure treatment of meat at 103 MPa and 40–60°C for 2.5 min improves the texture of meat and reduces cooking losses. Among the commercially produced items available, HPP-processed salted raw squid and fish sausages can be mentioned (Hayashi 1995). Some other interesting effects of HPP include improved microbiological safety and elimination of cooked flavors from sterilized meats and pâté (Johnston 1995), as well as tempering of chocolate, where the high pressures transform cocoa butter into the stable crystal form.

There are few reports on the effects on nutritional characteristics of HPP-treated foods. Elgasim and Kennick (1980) investigated pressure treatment at 103 MPa for 2 min of meat protein. The apparent digestibility was improved, and no adverse effects on the apparent biological value, net protein utilization, or protein efficiency ratio were observed. Butz et al. (2002) studied the influence of high pressure on the functional properties of a choice of vegetables at around 600 MPa and in combination with elevated temperatures. Carrots, tomatoes, and broccoli were investigated and the contents of health-promoting substances (e.g., vitamins, antioxidants, antimutagens) were assessed. In most cases high pressure did not induce loss of beneficial substances in the vegetable matrices but induced changes in the structure of the products, which resulted in altered physicochemical properties such as higher glucose retardation index and water retention or reduced extractability. Sancho et al. (1999) investigated the effect of HPP on selected hydrosoluble vitamins. Vitamin retention after pressurization was compared with that induced by several classic food processing treatments, such as pasteurization and sterilization. Minor variations were found among the vitamins after pressurization. Vitamins B_1 and B_6 underwent no significant losses after the treatments. Vitamin C levels, although significant, were not dependent on the intensity of the HPP. Naturally occurring vitamin C losses were analyzed in two representative foodstuffs (egg yolk and strawberry coulis) after several processes in order to validate the model system results. The results showed that HPP plays only a minor role in the degradation kinetics of vitamin C.

14.7   Equipment

The manufacturing of high-pressure machinery is a specialized and expensive procedure, but nowadays equipment for treatment of foods by HPP is readily available. The main components of a high-pressure system are (1) the pressure

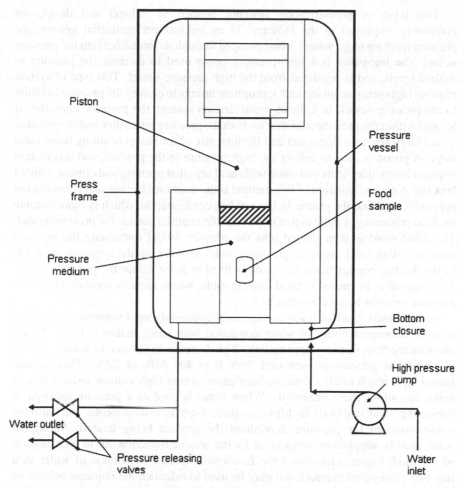

Fig. 14.3 Principle of ultrahigh hydrostatic pressure equipment: a direct pressure generation unit

vessel and its enclosure, (2) the pressure generation system, (3) the temperature control device, and (4) materials handling adaptation (Fig. 14.3).

The key component of HPP is the high-pressure vessel, in which products undergoing treatment are subjected to pressure. Such a vessel is generally made of low-alloy steel and is routinely used in the ceramic and metal industries. For food applications, however, a unique requirement for the high-pressure vessel is that it must undergo several thousand processing cycles per year to process large volumes of foods. The large number of pressurization and depressurization cycles required increases metal fatigue and reduces the life of the vessel. Furthermore, the vessel itself must be protected from any corrosion due to either the food material itself or any liquids used for cleaning, and must be easy to clean.

Two types of pressurization systems, defined as indirect and direct, are commonly employed in the industry. In an indirect pressurization system, the pressure medium (e.g., water) is first pumped through an intensifier into the pressure vessel. The intensifier is a high-pressure pump used to increase the pressure to desired levels, and is separated from the high-pressure vessel. This type of system requires high-pressure tubing and appropriate fittings to convey the pressure medium to the pressure vessel. In a direct pressurization system, the pressure intensifier is located within the pressure vessel, with both the pressure intensifier and the pressure vessel fabricated as a single unit and the total size of the vessel resulting being quite large. A piston is used to deliver the high pressure to the product, and this system requires heavy-duty seals that must withstand repeated opening and closure without leakage. A major limitation of this method is the need for efficient seals between the pressure vessel and the piston. In the wet bag configuration, which is more suitable for food processing, a mold is first filled with the material outside the pressure vessel. The filled mold is then moved into the pressure vessel containing the pressure medium. With cold isostatic pressure, water is used as the pressure medium. In the dry bag configuration, the mold is fixed in place within the pressure vessel. The material to be treated is filled into the mold, which remains separated from the pressure medium by an elastomer tool.

In HPP, inert gases or water are the most commonly used pressure media. The relative incompressibility of water compared with gases makes it the preferred pressure medium in many applications. The decrease in volume of water is about 5% when its pressure is increased from 0 to 400 MPa at 22°C. This volume reduction is much smaller than for inert gases, where high volume reductions can make operations more hazardous. When water is used as a pressure medium in subjecting food materials to high pressure, there is instantaneous and uniform transmission of the pressure throughout the product being treated. Additional water must be supplied to compensate for the volume reduction due to compression of the fluid. Figure 14.4 shows the fractional decrease in volume of water as a function of imposed pressure and may be used to calculate the required volume of water to be added. Also, small amounts of oil may be added to the water for anticorrosion and lubrication purposes.

A simplified mode of operation used in an HPP system is shown in Fig. 14.5. In a high-pressure process, the pressure vessel is filled with a food product and pressurized for the desired time, after which it is depressurized. The time required to pressurize the vessel is influenced by the compressibility of the pressure medium and the nature of the food material. If water is used as the pressure-transmitting medium for most food materials, the compressibility is similar to that of the pressure medium. Typically, the pressurization time for foods is independent of the quantity of the food placed in the pressure vessel. The presence of air in the food increases the pressurization time, since air is considerably more compressible than water. After pressurization, the food is kept under high pressure for the required process time, which may be several minutes. Upon completion of the exposure to pressure, depressurization can be done quite rapidly.

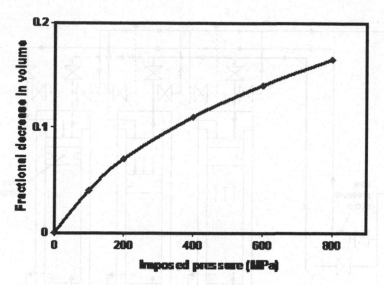

Fig. 14.4 Effect of applied pressure on fractional decrease of water volume

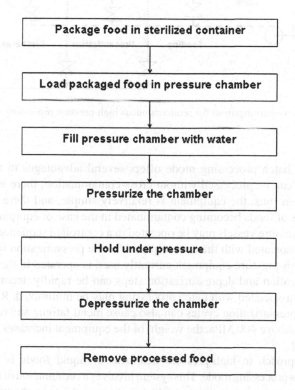

Fig. 14.5 Flow diagram of a high-pressure processing operation

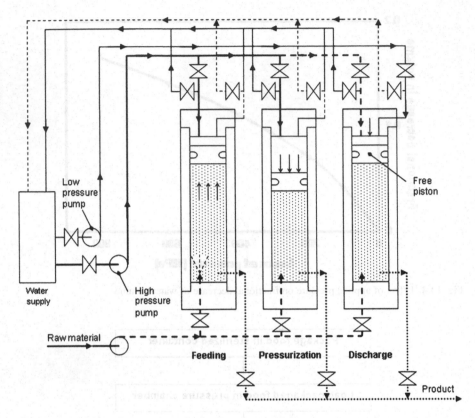

Fig. 14.6 Multivessel arrangement for semicontinuous high-pressure processing

HPP in the batch processing mode offers several advantages in that different types of foods can be processed without cross-contamination, there is no need for cleanup between runs, the equipment is relatively simple, and there is no risk of large quantities of foods becoming contaminated in the case of equipment malfunction. Several pressure vessels may be operated in a controlled sequence to minimize any time lag associated with the time required for the pressurization of the vessels. Most of the high-pressure equipment currently used is operated in the batch mode. Since pressurization and depressurization steps can be rapidly accomplished, the low efficiency associated with batch processing may be minimized. Rapid pressurization and depressurization cycles can also cause metal fatigue and reduce the life of equipment. Above 400 MPa, the weight of the equipment increases significantly, as does its cost.

Another approach to high-pressure treatment of liquid foods is the use of a semicontinuous processing mode. This system involves a combination of multiple pressure vessels that are sequenced to provide a continuous flow. As shown in Fig. 14.6, while one vessel is being pressurized, another may be in a decompression mode.

The compression process is done with water up to a maximum pressure of 400 MPa. Programmable pressure controllers are used to adjust pressurization and decompression rates. Appropriate temperature controls are used to maintain the temperatures between $-20°C$ and $80°C$. The reduction in cost of a semicontinuous process is about 27% over a batch process for 500 l/h production. Another continuous high-pressure system involves 5-m-long stainless steel pipes that are wound like a coil with a pressure resistance of 700 MPa. An air-driven hydraulic pump is used to introduce liquid product into the pipes. With the outlet valve closed, the liquid is subjected to pressure. The coiled pipes are placed in thermostatically controlled water baths, in which the temperature is maintained between $5°C$ and $80°C$. The outlet valve is gradually opened to release the pressurized product in a continuous manner.

Other innovations in high-pressure system design include the use of pulsating high pressures. The pressure vessel is similar to the vessels used in cold isostatic pressing. A unique feature of this new system is an air-driven pressure-increasing device that allows an instantaneous change in pressure. Additionally, a pressure-reducing valve attached to the pressure vessel is useful for releasing pressurized water. By manipulating the pressure-reducing valve, on obtains the desired pulsations. The pressure vessel is contained in a thermostatically controlled water bath. Reduced process times at high pressures were obtained when the system was used in combination with higher temperatures. These studies emphasize the synergistic benefit of pressure and temperature in selected food applications.

The cost of HPP is dependent upon the combination of the pressure, the pressure holding time, and the temperature at which the product is processed. These variables must therefore be selected carefully. The cost per unit of production is lower for a single large production unit than for several small pressure units in parallel. This cost saving is possible because the capital cost of manufacturing a large pressure unit is lower than that of manufacturing several small units.

14.8 Applications and Prospects for the Future

As previously pointed out, HPP represents one of the major recent advances in food technology and, as such, its potential for use in a number of applications in the food industry is great. Relevant applications of HPP include decontamination of raw milk and some curds and cheeses made from raw milk, reduction of the intensity of heat processing for prepared chilled meals containing thermosensitive food constituents, and sanitation and increase of the refrigerated shelf life of spreads, emulsified sauces, essential oils, aromatic extracts, and herbs (Cheftel 1995). HPP offers a unique opportunity to develop new foods with high nutritional and sensory quality, improved texture, more functionality, and increased shelf life (Mertens 1995). High-pressure treatment has the potential to improve the microbiological safety and quality of foods such as meat, milk, and dairy products. High pressure depresses the freezing point of water and the melting point of ice, allowing various high-density forms of ice to be obtained. These effects of pressure on the solid–liquid phase diagram of water have several potential applications, including pressure-assisted thawing and non-frozen storage at low temperature under pressure (Kalichevsky et al. 1995).

The perspectives for extending the range of HPP in applications in the food industry depend on further research, by both academia and industry. Research and development work is likely to include shelf life testing and appropriate microbial challenge testing. Incorporation of good manufacturing practices to guarantee the efficiency of HPP operations may also be a priority. More studies dealing with interactions between applied pressure and food constituents, focusing mainly on nutritive features, are necessary.

References

Amati A, Castellari M, Matricardi L, Arfelli G, Carpi G (1996) Modificazione indotte in mosti d'uva da trattamenti con alte pressioni idrostatiche. Ind Bevande 25: 324–328.

Anese M, Nicoli MC, Dallaglio G, Lerici CR (1995) Effect of high pressure treatments on peroxidase and polyphenoloxidase activities. J Food Biochem 18: 285–293.

Asaka M, Aoyama Y, Ritsuko N, Hayashi R (1994 Purification of a latent form of polyphenoloxidase from La France pear fruit and its pressure-activation. Biosci Biotechnol Biochem 58: 1486–1589.

Balogh T, Smout C, Ly Nguyen B, van Loey A, Hendrickx ME (2004) Thermal and high pressure inactivation kinetics of carrot pectinmethylesterase (PME): From model systems to real foods. Innov Food Sci Emerg Technol 5: 429–436.

Bauer BA, Knorr D (2005) The impact of pressure, temperature and treatment time on starches: pressure-induced starch gelatinisation as pressure time temperature indicator for high hydrostatic pressure processing. J Food Eng 68: 329–334.

Berlin DL, Herson DS, Hicks DT, Hoover DG (1999) Response of pathogenic Vibrio species to high hydrostatic pressure. Appl Environ Microbiol 65: 2776–2780.

Blaszczak W, Fornal J, Kiseleva VI, Yuryev VP, Sergeev AI, Sadowska J (2007) Effect of high pressure on thermal, structural and osmotic properties of waxy maize and Hylon VII starch blends. Carbohydr Polym 68: 387–396.

Butz P, Edenharder R, Fernández García A, Fister H, Merkel C, Tauscher B (2002) Changes in functional properties of vegetables induced by high pressure treatment. Food Res Int 35: 295–300.

Butz P, Koller D, Tauscher B (1994) Ultra-high pressure processing of onions: chemical and sensory changes. LWT-Food Sci Technol 27: 463–467.

Cano MP, Hernandez A, de Ancos B (1997) High pressure and temperature effects on enzyme inactivation in strawberry and orange products. J Food Sci 62: 85–88.

Castellari M, Matricardi L, Arfelli G, Rovereb P, Amati A (1997) Effects of high pressure processing on polyphenoloxidase enzyme activity of grape musts. Food Chem 60: 647–649.

Castro SM, Saraiva JA, Lopes-Da-Silva JA et al (2008) Effect of thermal blanching and of high pressure treatments on sweet green and red bell pepper fruits (Capsicum annuum L.). Food Chem 107: 1436–1449.

Cheftel JC (1992) Effects of high hydrostatic pressure on food constituents: an overview. In: Balny C, Hayashi R, Heremans K, Masson P (eds) High Pressure and Biotechnology, pp 211–218. John Libbey Eurotext, Montrouge.

Cheftel JC (1995) High pressure, microbial inactivation and food preservation. Food Sci Technol Int 1: 75–90.

Cheftel JC, Culioli J (1997) Effects of high pressure on meat: a review. Meat Sci 46: 211–236.

Crawford YJ, Murano EA, Olson DG, Shenoy K (1996) Use of high hydrostatic pressure and irradiation to eliminate *Clostridium sporogenes* in chicken breasts. J Food Prot 59: 711–715.

Earnshaw RG (1995) Understanding physical inactivation processes: combined preservation opportunities using heat, ultrasound and pressure. Int J Food Microbiol 28: 197–219.

Elgasim EA, Kennick WH (1980) Effect of pressurization of pre-rigor beef muscles on protein quality. J Food Sci 45: 1122–1124.

Farr D (1990) High-pressure technology in the food industry. Trends Food Sci Technol 1: 14–16.

Galazka VB, Leward DA (1995) Developments in high pressure food processing. In: Turner A (ed) Food Technology International Europe, pp 123–125. Sterling Publications International, London.

Gomes MRA, Ledward DA (1996) Effect of high-pressure treatment on the activity of some polyphenoloxidase. Food Chem 56: 1–5.

Gould GW, Sale AJH (1970) Initiation of germination of bacterial spores by hydrostatic pressure. J Gen Microbiol 60: 355–346.

Hayashi R (1995) Advances in high pressure food processing technology in Japan. In: Gaonkar AG (ed) Food Processing: Recent Developments, pp 185–195. Elsevier Science, Amsterdam.

Heinz V, Knorr D (2001) Effects of high pressure on spores. In: Hendrickx MEG, Knorr D (eds) Ultra-high Pressure Treatment of Foods, pp 77–113. Kluwer Academic/Plenum Publishers, New York.

Hendrickx ME, Ludikhuyze LR, van den Broeck I, Weemaes CA (1998) Effects of high-pressure on enzymes related to food quality. Trends Food Sci Technology 9:197–203.

Heremans K (1982) High pressure effects on proteins and other biomolecules. Annu Rev Biophys Bioeng 11: 1–21.

Heremans K (1995) High-pressure effects on biomolecules. In: Ledward DA, Johnston DE, Earnshaw RG, Hasting APM (eds) High-Pressure Processing of Foods, pp 81–97. Nottingham University Press, Nottingham.

Hite BH (1899) The effects of pressure in the preservation of milk. Bull WV Univ Agric Exp Stn Morgantown 58: 15–35.

Hite BH, Giddings NJ, Weakly CE (1914) The effects of pressure on certain microorganisms encountered in the preservation of fruits and vegetables. Bull WV Univ Agric Exp Stn Morgantown 146: 1–67.

Hoover DG (1993) Pressure effects on biological systems. Food Technol 47 (6): 150–155.

Irwe S, Olson I (1994) Reduction of pectinarase activity in orange juice by high pressure treatment. In: Singh RP, Oliveira FAR (eds) Minimal Processing of Foods and Process Optimization, pp 35–42. CRC Press Inc, Boca Raton, FL.

Johnston DE (1995) High pressure effects on milk and meat. In: Ledward DA, Johnston DE, Earnshaw RG, Hasting APM (eds) High-Pressure Processing of Foods, pp 99–122. Nottingham University Press, Nottingham.

Jolibert F, Tonello C, Sagegh P, Raymond J (1994) Les effets des hautes pressions sur la polyphenol oxydase des fruits. Bios Boissons 251: 27–35.

Kalichevsky MT, Knorr D, Lillford PJ (1995) Potential food applications of high-pressure effects on ice-water interactions. Trends Food Sci Technol 6: 253–259.

Knorr D (1994) Hydrostatic pressure treatment of food: microbiology. In: Gould GW (ed) New Methods of Food Preservation, pp 159–175. Blackie Academic and Professional, London.

Knorr D (1995) High pressure effects on plant derived foods. In: Ledward DA, Johnston DE, Earnshaw RG, Hasting APM (eds) High-Pressure Processing of Foods, pp 123–136. Nottingham University Press, Nottingham.

Knorr D, Heinz V, Buckow R (2006) High pressure application for food biopolymers. Biochim Biophys Acta 1764: 619–631.

Lakshmanan R, Patterson MF, Piggott JR (2005) Effects of high-pressure processing on proteolytic enzymes and proteins in cold-smoked salmon during refrigerated storage. Food Chem 90: 541–548.

Linton M, Patterson MF (2000) High pressure processing of foods for microbiological safety and quality. Acta Microbiol Immunol Hung 47: 175–182.

Ludikhuyze L, Claeys W, Hendrickx ME (2000) Combined pressure-temperature inactivation of alkaline phosphatase in bovine milk: a kinetic study. J Food Sci 65: 155–160.

Ludikhuyze L, Van Loey A, Indrawati, Hendrickx M (2002) High pressure processing of fruits and vegetables. In: Jongen W (ed) Fruit and Vegetable Processing: Improving Quality, pp 346–362. Woodhead Publishing Ltd, Cambridge, UK.

Macheboeuf MA, Basset J (1934) Die Wirkung sehr hoher Drucke auf Enzyme. Ergeb Enzymforsch 3: 303–308.

Mackey BM, Forestiere K, Isaacs NS (1995) factors affecting the resistance of *Listeria monocytogenes* to high hydrostatic pressure. Food Biotechnol 9: 1–11.

Mertens B (1995) Hydrostatic pressure treatment of food: equipment and processing. In: Gould GW (ed) New Methods in Food Preservation, pp 135–158. Aspen Publishers Inc, Gaithersburg, MD.

Messens W, Van Camp J, Huyghbaert A (1997) The use of high pressure to modify the functionality of food proteins. Trends Food Sci Technol 8: 107–112.

Meyer RS, Cooper KL, Knorr D, Lelieveld HLM (2000) High-pressure sterilization of foods. Food Technol 54 (11): 67–72.

Mills G, Earnshaw R, Patterson MF (1998) Effects of high hydrostatic pressure on *Clostridium sporogenes* spores. Lett Appl Microbiol 26: 227–230.

Morild E (1981) The theory of pressure effects on proteins. Adv Protein Chem 35: 93–166.

Nakayama A, Yano Y, Kobayashi S, Ishikawa M, Sakai K (1996) Comparison of pressure resistances of spores of six Bacillus strains with their heat resistances. Appl Environ Microbiol 62: 3897–900.

Oey I, Verlinde P, Hendrickx M, van Loey A (2006) Temperature and pressure stability of L-ascorbic acid and/or [6s] 5-methyltetrahydrofolic acid: a kinetic study. Eur Food.

Ogawa H, Fukuhisa K, Kubo Y, Fukumoto H (1990) Pressure inactivation of yeast, mould and pectinesterase in satsuma mandarin juice: Effects of juice concentration, pH, and organic acids and comparison with heat sanitation. Agric Biol Chem 5: 1219–1225.

Otero L, Ramos AM, de Elvira C, Sanz PD (2007) A model to design high-pressure processes towards a uniform temperature distribution. J Food Eng 78: 1463–1470.

Palou E, Hernandez-Salgado C, Lopez-Malo A, Barbosa-Cánovas GV, Swanson BG, Welti J (2000) High pressure-processed guacamole. Innov Food Sci Emerg Technol 1: 69–75.

Palou E, Lopez-Malo A, Barbosa-Cánovas GV, Welti-Chanes J, Swanson BG (1999) Polyphenoloxidase activity and color of blanched and high hydrostatic pressure treated banana puree. J Food Sci 64: 42–45.

Papineau AM, Hoover DG, Knorr D, Farkas DF (1991) Antimicrobial effect of water-soluble chitosans with high hydrostatic pressure. Food Biotechnol 5: 45–57.

Patterson MF, Quinn M, Simpson R, Gilmour A (1995) Sensitivity of vegetative pathogens to high hydrostatic pressure treatment in phosphate-buffer saline and foods. J Food Prot 58: 524–529.

Pauling L (1964) College Chemistry: An Introductory Textbook of General Chemistry. Freeman and Company, San Francisco, CA.

Quaglia RB, Gravina R, Paperi F, Paoletti F (1996) Effect of high pressure treatments on peroxidase activity, ascorbic acid content and texture in green peas. LWT-Food Sci Technol 29: 552–555.

Quiros A, Chichon A, Recio I, Lopez-Fandino R (2007) Analytical, nutritional and clinical methods: the use of high hydrostatic pressure to promote the proteolysis and release of bioactive peptides from ovalbumin. Food Chem 104: 1734–1739.

Rasanayagam V, Balasubramaniam VM, Ting E, Sizer CE, Bush C, Anderson C (2003) Compression heating of selected fatty food materials during high-pressure processing. J Food Sci 68: 254–259.

Rastogi NK, Raghavarao KSMS, Balasubramaniam VM, Niranjan K, Knorr D (2007) Opportunities and challenges in high pressure processing of foods. Crit Rev Food Sci Nutr 47: 69–112.

Reddy NR, Solomon HM, Tetzloff RC, Rhodehamel EJ (2003) Inactivation of *Clostridium botulinum* type A spores by high-pressure processing at elevated temperatures. J Food Prot 66: 1402–1407.

Reps A, Kolakowski P, Dajnowiec F (1998) The effect of high pressure on microorganisms and enzymes of ripening cheeses. In: Isaacs NS (ed) High Pressure Food Science, Bioscience and Chemistry, pp 265–270. Woodhead Publishing, Cambridge, UK.

Sale AJH, Gould GW, Hamilton WA (1970) Inactivation of bacterial spores by hydrostatic pressure. J Gen Microbiol 60: 323–334.

Sancho F, Lambert Y, Demazeau G, Largeteau A, Bouvier JM, Narbonne JF (1999) Effect of ultra-high hydrostatic pressure on hydrosoluble vitamins. J Food Eng 39: 247–253.

Seyderhelm I, Boguslawski S, Michaelis G, Knorr D (1996) Pressure induced inactivation of selected food enzymes. J Food Sci 61: 308–310.

Sila DN, Smout C, Satara Y, Truong V, van Loey A, Hendrickx ME (2007) Combined thermal and high pressure effect on carrot pectinmethylesterase stability and catalytic activity. J Food Eng 78: 755–764.

Stewart CM, Dunne CP, Sikes A, Hoover DG (2000) Sensitivity of spores of *Bacillus subtilis* and *Clostridium sporogenes* PA 3679 to combinations of high hydrostatic pressure and other processing parameters. Innov Food Sci Emerg Technol 1: 49–56.

Stolt M, Oinonen S, Autio K (2000) Effect of high pressure on the physical properties of barley starch. Innov Food Sci Emerg Technol 1: 167–175.

Sun N, Seunghwan L, Kyung BS (2002) Effect of high-pressure treatment on the molecular properties of mushroom polyphenoloxidase. LWT-Food Sci Technol 35: 315–318.

Tamaoka T, Hayashi R (1991) High-pressure effect on Millard reaction. Agric Biol Chem 55: 2071–2074.

Tauscher B (1995) Pasteurization of food by hydrostatic high pressure: chemical aspects. Lebensm-Unters-Forsch 200: 3–13.

Timson WJ, Short AJ (1965) Resistance of microorganisms to hydrostatic pressure. Biotechnol Bioeng 7: 139–59.

Toepfl S, Mathys A, Heinz V, Knorr D (2006) Potential of high hydrostatic pressure and pulsed electric fields for energy efficient and environmentally friendly food processing. Food Rev Int 22: 405–423.

Torres JA, Velazquez G (2008) Hydrostatic pressure processing of foods. In: Jun S, Irudayaraj J (eds) Food Processing Operations Modeling: Design and Analysis, pp 173–212. CRC Press, Boca Raton, FL.

van den Broeck I, Ludikhuyze LR, van Loey AM, Hendrickx ME (2000) Inactivation of orange pectinesterase by combined high-pressure and temperature treatments: a kinetic study. J Agric Food Chem 48: 1960–1970.

Weemaes CA, de Cordt SV, Ludikhuyze LR, van den Broeck I, Hendrickx ME, Tobback PP (1997) Influence of pH, benzoic acid, EDTA, and glutathione on the pressure and/or temperature inactivation kinetics of mushroom polyphenoloxidase. Biotechnol Prog 13: 25–32.

Weemaes CA, Ludikhuyze L, van den Broeck I, Hendrickx ME (1998) High pressure inactivation of polyphenoloxidases. J Food Sci 63: 873–877.

Weemaes CA, Ludikhuyze L, van den Broeck I, Hendrickx ME (1999) Kinetic study of antibrowning agents and pressure inactivation of avocado polyphenoloxidase. J Food Sci 64: 823–827.

Wicker L, Temelli F (1988) Heat inactivation of pectinarase in orange juice pulp. J Food Sci 53: 162–164.

Yemenicioglu A, Ozkan M, Cemeroglu B (1997) Heat inactivation kinetics of apple polyphenoloxidase and activation of its latent form. J Food Sci 62: 508–510.

Chapter 15
Protective and Preserving Food Packaging

15.1 Concepts and Definitions

Packaging of food materials serves different purposes, such as acting like a material handling tool and performing a food processing aid role. As a handling tool, packaging contains the desired unit amount of food and may facilitate the assembly of individual items into aggregates. For example, some fluid foods are package into bottles, which may be placed into boxes and, finally, these boxes may be arranged into easily handled pallets. In the case of packaging as a processing aid, an example is that of metal containers used in food sterilization, which serve not only a protective function but also ensure dimensional stability for the food to maintain a certain shape and location to facilitate heat-penetration calculations. Protection of the food product from physical, microbiological, or any other sort of damage is the most important aspect of packaging. The protective function of packaging takes into account the aspects of the quality of the food being protected. The quality of the food products, as they reach the consumer, depends on the conditions of the raw material, the processing method, the severity of the process, and the conditions of storage. The chemical, physical, and biological mechanisms involved in food deterioration are a direct function of diverse environmental factors, and the most pertinent barrier property of the food package differs for each product. Some of the environmental factors and the corresponding relevant packing properties are listed in Table 15.1.

The functionality of food packaging should include aesthetic aspects and its suitability to act as a barrier to, mainly, light and moisture. Light transmission is required to display package contents, but is restricted when foods may be deteriorated by light causing reactions such as the oxidation of lipids and color changes. Selection of the insulation properties of packaging materials is performed as a means of controlling the rate of heat transfer to foods. The rates of moisture or oxygen migration are a direct function of the shelf life of dehydrated foods, being considered the most important properties of food packaging materials.

E. Ortega-Rivas, *Non-thermal Food Engineering Operations*,
Food Engineering Series, DOI 10.1007/978-1-4614-2038-5_15,
© Springer Science+Business Media, LLC 2012

Table 15.1 Important
interactions between
package properties and
environmental aspects

Environmental aspect	Relevant package property
Mechanical shocks or stresses	Strength
Temperature	Thermal conductivity
Light intensity	Light transmission
Pressure of oxygen or water vapor	Permeability
Biological agents	Penetrability

Vapor or gases, such as oxygen, nitrogen, and carbon dioxide, can penetrate the packaging materials through microscopic pores or activated diffusion owing to concentration gradients. The gas or vapor permeability for one-dimensional diffusion can be calculated by Fick's law of diffusion:

$$J = -D_g A \frac{dC}{dX}, \tag{15.1}$$

where J is the rate of diffusion, D_g is the gas diffusivity, A is the surface area, C is the gas concentration, and X is the distance in the direction of diffusion. The gas concentration can be defined by Henry's law as

$$C = SP, \tag{15.2}$$

where S is the gas solubility and P is the gas partial pressure.

Substituting Eq. 15.2 into Eq. 15.1, we obtain

$$J = -D_g SA \frac{dP}{dX}. \tag{15.3}$$

A permeability coefficient B can be defined as the product of the gas diffusivity coefficient and the gas solubility ($B = D_g S$), and describes the amount of gas permeating through a unit film thickness per unit time, per unit packaging surface, and per unit pressure difference between the environment and the packaged material. Equation 15.3 can, therefore, be simplified to the form

$$J = -BA \frac{\Delta P}{\Delta X}. \tag{15.4}$$

Intact packaging materials are a barrier against microorganisms, but seals are a potential source of contamination. Contaminated air or water may sometimes be drawn through pinholes in hermetically sealed containers as the head space vacuum forms. The ability of packages to protect foods from mechanical damage can be evaluated by tensile strength, Young's modulus of elasticity, or yield strength.

The water vapor transmission characteristics of a packaging material are described by the water transmission rate, the water vapor permeance (WVP), and the water vapor permeability. The water vapor transmission rate (WVTR) can be defined as 1 g of water vapor transmitted from 1 m^2 of film area per day. In mathematical form, such a rate may be represented by

$$WVTR = \frac{24m_v}{tA},$$ (15.5)

where m_v is the mass gain or loss, t is the time, and A is the surface area of the packaging material.

Taking into account the pressure factor, we can express Eq. 15.5 for the conditions referred to previously when a vapor pressure difference of 1 mmHg is maintained. The above expression is, thus, transformed into

$$WVP = \frac{WVTR}{\Delta P} = \frac{WVTR}{(P_1 - P_2)} = \frac{WVTR}{P_s(RH_1 - RH_2)},$$ (15.6)

where ΔP is the vapor pressure difference, $RH_1 = P_1/P_s$ and $RH_2 = P_2/P_s$ are the relative humidity on each side of the film specimen, and P_s is the saturation vapor pressure.

An Arrhenius-type equation can be used to measure the effect of temperature on the permeability coefficient B, as follows:

$$B = B_0 e^{-E_a/RT},$$ (15.7)

where B_0 is a constant, E_a is the activation energy, R is the universal gas constant, and T is the absolute temperature.

15.2 Packaging Materials

Natural materials, such as skins, leaves, and bark, were the original sources for food packaging. Throughout the years, the sources have been diversified and now include natural and synthetic materials. Packaging materials are commonly categorized into rigid and flexible structures. Plastic film, foil, paper, and textiles are flexible materials, whereas glass, metals, and hard plastics constitute rigid materials.

Plastic materials have been used as food packaging materials practically since their appearance in the 1960s. Plastics for food packaging are used for both flexible and rigid applications and the examples are varied: poly(ethylene terephthalate), high- or low-density polyethylene, oriented polypropylene, poly(vinyl chloride), poly(vinylidene chloride), ethylene vinyl alcohol, polypropylene, and nylon. Advances in polymer science and technology have allowed the original limitations of some plastic materials, such as permeability and heat resistance, to be overcome, principally by improvement of barrier properties. High-nitrile resins, for instance, are used extensively in packaging because of their gas barrier and chemical resistance attributes. Food products currently packaged in high-nitrile resins include processed meats, cheeses, bakery products, sauces, peanut butter, and cooking oil. Ethylene vinyl alcohol, which offers a superior barrier for gases, odors, and aromas, was the first material used for orange juice distribution in 1.89-l containers because its structure provided an excellent barrier to oxygen and prevented flavor scalping. Poly(vinylidene chloride), commercialized under the trade name Saran®, is the oldest material to be used with appropriate barrier characteristics for liquids and gases. It has been widely used in the form of coating on a polyester, cellophane, and polypropylene for packaging of snack foods, processed meats, and cheese.

Metalizing can be used to improve the barrier properties of a clean film. Metal adhesion is enhanced after a special treatment of the film and then it is plated with a thin coating of metal, normally aluminum. The layer of aluminum is about 30 nm thick, and provides barrier properties that are difficult to achieve by other methods. Metalized films improve greatly the moisture and gas barrier properties, and also keep out of light that may otherwise cause rancidity of most snack products.

Metal food packaging foils are generally fabricated from pure aluminum, with the use of some other metals, such as lead, tin, zinc, and steel, not being able to compete against use of aluminum. The thickness of aluminum foils ranges from 4 to 150 µm, and practically determines their protective properties. Low-thickness foils normally have microscopic discontinuities or pinholes in order to allow limited diffusion of gases and vapors. The properties of thinner foils can be improved by combination with one or more plastics in the form of coatings or laminations.

Paper is also used in a great proportion for packaging and distribution of different food items, mainly because of its low cost, versatility, and availability. The packaging properties of paper differ considerably depending on the fabrication process and the additional treatments to which it may be subjected. The strength and mechanical properties may be improved with inclusion of fillers and binding materials. Some physicochemical properties, such as permeability to liquids, vapors, and gases, can be modified by impregnating, coating, or laminating paper with different materials, such as waxes, plastics, resins, gums, and adhesives. A good example of this type of treatment is the use of paperboard layered with plastic film and aluminum foil in aseptic processing packages. The typical container used in such a process is manufactured with paperboard, which comprises five to seven plies including polyethylene, paper, and aluminum foil.

Similarly to the case of plastic materials, paper can be used as a flexible or rigid raw material for food packaging. Flexible packaging applications include the use of papers for wrapping materials, as well as for fabrication of bags, envelopes, liners, and overwraps. Examples of types of papers for this purpose are kraft paper, greaseproof papers, glassine papers, and waxed papers. Rigid containers comprise a great variety of types, including the above-mentioned example of containers for aseptic processing, as well as cartons, boxes, drums, and liquid-tight cups. They can be manufactured from paperboard, laminated papers, or corrugated board, in addition to the previously mentioned laminated and coated paperboards.

Glass, in the form of bottles, jars, jugs, and some other types of containers, has been used as a food packaging material for centuries. The most important valued attributes of glass are its moldability, inertness, and strength. A major disadvantage of glass is its fragility. Minor modifications in the composition of glass may improve its properties. For example, color can be controlled by inclusion of small amount of oxides of metals such as chromium, cobalt and iron, whereas semiopacity can be provided by addition of fluorinated compounds. In terms of mechanical properties, since glass is brittle, its tensile strength depends greatly on the condition of its surface. Commercial glass containers have only a small fraction of the potentially possible tensile strength, as their surface conditions cannot be kept at the perfect level required for maximum strength. Consequently, glass containers have to be made out of fairly thick glass, meaning they have considerable weight. An alternative is the use of very thin glass encased in a plastic exterior in order to obtain a container combining the advantage of an inert inner surface and a resistant external surface.

The widest use of a metal in food packaging is represented by the so-called tin can, which has been traditionally employed for contained heat sterilization of many foodstuffs. The tin can is made of tinplate, which consists of a base sheet of steel with a coating of tin applied by "hot dipping" or by an electrolytic process. The amount of tin depends on the process and the type of can. The thickness of a thin plate is typically approximately 0.25 mm, with the contribution of the tin coating being only about 1.3 μm of such thickness. To make the can more suitable for specific applications, enamels or linings can be applied to either the coated tinplate or the steel sheet without the coating. The composition of these linings depends on the specific application. Plastics, shellacs, resins, glass, and inorganic oxides have all been used as lining materials for tinplates. Apart from the typical tin can, developments in the can industry have led to the use of metals other than tin for the manufacture of metal containers. The most important case is that of the aluminum can, used mainly for beer, carbonated beverages, and some fruit juices. Other innovations include tin-free steel utilizing direct chromium-containing coatings or thin thermoplastic coatings, as well as the use of cans with cemented or welded, rather than soldered, side seams.

15.3   Methods of Packaging: Modified and Controlled Atmosphere Packaging

A number of food materials are advantageously stored in atmospheres different from normal air. For example, the fresh fruit respiration process takes up oxygen and evolves carbon dioxide. The rate of respiration can be reduced by cooling, thus extending the shelf life, but can be reduced even further by storing the fruit in an atmosphere richer in carbon dioxide and poorer in oxygen than normal air. This technique, known as controlled atmosphere or gas storage, is in extensive commercial use. The oxygen and CO_2 levels used differ markedly between varieties and are controlled to optimum values, since too great a modification can lead to secondary spoilage (Salunkhe 1974). Storage chambers for controlled atmosphere storage are not large, since it is desirable to fill the stores and close them within 7 days to take full advantage of the artificially generated atmospheres. The storage chambers can be leak-tight, to rely solely on the respiratory activity of the fruit to control the atmosphere, or can accept some degree of leakage and inert gases to modify the atmosphere. Nitrogen gas or exhaust gases produced from burning hydrocarbon fuels in air can be used to aid in controlling the atmosphere, so as to create the desired atmosphere quicker than could be achieved by relying on the respiration rate of the fruit alone. Fruit packed in perforated or partly closed plastic containers can develop local atmospheres differing from air. Such storage has been called modified atmosphere storage, but it has the disadvantage that the composition of the atmosphere cannot be tightly controlled. Another method for modifying the storage atmosphere is to reduce the pressure of the air surrounding the fruit, but the effect appears to be largely one of mechanically reducing the oxygen concentration (Salunkhe 1974).

In terms strictly of packaging, minimal food preservation methods aimed at maintaining quality and extending storage life are known as modified atmosphere packaging (MAP) and controlled atmosphere packaging. The term "modified atmosphere technology" includes controlled atmosphere storage, ultralow oxygen storage, superatmospheric oxygen storage, gas packaging, vacuum packaging, passive MAP, and active packaging (Hertog and Banks 2003). The common principle of all these techniques is the manipulation or control of the composition of the atmosphere that surrounds a product in order to maintain quality during the shelf life. Although MAP requires strict control and adherence to quality assurance programs in order to be successful, the technology has shown rapid growth, with the MAP market exceeding other food processing categories such aseptic, retort pouch, and canned or frozen foods (Forcinio 1997).

"Modified atmosphere" and "controlled atmosphere" have been used indistinctly but they differ considerably. "Controlled atmosphere" implies a precise control of the atmosphere in specific concentrations, whereas "modified atmosphere" means such a control is not applied. "Modified atmosphere packaging" (MAP) has been defined as the enclosure of food products in gas-barrier materials, in which the gaseous environment has been changed to inhibit spoilage agents so as to maintain a higher quality of perishable foods during their natural life, or to extend the shelf life (Church and Parsons 1995). In controlled atmosphere storage, the composition of the atmosphere is controlled throughout the storage period, whereas in MAP, the composition is modified at the beginning and there is additional modification caused by food characteristics (respiration rate and biochemical changes), the properties of the package (permeability), and the storage conditions (temperature). "Modified atmosphere packaging" (MAP) is defined as a form of packaging that involves the removal of air from the pack and its replacement with a single gas of a mixture of gases (Blakistone 2000). Depending on the type of product, a specific mixture of gas will be used. Two types of MAP have been defined: vacuum packaging and gas, and gas flush or gas exchange packaging (Church and Parsons 1995). The modification of the atmosphere can be achieved using different methods, such as vacuum packaging, gas packaging, a compensated vacuum, passive atmosphere modification, or active packaging. Oxygen, CO_2, and ethylene can be used as absorbers (Blakistone 2000). Vacuum packaging is used to inhibit aerobic bacteria, oxidative reactions, and spoilage agents by means of making oxygen unavailable in order to achieve a longer shelf life. Oxygen is reduced to less than 1%, but some anaerobic/microaerophilic organisms can survive and non-oxidative reactions can occur, contributing to some sort of deterioration. Deformation through product compression occurs, which limits vacuum packaging of soft products such as bakery items. Gas packaging, on the other hand, tries to overcome the problems related to the use of a vacuum. The modification of the headspace atmosphere inside a package can be obtained by mechanical replacement of air with a gas mixture or by generating the atmosphere in two ways: actively or passively. The mechanical replacement of the atmosphere is obtained through gas flushing and a compensated vacuum.

The atmosphere generated by the passive method involves the packaging of the product under ambient conditions using a permeable film (Kader et al. 1998) and is commonly used in packaging of fruits and vegetables. During the storage, the respiration of the product consumes the surrounding oxygen, producing CO_2, which changes the atmosphere. Passive atmosphere modification is a complex process where the interplay of many elements conjugates to modify the atmosphere, until a steady state is reached (Al-Ati and Hotchkiss 2002). The active method involves the addition of certain additives to the packaging film or inside the packaging containers to modify the headspace atmosphere. Some of the additives used include oxygen absorbents, carbon dioxide absorbents/emitters, and ethylene absorbents (Kader et al. 1998).

"Equilibrium modified atmosphere packaging" is yet another term describing a systematic approach to alter of the air around the commodity to a gas combination of 1–5% O_2 and 3–10% CO_2. If the respiration rate of the produce is known at the desired storage temperature, a packaging configuration can be designed to keep the gas concentrations constant during the storage time of the produce. Packages are designed by selecting the appropriate film permeability at the storage temperature, the amount of minimally processed vegetables in the package, and the package dimensions.

A key aspect of MAP technology is the storage temperature, considered the single most important element for this technology, even more important than the modified atmosphere (Brody 1996). Temperature control, in conjunction with good sanitation practices and high initial product quality, ensure the success of MAP. Programs such as good manufacturing practices and hazard analysis critical control points should be maintained when processing foods under MAP. A common misconception is that MAP can overcome temperature and microbiological abuse, which is wrong, since MAP can help extend shelf life but will not improve quality, although it can slow the loss of quality (Brody 1999; Day 1991; Zagory 2000).

15.4   Shelf Life and Quality of Packaged Foods

The shelf life of packaged foods is controlled by the properties of the foodstuff, such as water activity, pH, and potential microbial/enzymic deterioration, along with the barrier properties of the packaging material with regard to oxygen, light, moisture, and carbon dioxide. Moisture loss or uptake is one of the most important factors determining the shelf life of foods. The microclimate within a package is controlled by the vapor pressure of moisture in the food at the storage temperature. The changes in moisture content depend on the WVTR of the package. Control of moisture exchange is also necessary to prevent condensation inside the package, which may result in mold growth. To control the moisture content inside the package, the water vapor permeability of the packaging material, its surface area, and its thickness should be selected on the basis of the required storage time or shelf life.

On the basis of moisture sorption isotherms, simple equations can be used to estimate the gain or loss of moisture held in a semipermeable membrane. Labuza and Contreras Medellin (1981) derived the following expression to predict the change in weight in packaged dried foods:

$$\ln\left(\frac{m_e - m_i}{m_e - m}\right) = \frac{B}{x}\frac{A}{W_s}\frac{P_0}{b}t, \tag{15.8}$$

where m_e is the moisture content of the isotherm based on a straight line approximation, m_i is the initial moisture content, m is the moisture content at time t, B/x is the film permeability, A is the film packaging area, W_s is the weight of dry solids in the package, P_0 is the vapor pressure of pure water at temperature T, and b is the isotherm slope.

Some foods are prone to oxidation, so they should be stored using packaging materials with low oxygen permeability. Sometimes, desiccants or oxygen scavengers can be placed inside a package as a means of controlling water vapor and oxygen contents. Packaging should also retain desirable odors, and may prevent odor pickup from plasticizers, printing inks, adhesives, or solvents used in the fabrication of the packaging material. Most foods deteriorate more quickly at higher temperatures, so the storage conditions should be controlled to minimize temperature fluctuations.

MAP and controlled atmosphere packaging are used extensively in most food categories, such as fresh meats, cured and processed meats, fruits and vegetables, bakery products, dairy products, and seafood. More recent applications are for precooked pastas, prepared salads, and cottage cheese. MAP is part of the food industry around the world and new applications are evaluated daily, with several publications dealing with MAP and intelligent and active packaging describing recent advances in these technologies. Research on this topic has been conducted for many years (Blakistone 1999). MAP and controlled atmosphere storage have been the subject of numerous reviews and analyses from different points of view: microbiological safety (Brody 1989; Farber et al. 2003), processing of commodities

(Jayas and Jeyamkondan 2002; Young et al. 1988), packaging of fresh produce (Farber et al. 2003; Soliva-Fortuny and Martin-Belloso 2003; Zagory and Kader 1988), processing of fish and seafood products (Sivertsvik et al. 2002; Skura 1991), general approaches (Church and Parsons 1995; Day 1989), quality and shelf life (Rai et al. 2002), processing of grains (Jayas and Jeyamkondan 2002), and mathematical modeling (Al-Ati and Hotchkiss 2002).

There has been an attempt to break from the traditional MAP process based on the use of gas mixtures and films, by using an alternative approach incorporating intrinsic and extrinsic elements to ensure food safety proactively. This approach is denominated "novel MAP," and the intrinsic elements are intelligent or active packaging, whereas the extrinsic elements refer to the processing environment (Yuan 2003). Intelligent or active packaging technologies are denominated "interactive packaging," a concept used to describe those technologies incorporated into the package or the packaging film to maintain or monitor the quality or safety of the product. In terms of the extrinsic elements, the quality of the product and the storage temperature are considered the most important. "Active packaging" refers to those technologies intended to interact with the internal gas environment and/or directly with the product in order to improve quality or shelf life. A modification of the gas environment occurs by removing (absorbing or scavenging) gases from or adding gases or vapors (emit) to the package. Some of the additives used include oxygen absorbents, carbon dioxide scavengers, humidity absorbers, controllers, enzymically active systems, and antimicrobial agents (Kader et al. 1998). Numerous studies have been conducted on active packaging. Perry et al. (2006) presented an overview of the utilization of active and intelligent packaging systems for meat- and muscle-based products. Gill (2003) presented some of the applications of MAP technology on meats. The main route of gas exchange in MAP is through gas-permeable films, perforations in films, or both. In active or intelligent packaging, gas can be modified by flushing it with a specific mixture to get a precise initial composition of the atmosphere. Gases may be also actively released or scavenged in the package, a partial vacuum can be imposed, biologically active materials can be incorporated in the package, or sensors may be used to respond to the product or package conditions (Beaudry 2007). A description of the additives used in active packaging for a variety of foods was presented by Lopez-Rubio et al. (2005).

Numerous factors influence MAP design. Mathematical models are useful tools for defining the package requirements for MAP, and the development of models for MAP has been very productive. Those models are based, mainly, on mass balances for O_2 and CO_2 to describe the interactions among the product respiration, the film permeability, and the environment. In many cases, it takes a long time to reach gas equilibrium, or there is a transient period during which the product is exposed to an unsuitable gas composition, thus, preventing the positive effects of the steady-state atmosphere. In a MAP system, gas exchange depends on the food product type, freshness, microbial load, lipid content, package permeability, temperature, and storage time. Gas transfer in MAP systems has been described by different models. A summary of the major models developed for respiration was presented by Fonseca et al. (2002), where their strong points and limitations are discussed.

One of the most suitable models based on enzyme kinetics was proposed by Lee et al. (1991) for predicting respiration rates of fresh produce as a function of O_2 and CO_2 concentrations. The model describes the dependence of respiration on O_2 as assumed to follow a Michaelis–Menten-type equation and the effect of CO_2 is considered to be one of uncompetitive inhibition. In different studies, experimental data have provided a good fitting to this proposed model and even, recently, one study conducted with mushrooms confirmed that the respiration rates of commodities stored under MAP can be predicted using the enzyme kinetics approach (Deepak and Shashi 2007). Modeling for specific applications such as for microperforated or macroperforated films has been reported (Techavises and Hikida 2008). Some of them include O_2, CO_2, and N_2 exchanges in MAP, whereas others include atmospheric gas and water vapor exchanges. The model is composed of an empirical equation for effective permeability and is based on Fick's law for prediction of gas and water vapor exchange in MAP films with macroperforations. Another model on microperforated films for determination of gas transmission rates (O_2 and CO_2), agrees with the modified Fick's law. This model allows the estimation of the gas flow through the film by a simple measurement of the perforation size with an ocular microscope and can be used for a wide range of conditions, given the range of the dimensions of the microperforations used, a feature that establishes the difference from other models (Gonzalez et al. 2008).

A model that predicts the respiration rate under superatmospheric oxygen, different from the Michaelis–Menten-type enzyme kinetics approach, is proposed for fresh cut melon, where the experimental data showed that the best fit to concentrations inside the package was achieved with the Weibull model for O_2 concentration and the logistic model for CO_2 concentration. The respiratory activity of fresh cut melon stored under 70 kPa O_2 atmospheres was adequately described through CO_2 production rates (Oms-Oliu et al. 2008). The respiration rate of bananas has been described by a mathematical model derived from data obtained by storing such fruit at 10, 15, 20, 25 and 30°C. The generated data was used to develop two different models based on regression analysis and enzyme kinetics, respectively. Both models showed good agreement with the experimentally estimated respiration rate (Bhande et al. 2008). Another example of the advances in modeling of MAP is the model proposed by Charles et al. (2005) consisting of modeling an active package with an oxygen absorber for respiring products that allows prediction of the gas exchange dynamics in an active MAP associated with an oxygen absorber. The model was based on the following parameters: vegetable respiration (tomato as a model), film permeability, and oxygen absorption kinetics of the absorber.

References

Al-Ati T, Hotchkiss, JH (2002) Application of packaging and modified atmosphere to fresh-cut fruits and vegetables. In: Lamikanra O (ed) Fresh Cut Fruits and Vegetables, pp 305–338. CRC Press, Boca Raton, FL.

Beaudry R (2007) MAP as a basis for active packaging. In: Wilson CH (ed) Intelligent and Active Packaging for Fruits and Vegetables, pp 31–56. CRC Press, Boca Raton FL.

Bhande SD, Ravindra MR, Goswami TK (2008) Respiration rate of banana fruit under aerobic conditions at different storage temperatures. J Food Eng 87: 116–123.

Blakistone BA (1999) Introduction. In: Blakistone BA (ed) Principles and Applications of Modified Atmosphere Packaging of Foods, pp 1–13. Aspen Publishers Inc, Gaithersburg, MD.

Blakistone BA (2000) Preface. In: Blakistone BA (ed) Principles and Applications of Modified Atmosphere Packaging of Foods, pp 1–4. Aspen Publishers Inc, Gaithersburg MD.

Brody AL (1989) Controlled/Modified Atmosphere/Vacuum Packaging of Foods. Food & Nutrition Press, Trumbull CT.

Brody AL (1996) Fundamentals of modified atmosphere packaging. Paper presented at the Conference of the Society of Manufacturing Engineers, Dearborn, MI, December, 1996.

Brody AL (1999) Markets for MAP foods. In: Blakistone BA (ed) Principles and Applications of Modified Atmosphere Packaging of Foods, pp 14–38. Aspen Publishers Inc Gaithersburg, MD.

Charles F, Sanchez J, Gontard N (2005) Modeling of active modified atmosphere packaging of endives exposed to several postharvest temperatures. J Food Sci 70: E443–E449.

Church I, Parsons AL (1995) Modified atmosphere packaging technology: a review. J Sci Food Agric 67: 143–152.

Day BPF (1989) Modified atmosphere packaging. Principles of production and distribution of chilled foods. In: Proceedings International Conference on Modified Atmosphere Packaging, pp 17–20. Campden and Chorleywood Food Research Association, Chipping Campden

Day BPF (1991) A perspective of modified atmosphere packaging of fresh produce in Western Europe. Food Sci Technol Today 4: 215–221.

Deepak RR, Shashi P (2007) Transient state in-pack respiration rates of mushroom under modified atmosphere packaging based on enzyme kinetics. Biosyst Eng 98: 319–326.

Farber JN, Harris LJ, Parish ME, et al (2003) Microbiological safety of controlled and modified atmosphere packaging of fresh and fresh-cut produce. Compr Rev Food Sci Food Saf 2: 142–160.

Fonseca SC, Oliveira FAR, Brecht JK (2002) Modelling respiration rate of fresh fruits and vegetables for modified atmosphere packaging: a review. J Food Eng 52: 99–119.

Forcinio H (1997) Mapping out a fresh approach. Prep Foods 4: 62–64.

Gill., C.O. 2003. Active packaging in practice: meat. In: Ahvenainen R (ed) Novel Food Packaging Techniques, pp 365–383. Woodhead Publishing Ltd. Cambridge UK.

Gonzalez J, Ferrer A, Oria R, Salvador ML (2008) Determination of O_2 and CO_2 transmission rates through microperforated films for modified atmosphere packaging of fresh fruits and vegetables. J Food Eng 86: 194–201.

Hertog MLATM, Banks NH (2003) Improving MAP through conceptual models. In: Ahvenainen R (ed) Novel Food Packaging Techniques, pp 337–362. Woodhead Publishing Ltd, Cambridge, UK.

Jayas DS, Jeyamkondan S (2002) Modified atmosphere storage of grains, meats, fruits and vegetables. Biosyst Eng 82: 235–251.

Kader AA, Singh RP, Mannapperuma JD (1998) Technologies to extend the refrigerated shelf life of fresh fruits. In: Taub IA, Singh RP (eds) Food Storage Stability, pp 419–434. CRC Press Boca Raton, FL.

Labuza TP, Contreras Medellin R (1981) Prediction of moisture protection requirements for foods. Cereal Foods World 26: 335–343.

Lee DS, Haggar P, Lee J, Yam KL (1991) Model for fresh produce respiration in modified atmospheres based on principles of enzyme kinetics. J Food Sci 56: 1580–1585.

Lopez-Rubio A, Almenar E, Hernández-Muñoz P, Lagarón JM, Catalá R, Gavara R (2005) Overview of active polymer-based packaging technologies for food applications. Food Rev Int 20: 357–387.

Oms-Oliu G, Odriozola-Serrano I, Soliva-Fortuny R, Martín-Belloso O (2008) The role of peroxidase on the antioxidant potential of fresh-cut 'Piel de Sapo' melon packaged under different modified atmospheres. Food Chem 106: 1085–1092.

Perry JP, Grady MN, Hogan SA (2006) Past, current and potential utilization of active and intelligent packaging systems for meat and muscle based products: a review. Meat Sci 74: 113–130.

Rai DR, Oberoi HS, Baboo B (2002) Modified atmosphere Packaging and its effect on quality and shelf life of fruits and vegetables-an overview. J Food Sci Technol 39:199–207

Salunkhe DK (1974) Storage Processing and Nutritional Quality of Fruits and Vegetables. CRC Press, Boca Raton, FL.

Sivertsvik M, Jeksrud WK, Rosnes JT (2002) A review of modified atmosphere packaging of fish and fishery products-significance of microbial growth, activities and safety. Int J Food Sci Technol 37: 107–127.

Skura BJ (1991) Modified atmosphere packaging of fish and fish products. In: Ooraikul B, Stiles ME (eds) Modified Atmosphere Packaging of Food, pp 148–168. Ellis Horwood, New York.

Soliva-Fortuny RC, Martin-Belloso O (2003) New advances in extending the shelf lie of fresh-cut fruits: a review. Trends Food Sci Technol 14: 341–353.

Techavises N, Hikida Y (2008) Development of a mathematical model for simulating gas an water vapor exchanges ion modified atmosphere packaging with macroscopic perforations. J Food Eng 85: 94–104.

Young LL, Reviere RD, Cole AB (1988) Fresh red meats: a place to apply modified atmospheres. Food Technol 42(9): 65–66, 68–69.

Yuan JTC (2003) Modified atmosphere packaging for shelf-life extension. In: Novak JS, Sapers GM, Juneja VK (eds) Microbial Safety of Minimally Processed Foods, pp 205–220. CRC Press, Boca Raton FL.

Zagory D (2000) What modified atmospheres packaging can and can't do for you. Paper presented at the 16th Annual Postharvest Conference, Yakima, WA, 14–15 March, 2000.

Zagory D, Kader AA (1988) Modified atmosphere packaging of fresh produce. Food Technol 42(9): 70–77.

Chapter 16
Other Methods

16.1 Non-conventional Chemical Reagents

The use of chemicals as preservatives is an everyday practice in food processing to achieve sufficiently long shelf life for foods, as well as a high degree of safety with respect to food-borne pathogenic microorganisms. In spite of this, consumers increasingly refuse foods prepared with preservatives of chemical origin owing, mainly, to justified or unjustified concerns regarding the various negative side effects that chemicals in foods may cause. Since, as described in previous chapters, the search is still ongoing for milder food preservation techniques to produce foods with a more natural appearance and nutritious quality than can be achieved by traditional food preservation methods, the need for alternative chemical compounds may also be considered a priority for food processors. Natural antimicrobial chemicals from different sources have been used as an alternative to conventional chemicals used in food processing, and they may be generally termed "non-conventional chemical reagents."

Natural antimicrobial chemicals of plant origin can be used efficiently in a number of food preservation applications. Most of these chemicals are identified as secondary metabolites, mainly being of terpenoid or phenolic origin. The rest are hydrolytic enzymes, such as glucanases and chitinases, and proteins with antimicrobial activity acting specifically on membranes of invading microorganisms (Bowles 1990). The main antimicrobials of plant origin are phytoalexins, organic acids, phenolic compounds, and essential oils.

Phytoalexins are defined as host-synthesized, low molecular weight, wide-spectrum antimicrobial compounds, whose synthesis from distant precursors is induced by plants in response to microbial infection. More than 200 different phytoalexins have been identified in more than 20 plant families. Phytoalexins are classified as antibiotics, and are generally effective against phytopathogenic fungi (Van Etten 1976), although activity has also been reported against Gram-positive bacteria (Lund and Lyon 1975). The use of phytoalexins as food preservatives has been suggested from some time (Gould 1996), but there are still very few

E. Ortega-Rivas, *Non-thermal Food Engineering Operations*,
Food Engineering Series, DOI 10.1007/978-1-4614-2038-5_16,
© Springer Science+Business Media, LLC 2012

examples of the actual use of these compounds in food preservation, possibly because they only show adequate antimicrobial effects at relatively high concentrations.

Organic acids, such as citric, succinic, malic, and tartaric acids, which are commonly found in fruits and vegetables, also show some degree of antimicrobial properties. Lactic and propionic acids do not occur naturally in foods, other than in trace amounts, but are readily formed during natural fermentation. The antimicrobial activity of various organic acids has been well documented (Stratford and Eklund 2003) and they are supposed to target cell walls, cell membranes, metabolic enzymes, protein synthesis systems, and genetic material. They are, thus, active against a wide range of microorganisms. The organic acids naturally contained in some crops may well contribute to their natural resistance against microbial damage. Many organic acids and their derivatives have been used as food preservatives.

Phenolic acids are plant metabolites that are widely spread throughout the plant kingdom. Recent interest in phenolic acids stems from their potential protective role, through ingestion of fruits and vegetables, against oxidative damage diseases (coronary heart disease, stroke, and cancers). Phenolic compounds are essential for the growth and reproduction of plants, and are produced as a response to defend injured plants against pathogens. Phenols, sometimes called phenolics, are a class of chemical compounds consisting of a hydroxyl functional group ($-OH$) attached to an aromatic hydrocarbon group. The simplest of the class is phenol (C_6H_5OH). Some phenols are germicidal and are used in formulating disinfectants. The role of plant phenolics in the chemical defense of plants against microorganisms is still unclear. Nonetheless, it has been observed that a vast range of phenolic compounds contribute to the defense mechanisms of plant tissues and, also, to the sensory and nutritional qualities of fresh or processed plants. Phenolic compound are varied and abundant and include phenol, the parent compound, which is used as a disinfectant and for chemical synthesis, polyphenols such as flavonoids and tannins, capsaicin, the pungent compound of chilli peppers, and the amino acid tyrosine (Nychas 1995). The antimicrobial activities of the naturally occurring phenols from olives, tea, and coffee have been studied in more detail than those from other sources, which may be partly due to the high value of the products being processed. Phenolics from spices, such as gingerol, zingerone, and capsaicin, have been found to inhibit germination of bacterial spores. Native plant phenols exhibit important food preservation characteristics and have, as a group, an impressive antimicrobial spectrum. Their deliberate use as food preservatives has, however, rarely been exploited.

Essential oils are liquids that are generally distilled from the leaves, stems, flowers, bark, roots, or other elements of a plant. Essential oils, contrary to the use of this word, do not really have an oily feeling. Most essential oils are clear, but some oils such as patchouli, orange, and lemongrass are amber or yellow. Essential oils contain the true essence of the plant from which they were derived. Essential oils are highly concentrated and a little goes a long way. The chemical composition and aroma of essential oils can provide valuable psychological and physical therapeutic benefits. Essential oils are mostly soluble in alcohol and, to a limited extent, in water.

They consist of mixtures of esters, aldehydes, ketones, and terpenes (Hargreaves et al. 1975). The antimicrobial properties of essential oils have been known for many centuries. In recent years, a large number of essential oils and their constituents have been investigated for their antimicrobial properties against some bacteria and fungi in more than 500 reports (Kalemba and Kunicka 2003). Essential oil components with a broad spectrum of antimicrobial activity include thymol from thyme and oregano, cinnamaldehyde from cinnamon, and eugenol from clove. The antimicrobial properties of plant essential oils against important food-borne pathogens have been well investigated. For example, *Campylobacter jejuni*, *Salmonella enteritidis*, *Escherichia coli*, *Staphylococcus aureus*, and *Listeria monocytogenes* activities have been assayed against 21 different essential oils (Smith-Palmer et al. 1998). Essential oils of bay, cinnamon, clove, and thyme were most inhibitory, each having a bacteriostatic concentration of 0.075% or less against all five pathogens. In general, Gram-positive bacteria were more sensitive to inhibition by plant essential oils than the Gram-negative bacteria. Of the bacteria investigated, *Campylobacter jejuni* was the most resistant to plant essential oils, with only the oils of bay and thyme having a bacteriocidal concentration of less than 1%. At 35°C, *Listeria monocytogenes* was extremely sensitive to the oil of nutmeg. A concentration of less than 0.01% was bacteriostatic and 0.05% was bacteriocidal, but when the temperature was reduced to 4°C, the bacteriostatic concentration was increased to 0.5% and the bacteriocidal concentration was increased to more than 1%. Fungicidal and bactericidal compounds from natural sources, such as cinnamaldehyde, may offer interesting possibilities for disinfecting fresh and minimally processed fruits and vegetables. A crucial problem associated with the practical use of these compounds in such an application is not the efficacy but, rather, the specific odors associated with them at higher doses. To overcome this problem, antifungal plant metabolites should be selected for both efficacy and minimal interference with the natural odor of the treated food.

16.2 Biocontrol Cultures

The use of biocontrol cultures to cause competitive growth between pathogenic bacteria and such a culture, which is non-pathogenic, is yet another example of a non-thermal alternative food processing technology. Microorganisms produce a broad range of components having an influence on the growth of other microorganisms present in their environment. These developed components often increase the competitive edge of the producing organism, so they represent an important feature for their survival and proliferation. Lactic acid bacteria (LAB) constitute the most important single group of organisms to be considered as a source of biopreservatives with relevant applications in food preservation. For centuries, LAB have been used in food fermentations to produce stable foodstuffs, including cheeses, sausages, and vegetables such as sauerkraut. Fermented products naturally containing LAB and the antimicrobials they produce have been consumed without negative health effects, so this range of products have the generally recognized as safe (GRAS) food status. LAB may produce antimicrobial compounds with a broad range inhibition spectrum, such as organic acids and hydrogen peroxide, as well as compounds with a narrower inhibition spectrum such as bacteriocins. The use of LAB as food preservatives is possible via the application of the producing organism as a "protective culture" to the food product and relying on its proliferation and consequent competition with the microorganisms to be suppressed. Alternatively, the antimicrobial compounds may be prepared and then utilized directly in the food to be protected. This procedure represents the advantage of having an instant and more controllable antimicrobial effect. The use of protective cultures in most countries needs only be declared on the product labeling, but the use of antimicrobial metabolites, such as the bacteriocins, is subject to specific rules and regulations in different parts of the world.

Examples of biocontrol cultures are diverse. Protective LAB cultures in bacon were found to be more effective than 120 ppm sodium nitrite, which is the maximum amount permitted on cured meats (Tanaka et al. 1985). Other applications of LAB cultures as a biocontrol hurdle have been investigated in refrigerated meat products (Holzapfel et al. 1995). With regard to vegetables, Romick (1994) used a *Lactobacillus plantarum* strain as a protective culture to prevent growth of *Lactobacillus monocytogenes* in brined, non-acidified, refrigerated pickle products, and Vescovo et al. (1996) tested LAB strains, isolated from fresh vegetable salad ingredients, against a variety of pathogens in salad products. An advantage of biocontrol cultures is that the inhibiting hurdles are themselves microorganisms, so they tend to grow as conditions become more favorable to growth of bacteria. In this sense, the shelf life of products may be extended, since there is no need for additional hurdles such as refrigeration.

Bacteriocins are small proteins produced by many bacteria, including LAB. Bacteriocins generally produced by LAB inhibit the growth of other LAB, but some are bactericidal to a number food pathogens and food-spoilage bacteria of the Gram-positive kind. The LAB bacteriocins may not, therefore, be used as a unique food preservation factor, but may be used as a specific hurdle to suppress the growth of notorious Gram-positive pathogens, such as *Listeria monocytogenes*, *Clostridium botulinum*, and *Bacillus cereus*. Although many different bacteriocins have been identified as potential food preservatives over the years, in practice only two bacteriocins have been widely employed: nisin and pediocin.

Nisin is a protein consisting of 34 amino acids, and is stable on retorting operations and effectively inhibits the growth of Gram-positive food-borne pathogens, such as *Listeria monocytogenes* and *Staphylococcus aureus*, and it also prevents growth of spores of *Clostridium* and *Bacillus* spp. Nisin was considered, originally, for use as an antibiotic, but because of it limited range of inhibition, it was not considered suitable for any therapeutic use. Since nisin is completely degraded in the alimentary tract, it has been found most suitable for use as a food additive. In canned foods, such as vegetables, soups, and puddings, nisin has been applied in combination with heating to successfully neutralize heat-resistant spores of flat-sour thermophilic bacteria. Routine heating and nisin may be used in conjunction for milk preservation in regions where pasteurization, refrigeration, and transportation are not entirely adequate. Nisin used along with acetic, lactic, or citric acid improves blanching and pasteurization methods compared with the use of the organic acids alone. Nisin has been used frequently, in combination with nitrite products, for meat preservation. The combined application may allow less nitrite to exert an identical degree of inhibition of *Clostridium* spp as nitrite alone. Similar findings have been observed for *Clostridium* spp. suppression in bacon and sausages.

Pediocins are bacteriocins produced by LAB of the genus *Pediococcus*. For example, it has been known for some time that *Pediococcus pentosaceous* inhibits growth and acid production of *Lactobacillus plantarum*, an undesirable competitor in mixed-brine cucumber fermentation. Several applications of pediocins have been evaluated with regard to food safety. Pediocin PA-1, produced by a strain of *Pediococcus acidilactici*, has been shown to inhibit growth of *Listeria monocytogenes* added to cottage cheese, half-and-half cream, and cheese sauce for 1 week at 4°C. Although rapid growth to high cell densities was observed in the control samples, a synergistic action was noted between the effect of the bacteriocin and the lactic acid. Extensive tests have shown that this pediocin is non-toxic, non-immunogenic, and is readily hydrolyzed by gastric enzymes.

Although Gram-negative bacteria, yeasts, and molds are not generally sensitive to the action of bacteriocins produced by LAB, the presence of gelatin agents, surfactants, or osmotic shock may sensitize them. A combined preservation scheme may be advantageous, as shown by Stevens et al. (1991) for synergistic effects observed in treatments of nisin and other hurdles tested with *Salmonella* and other Gram-negative species. Other reports highlighted the benefits of combining nisin with EDTA (Blackburn et al. 1990), citrate (Blackburn et al. 1989), and lysozyme and citric acid (Anderson 1992).

16.3   Some Other Alternatives and Future Trends

Magnetic fields, with intensities between 5 and 50 T and either stable or oscillating at 5–50 kHz, are able to destroy vegetative cells of different microorganisms and have been investigated as a food preservation technology. As far as they have been studied, magnetic fields exert no effect on spores or enzymes, and some types of vegetative cells may be even stimulated to grow. The effects of magnetic fields are not well understood, but may involve translocation of free radicals and disruption of cell membranes. The above-mentioned field strengths can only be achieved using liquid-helium-cooled coils and there may be safety concerns over the use of such powerful magnetic fields in a local processing environment.

Although not well understood, the effects of magnetic fields on microbial populations in foods may depend on the magnetic field intensity, the number of pulses, the frequency of the pulses, and the properties of the food being treated, such as resistivity, electrical conductivity, and thickness. Since the underlying mechanisms of potential inactivation are also uncertain, additional research may be required to critically assess the potential of this technology in key issues such as elucidating destruction kinetics, determining crucial process factors, and validating the technology using indicator organisms or surrogates. It is unclear, however, if there would be a vested interest for research groups or established industries in having aggressive research programs on the technology, considering that some other thoroughly investigated options, such as pulsed electric fields, have not yet successfully completed a technology transfer process for wide industrialization and commercialization.

Apart from magnetic fields, some other non-thermal food processing alternatives include use of ozone and cold atmospheric plasma. Ozone is a potent antimicrobial agent. It can effectively kill viruses, bacteria, fungi, and parasites, including those causing food spoilage or human diseases. The efficacy of ozone, however, depends on the target microorganism and the treatment condition. Commercial applications of ozone include purification of drinking water, sterilization of containers for aseptic packaging, decontamination of fresh produce, and food preservation in cold storage. Ozone is commonly generated by electrical discharge, in which dry air or oxygen is passed between two parallel or concentric electrodes that are coated with a dielectric material. Oxygen molecules are broken down to charged oxygen atoms, which recombine to form ozone molecules. Depending on the feed gas, the ozone production rate varies. Ozone destroys microorganisms by reacting with oxidizable cellular components, particularly those containing double bonds, sulf-hydryl groups, and phenolic rings. Therefore, membrane phospholipids, intracellu-lar enzymes, and genomic material are targeted by ozone; these reactions result in cell damage and death of microorganisms. Ozone can be applied in an aqueous solution or the gas phase to decontaminate food-contact surfaces, sanitize equip-ment, recycle wastewater, and decrease pesticide levels on fresh produce. The microbiological quality and shelf life of vegetables, fruits, cheeses, eggs, nuts, and meats can be improved when these products are directly treated with ozone or stored in an ozone-containing environment. For processing of fresh produce, ozone can be used to sanitize the processing water or to decontaminate the product itself. The use of ozone in the gas phase helps in controlling mold and bacteria, both in the air and on the surface of the product.

Cold atmospheric plasma treatment has been used in microbial inactivation, particularly in *Salmonella* spp. Plasmas are created when gases are excited by externally applied energy sources, and they consist of a variety of highly energetic particles, which in combination are able to inactivate microorganisms. Exactly how this is done is not fully understood, but investigations have focused on applications to fruits, vegetables, and other foods with fragile surfaces. These foods are either not adequately sanitized or otherwise unsuitable for treatment with chemicals, heat, or other conventional food processing tools. Overall, plasma treatment has been shown to be highly efficient at inactivating *Salmonella* spp., killing cells in a very short period of time. The published literature does not clearly identify what factors are behind this inactivation, in part owing to the different and diverse experimental conditions used in the studies. A greater understanding of the active antimicrobial compounds in plasmas is needed to understand how they inactivate microorganisms. It is important to be certain that no harmful by-products are generated, and that the process does not adversely affect the quality and the shelf life of the treated food products.

Interest in non-thermal food processing technologies has increased appreciably in recent decades. These technologies promise to maintain the critical balance between safety and marketability of a new generation of foods. Some of these technologies have been optimized for the cold pasteurization of foods, and a limited number of such products are now commercially available. Non-thermal technologies can be used to produce safe fresh-like acid foods (e.g., fruit juices), but extensive research is needed to adapt these technologies for the production of shelf-stable low-acid foods. Current limitations of emerging non-thermal technologies can be overcome when they are combined with conventional preservation methods. The primary barrier to wider commercialization of many non-thermal food processing technologies can be attributed to their high capital cost compared with traditional pasteurization methods, partly due to the relatively small market for alternative pasteurization equipment. In this sense, many of these technologies are not different from some other alternative pasteurization methods. Irradiation, for instance, which has been a research topic for more than a century, does not represent a major factor in the current food processing environment. It is, therefore, likely that numerous visionary research schemes worldwide are needed to successfully position some of these alternative food processing technologies in diverse markets. Perhaps one of the cyclic economic recessions, which harshly affect different industrial sectors, may be an opportunity for investment in alternative food processing technologies because of the strategic position that the food industry has around the globe. In this context, only time will tell.

References

Anderson W (1992) Compositions having antibacterial properties and use of such compositions in suppressing growth of micro-organisms. Eur Patent 0466244A1.

Blackburn P, Polak J, Gusik SA, Rubino S (1989) Nisin composition for use as enhanced, broad range bactericides. US Patent WO 89/1239.

Blackburn P, Polak J, Gusik SA, Rubino S (1990) Novel bacteriocin compositions for use as enhanced broad range bactericides and methods of prevention and treating microbial infection. US Patent WO 90/09739.

Bowles DJ (1990) Defense-related proteins in higher plants. Ann Rev Biochem 59: 873–907.

Gould GW (1996) Industry perspective on the use of natural antimicrobials and inhibitors for food preservation. J Food Prot 59: 82–86.

Hargreaves LL, Jarvis B, Rawlinson AP, Wood JM (1975) The Antimicrobial Effects of Spices, Herbs and Extracts from These and Other Food Plants. Scientific and Technical Survey No 88. The British Food Manufacturing Industries Research Association.

Holzapfel WH, Geisen R, Schillinger U (1995) Biological preservation of foods with reference to protective cultures, bacterocins, and food-grade enzymes. Int J Food Microbiol 24: 343–362.

Kalemba D, Kunicka A (2003) Antibacterial and antifungal properties of essential oils. Curr Med Chem 10: 813–829.

Lund BM, Lyon GD (1975) Detection of inhibitors of *Erwinia cartovara* and *E. herbicola* on thin layer chromatograms. J Chromatogr 110: 193–196.

Nychas JGE (1995) Natural antimicrobials from plants. In: Gould GW (ed) New Methods of Food Preservation, pp 58–89. Aspen Publishers Inc, Gaithersburg MD.

Romick TL (1994) Biocontrol of *Listeria monocytogenes*, a psychrotrophic pathogen model in low salt, non-acified, refrigerated food products. PhD thesis, North Carolina State University, Raleigh, NC.

Smith-Palmer A, Steward J, Fyfe L (1998) Antimicrobial properties of plant essential oils and essences against five important food-borne pathogens. Lett Appl Microbiol 26: 118–122.

Stevens KA, Sheldon BW, Klapes NA, Klaenhammer TR (1991) Nisin treatment for inactivation of *Salmonella* species and other Gram-negative bacteria. Appl Environ Microbiol 57: 3613–3615.

Stratford M, Eklund T (2003) Organic acids and esters. In: Russell NJ, Gould GW (eds) Food Preservatives, pp 48–84. Kluwer Academic/Plenum Publishers, New York.

Tanaka N, Meske L, Doyle MP, Traisman E, Thayer DW, Johnston RW (1985) Plants trials of bacon made with lactic acid bacteria, sucrose and lowered sodium nitrite. J Food Prot 48: 679–686.

Van Etten HD (1976) Antifungal activity of pterocarpans and other elected isoflavonoids. Phytochemistry 15: 655–659.

Vescovo M, Torriani S, Orsi C, Macchiarlol F, Scolari G (1996) Application of antimicrobial-producing lactic acid bacteria to control pathogens in ready-to-use vegetables. J Appl Bacteriol 81: 113–119.

Appendix

Notation

(Dimensions given in terms of mass, M, length, L, time, T, and temperature, θ)

a	Exponent
A	Area (L^2), amplitude (L)
A_a	Arithmetic mean of the cake area in a filtering centrifuge (L^2)
A_1	Logarithmic mean of the cake area in a filtering centrifuge (L^2)
A_m	Area of the filter medium in a filtering centrifuge (L^2)
b	Exponent, constant, height of the bowl of a tubular centrifuge (L), isotherm slope
b_1	Empirical constant
B	Intercept with the y-axis in a plot of a constant-pressure filtration run (T/L^3), permeability coefficient
B_0	Constant
B'	Intercept with the y-axis in a plot of a constant-rate filtration run (M/LT^2)
c	Exponent, capillary viscosity constant, velocity (L/T), velocity of light (L/T)
c_p	Specific heat ($L^2/T^2\theta$)
C	Volume fraction of solids in suspension, gas concentration (M/L^3)
C_f	Volume fraction of solids in the feed
C_p	Heat capacity at constant pressure ($L^2/T^2\theta$)
C_u	Volume fraction of solids in the underflow
C_v	Heat capacity at constant volume ($L^2/T^2\theta$)
C_D	Drag coefficient
d	Exponent, diameter of the rod supporting rollers in chain conveyors (L), radius of the product in roller mills (L), diameter of the neck of a hydrometer (L), exponent, depth of a layer (L)
d_c	Cell diameter (L)
D	Diameter (L), screen aperture (L), dose of microbial inactivation energy
D_c	Cyclone and hydrocyclone diameter (L)
D_f	Diameter of the feed in roller mills (L)
D_g	Gas diffusivity (L^2/T)

(continued)

E. Ortega-Rivas, *Non-thermal Food Engineering Operations*,
Food Engineering Series, DOI 10.1007/978-1-4614-2038-5,
© Springer Science+Business Media, LLC 2012

D_i	Inlet diameter of a hydrocyclone (L)
D_o	Overflow pipe diameter of a hydrocyclone (L)
D_p	Diameter of the product in roller mills (L)
D_{pc}	Cut diameter (L)
D_r	Diameter of the roll in roller mills (L)
D_u	Underflow diameter of a hydrocyclone (L)
D_{UV}	Decimal reduction rate in UV treatment
D_0	Lethality constant
D_{10}	Decimal reduction time, decimal reduction dose
e	Exponent
E	Energy (ML^2/T^2), overall screen efficiency, Young's modulus (M/LT^2)
E_a	Activation energy (ML^2/T^2)
E_c	Electric field intensity threshold value (V^a/L)
E_d	Energy absorbed (ML^2/T^2)
E_i	Bond work index (ML^2/T^2)
E_O	Oversize screen efficiency
E_p	Partial efficiency
E_t	Total efficiency
Eu	Euler number
E_U	Undersize screen efficiency
E_0	Radiating light hitting the surface of a material
f	Frequency (dimensionless), frequency ($1/T$), fanning friction factor
f_a	Size fraction of one component of average weight w_a
$f(x)$	Frequency or occurrence related to size
$f_L(x)$	Size distribution function by length
$f_M(x)$	Size distribution function by mass (volume)
$f_N(x)$	Size distribution function by number
$f_S(x)$	Size distribution function by surface
F	Force (ML/T^2), mass flow rate of solids in the feed (M/T), cumulative percentage of the coarse fraction in the feed
F_f	Cumulative percentage of the fines fraction in the feed
F_A	Buoyancy force (ML/T^2)
$F(x)$	Cumulative frequency
F_D	Drag force (ML/T^2)
$F(x)$	Cumulative percentage oversize of feed solids
Fr	Froude number
g	Acceleration due to gravity (L/T^2)
G_c	Critical flux (M/L^2T)
$G(x)$	Grade efficiency
$G'(x)$	Reduced grade efficiency
h	Height (L), length of the hydrometer not submerged in the liquid (L), Planck's constant (L^2M/T)
H	Height or depth (L), relative humidity
I	Intensity of radiation at a given distance (L^2/T^2)
I_0	Intensity of radiation at the surface of an absorber (L^2/T^2)
J	Rate of diffusion (M/T)

(continued)

k	Constant, coordination number
k_p	Constant for a family of geometrically similar hydrocyclones
k_1	Constant
k_2	Constant
K	Constant, correlation constant of the power law, slope of the line in a plot of a constant-pressure filtration run (T/L^6)
K_c	Constant depending on the filter medium in air filtration
K_1	Cake resistance factor in air filtration
K'	Fluid consistency index (MT^n/L^2), slope of the line in a plot of a constant-rate filtration run (M/L^4T^2)
K''	Constant in constant-rate filtration
L	Length (L)
m	Mass (M), weight of a hydrometer (M), constant, moisture content
m_e	Moisture content of an isotherm based on a straight-line approximation
m_f	Weight of a pycnometer filled with a sample (M)
m_i	Initial moisture content
m_l	Weight of a pycnometer filled with liquid (M)
m_s	Weight of a pycnometer filled with a solid (M)
m_{sl}	Weight of a pycnometer filled with liquid and a solid (M)
m_{LC}	Weight of a container partially filled with liquid (M)
m_{LCS}	Weight of a container with liquid and a submerged solid (M)
m_0	Weight of an empty pycnometer (M)
M	Mesh size, mass flow rate of solids in suspension (M/T), molar concentration
M_c	Mass flow rate of separated solids (M/T), the mass of the solid cake in the bowl of a filtering centrifuge (M)
M_f	Mass flow rate of unseparated solids (M/T)
M_s	Weight of a specimen (M)
M_1	Mixing index
M_2	Mixing index
M_3	Mixing index
N	Rotation speed (1/T), sample size, initial microbial load
N_c	Critical rotation speed (1/T)
N_t	Microbial load after treatment time t
n	Exponent, shear index, number of particles in a sample, slope
n_p	Constant for a family of geometrically similar hydrocyclones
n'	Flow behavior index
O	Mass flow rate of solids in the overflow (M/T)
p	Proportion by weight of the component within a total sample weight w, agglomerating pressure (M/LT^2)
P	Pressure (M/LT^2), power (ML^2/T^3)
P_R	Pressure at reference pressure conditions (M/LT^2)
P_0	Vapor pressure of pure water at a given temperature (M/LT^2)
q	Proportion by weight of a component within a total sample weight w
Q	Volumetric flow rate (L^3/T)
r	Radius of a particle in roller mills, radius of a feed particle in roller mills (L), radius of a ball in tumbling mills (L), radius (L)

(continued)

R	Radius (L), radius of the roll in roller mills (L), radius of the drum in tumbling mills (L), radius of rotation (L), universal gas constant
Re	Reynolds number
Re_p	Particle Reynolds number
Re^*	Reynolds number for power-law fluids
R_f	Underflow-to-throughput ratio
R_i	Neutral zone in a centrifuge (L)
R_m	Resistance of the filter medium (1/L)
R_x	Outer radii of a stack of discs in a disc-bowl centrifuge (L)
R_y	Inner radii of a stack of discs in a disc-bowl centrifuge (L)
R_A	Weir radius for the denser phase in a centrifuge (L)
R_B	Weir radius for the less dense phase in a centrifuge (L)
s	Standard deviation of the analyses of the average value of the fraction of a specific powder, specific surface of a solid (L^2), empirical constant, compressibility coefficient
s_p	Surface area of a particle (L^2)
s_r	Estimate of the standard deviation of the fraction of a specific powder under complete randomization
s_0	Estimate of the standard deviation of the fraction of a specific powder under complete segregation
S	Surface (L^2), number of discs in the stack in a disc-bowl centrifuge
SG	Specific gravity
Stk_{50}	Stokes number
Stk'_{50}	Stokes number including the reduced cut size
Stk^*_{50}	Stokes number for power-law fluids
$Stk^*_{50}(r)$	Stokes number for power-law fluids including the reduced cut size
t	Time (T)
t_c	Treatment time threshold value in pulsed electric field technology (T)
T	Temperature (θ), tensile stress (M/LT^2)
TMP	Transmembrane pressure (M/LT^2)
u	Linear velocity in the vertical direction (L/T), fluid–particle relative velocity (L/T), settling velocity at concentration C (L/T)
u_g	Terminal settling velocity under gravity (L/T)
u_t	Terminal settling velocity of particles (L/T), channel velocity calculated from the cross section (L/T)
U	Mass flow rate of solids in the underflow (M/T), volumetric flow rate of the underflow (L^3/T)
v	Linear velocity in the horizontal direction (L/T)
v_g	Terminal settling velocity under gravity (L/T)
v_r	Radial settling velocity (L/T)
v_t	Terminal settling velocity (L/T)
v_{tan}	Tangential velocity (L/T)
V	Volume (L^3), compacted volume of solids at a given pressure (L^3), linear velocity (L/T)
V_f	Gas superficial velocity through the filter medium in air filtration (L/T)
V_p	Volume of a particle (L^3)
V_K	Volume of a test body (L^3)
V_R	Relation of volumes (V/V_s)

(continued)

V_s	Volume of solid material (L^3)
V_0	Sliding velocity at a reference location defined by Eq. 3.16 (L/T), initial volume (L^3), mean gas velocity (L/T)
V'	Activation volume constant
w	Weight of a sample (M), mass of solids deposited on the medium per unit volume of filtrate (M/L^3)
w_a	Average weight of a sample (M)
W_s	Weight of dry solids in a package (M)
WVP	Water vapor permeance (T/L)
WVTR	Water vapor transmission rate (M/L^2T)
x	Particle size
x_c	Cut point (L)
x_i	Every measured value of the fraction of one powder
x_p	Diameter of a particle (L)
x_{sv}	Equivalent surface diameter of a particle (L)
x_{50}	Cut size (L)
x'_{50}	Reduced cut size (L)
X	Fraction, distance (L)
X_F	Mass fraction of coarse particles in the feed
X_O	Mass fraction of coarse particles in the overflow
X_U	Mass fraction of coarse particles in the underflow
$Z(P)$	Pressure increase required to accomplish a one log reduction cycle in D_{10}

Greek letters

α	Half the angle of nip in roller mills, angle to the vertical in tumbling mills, surface tension of liquid (M/LT^2), specific cake resistance (L/M)
α_0	Empirical constant in Eq. 7.71
α'_0	Empirical constant in Eq. 7.71
β	Empirical constant in Eq. 7.72
γ	Viscosity coefficient for power-law fluids (M/LT^{2-n})
$\dot{\gamma}$	Shear rate (1/T)
δ	Membrane thickness (L)
ΔP	Pressure drop (M/LT^2)
Δm	Apparent increase in weight (M)
ΔG	Apparent weight force (ML/T^2)
ΔP_c	Pressure drop through the powder layer in air filtration (M/LT^2)
ΔP_f	Pressure drop across the filter medium of an air filter (M/LT^2)
$-\Delta P$	Total pressure drop across the filter (M/LT^2)
$-\Delta P_c$	Pressure drop across the cake (M/LT^2)
$-\Delta P_m$	Pressure drop across the medium (M/LT^2)
ΔV^+	Activation volume (L^3)
ε	Porosity or voidage, strain
ε_A	Axial strain
ε_L	Lateral strain
η	Apparent viscosity (M/LT)
θ	Angle between the particle and the screen aperture, hydrocyclone cone angle
λ	Wavelength (L)
μ	Liquid absolute viscosity (M/LT), Poisson's ratio, friction coefficient

(continued)

μ_g	Gas viscosity (M/LT)
μ_o	Viscosity of a pure solvent (M/LT)
μ'	Friction coefficient in roller mills
ν	Kinematic viscosity (L^2/T)
π	Osmotic pressure (M/LT^2)
ρ	Liquid density (M/L^3), frequency
ρ_b	Bulk density (M/L^3)
ρ_g	Gas density (M/L^3)
ρ_s	Solids density (M/L^3)
ρ_A	Density of the denser phase in a centrifuge (M/L^3)
ρ_B	Density of the less phase in a centrifuge (M/L^3)
σ	Interfacial tension (M/T^2), stress (M/LT^2)
σ_t	Strength of agglomerates (M/LT^2)
Σ	Operator, meaning "algebraic sum of," characteristic geometrical features of a centrifuge equivalent to the area of a gravity settling tank with settling characteristics similar to those of a centrifuge
τ	Shear stress (M/LT^2)
τ_0	Intercept with the y-axis in plot of shear stress versus shear rate
ϕ	Volume fraction of spheres in suspension
Φ	Sphericity
ω	Angular velocity (1/T)
Ω	Conical half angle of the discs in a disc-bowl centrifuge

[a]Volt

Index